Antenna Technology for Terahertz Wireless Communication

Uri Nissanov • Ghanshyam Singh

Antenna Technology for Terahertz Wireless Communication

Springer

Uri Nissanov
University of Johannesburg
Johannesburg, South Africa

Ghanshyam Singh
University of Johannesburg
Johannesburg, South Africa

ISBN 978-3-031-35899-9 ISBN 978-3-031-35900-2 (eBook)
https://doi.org/10.1007/978-3-031-35900-2

This Springer imprint is published by the registered company Springer Nature Switzerland AG
The registered company address is: Gewerbestrasse 11, 6330 Cham, Switzerland

Preface

Terahertz wave is a relatively unexplored part of the electromagnetic spectrum and is known as the terahertz gap owing to the lack of efficient sources and detectors which could generate and detect the signals in this frequency regime. However, this spectrum regime is scientifically rich with emerging possibilities and attracting much attention due to its unique properties favorable to various applications such as future-generation communication, non-destructive testing, security scanning, and process control. Therefore, terahertz science and technology are progressing tremendously, as evidenced by the ever-increasing potential applications such as future-generation communication systems with imaging and sensing. The field is becoming large enough for us to subdivide it into various specializations. This technology, although rapidly progressing, is still relatively immature. Therefore, it is highly appropriate that this book focuses on emergent applications and technology in order to inspire and motivate further progress in the engineering domain and has become an essential tool for the non-invasive sensing and imaging of various materials and structures, with key applications in defense, security, and healthcare industry (biomedical imaging and sensing).

This book emphasizes terahertz wireless communication, particularly for 6G, with its enabling technologies, including antenna design and channel modeling with channel characteristics for the success of reliable 6G wireless communication. This book presents the THz antennas technologies, highlights the microstrip-based substrates antennas technologies, and proposes some useful substrates to reduce the conductor and substrate losses at the THz regime. Furthermore, this book shows the design consideration of the THz unit-cell microstrip antenna and the techniques to boost the microstrip antennas' gain and impedance bandwidth (BW), which influence the transmit and receive signal distance propagation transmit, due to high path losses of the humid atmospheric conditions, and transmit and receive data rates, respectively. Moreover, this book discusses the multi-beam and beamforming THz antennas with the UM-MIMO feature. Besides that, this book describes the reconfigurable capabilities, AI, ML, and DL technologies that will influence the success of 6G wireless communication and suggests a remedy for integrating multiple radios into SoC design. Integrating numerous reconfigurable radiators with AI, ML, and DL technologies has become a desired and fundamental property of modern RF systems for the future 6G wireless communication. This book offers the full-wave 3D commercial solver comparison for validating the 6G wireless antennas, besides the experimental verification with the fabrication of the THz prototype antenna, which is extremely expensive.

- Overview of wireless communication evolutions, THz channel modeling, and 6G wireless communication properties.
- This book shows the design consideration of the THz unit-cell microstrip antenna and the techniques to boost the microstrip antenna's gain and impedance bandwidth simultaneously.
- This book discusses the ultra-massive multiple-input, multiple-output (UM-MIMO) antenna systems' features, such as multi-beam and beamforming at terahertz frequencies, that will influence the success of 6G wireless communication.
- This book describes the reconfigurable capabilities, AI, ML, and DL technologies that will influence the success of 6G wireless communication.
- Pillar of solver compression to THz antenna, so it should be as close as it can be to THz antenna experimental verification.

The topics covered in the book are concerned with future-generation wireless communication systems and their wireless channel modeling for required error-free data rates. In this way, it will attract various stages of readers as given below:

- The book will help academia and industry conceptualize the development of future 6G wireless communication systems with suitable channel modeling.
- This book will present an overview of wireless communication evolutions and channel estimation for THz-based future 6G wireless communication systems.
- This book will present antenna technologies for 6G wireless communication in the THz regime.
- For the undergraduate and postgraduate students, it will assist in learning new technology for designing high gain and broadband THz antennas for 6G wireless communication.
- This book will present an ideal reference that not only helps researchers in academia and industry learn more in this evolving field and inspires future 6G wireless communication researchers toward further innovation.

This book comprises ten chapters. Chapter 1 provides that wireless communication technology has evolved permanently to meet the increasing demands of upper specifications, such as higher data rates and connections worldwide. The 6G enabling technologies are mmWave and THz technologies because of the lack of available spectrum below 100 GHz for the bandwidth requirements to support data rates beyond 1 Tb/s. The 6G will also use free-space optics and visible light communication (VLC) to supply incredibly high data rate communications via short-to-medium ranges as a line-of-sight (LoS) and non-line-of-sight (NLoS) propagations. In addition, the 6G will use programmable metasurfaces technologies to manipulate the propagation of electromagnetic waves in the communication link. The THz challenges will be that the bandwidth of at least 10 GHz, which is needed to supply the needed data rate beyond 1 Tb/s, so mmWave, and THz frequencies should be used, so the high frequencies enhance the Doppler rate when the broad bandwidth causes hardware deteriorations such as non-linearities and phase noise that must be offset. One of the crucial drawbacks of the THz waves is the extreme atmospheric absorption losses caused by water vapor molecular absorption and other atmospheric conditions, such as rain. So, to compensate for these atmospheric absorption losses, we need to design 6G wireless communication networks in a frequency band in which atmospheric absorption losses are less than 0.1 dB/m.

Furthermore, to allow medium-range (50–200 m) LoS outdoor 6G wireless communication networks, high gain antennas must be designed due to the lack of THz emitters sources beyond 100 mW to compensate for these atmospheric absorption losses. They are diminishing its resources, energy, and emissions footstep and enhancing sustainability in different industries and society. One of the most significant issues regarding using the THz regime is designing a novel transceiver module because the present designs do not fit THz frequencies.

Chapter 2 mainly concerns the ways to generate THz electromagnetic radiation, which can be with THz solid-state sources, THz photonic solid-state sources, and THz VTSs. Then, it further presents the major design concepts of the THz wireless communication modules in transmitter and receiver. In the end, the design concepts of the THz high-rate PHY layer are presented. To fulfill the THz high-rate physical (PHY) layer in 6G wireless networks design challenges, all of the described THz transmitter and receiver wireless communication modules should work in synergy and optimally because the 6G wireless communication network will not provide the demands when there is a failure in a specific module.

Chapter 3 presents the THz wireless communication standardization, potential applications, channel properties, modeling, and link budget. The THz potential communication applications can be as chip-to-chip THz wireless communication, THz-WLAN, THz-WPAN, THz-kiosk downloading, THz wireless connections via servers inside a data center, THz backhauling and fronthaul for macro and small cells wireless communications, THz airborne and inter-satellite wireless communications, THz autonomous vehicles and UAV's wireless communications, and finally THz nano-cells wireless communication. The communication propagation distance can vary between a few cm indoors and several km outdoors. At the same time, the THz has some channel properties difficulties such as the FSPL, e.g., spreading loss and atmospheric absorption of the water vapor loss. For communication proposes, we need to operate at frequency bands below 1000 GHz so that atmosphere absorption loss needs to be below 100 dB/km. Therefore, the frequency windows are approximate: 125–175 GHz, 190–320 GHz, 335–376 GHz, 385–443 GHz, 453–515 GHz, 627–711 GHz, and 808–902 GHz.

Furthermore, the THz channel properties are snow and rain, fog and cloud, blockage, diffraction, scattering, reflection, Doppler frequency shift effect, and DS, where all of these properties attenuate the THz wave signals. So, to make a reliable THz wireless backhauling and fronthaul communication and based on the THz link budget and THz channel properties, there is a need to develop THz high-gain antennas with gains of at least 31.8–53.6 dB to facilitate reliable 6G LoS wireless communication with a data rate of 20–100 Gb/s and propagation distance of 100–300 m for 100–500 GHz. The THz wireless communication channel modeling can be a deterministic or stochastic model based on the RT scheme. Researchers are using THz channel simulators because setting up a standard THz wireless communication channel model is complex for the reasons described in this chapter and above.

Chapter 4 concerns the THz antenna technologies for 6G communication system networks. The THz antenna technologies are classified into three classes depending on their content processing: metallic antennas, dielectric antennas, and state-of-the-art materials antennas. We can use the THz horn antenna, THz dielectric lens antennas, THz nano- and plasmonic antennas based on graphene, and THz planar microstrip for the THz wireless communication antennas. In addition, the mathematical modeling of THz rectangular microstrip antenna design, feeding techniques, and design considerations parameters of THz microstrip antennas are presented. Furthermore, the THz microstrip antenna lamination with potential examples is presented. The THz antennas can be designed and simulated within full-wave electromagnetic commercial simulators such as the CST MWS, Ansys HFSS, and multi-physics COMSOL simulators. At the end of this chapter, the fabrication of the THz antennas with testbeds is discussed.

Chapter 5 explores the THz passive components for 6G communication networks such as THz Wilkinson power divider, THz T-junction power divider, and THz silicon integrated waveguide (SIW) with their design considerations. In this chapter, several models are designed, and their simulation results are presented as follows. (1) A THz Wilkinson 1 to 2 equal power divider, and the simulation results of resonance frequency, impedance bandwidth, isolation, and insertion loss were 123.8 GHz, >48.24 GHz, −12.4 dB, and 3.76–5.56 dB, respectively. (2) A THz Wilkinson 1 to 32 equal power divider, and the simulation results of resonance frequency, impedance bandwidth, and insertion loss were 119.56 GHz, 29 GHz, and 21.34–27.18 dB, respectively. (3) A THz T-junction 1 to 2 equal power divider, and the simulation results of resonance frequency, impedance bandwidth, and insertion loss were 113.28 GHz, >70 GHz, and 3.46–7.04 dB, respectively. (4) A THz T-junction 1 to 8 equal power divider, and the simulation results of resonance frequency, impedance bandwidth, and insertion loss were 118.78 GHz, 21.5 GHz, and 13.38–18.36 dB, respectively. (5) A THz T-junction 1 to 8 unequal power divider, and the resonance frequency, impedance bandwidth, and insertion loss simulation results were 116.9 GHz, 13.56 GHz, and 10.33–17.09 dB, respectively. (6) A wideband THz SIW, and the resonance frequency, impedance bandwidth, and insertion loss simulation results were 108.1 GHz, 12.06 GHz, and 2–2.4 dB, respectively. (7) Finally, a THz SIW with dual outputs and the resonance frequency, impedance bandwidth, and insertion loss simulation results were 109 GHz, 5.8 GHz, and 4.7–4.76 dB, respectively. The design and simulation are performed using CST MWS based on the FIT solver, and the results are compared with CST MWS based on the FEM solver or by Ansys HFSS based on the FEM solver, and an acceptable agreement is achieved.

Therefore, Chap. 6 exploited the THz unit-cell microstrip antenna and THz unit-cell frequency selective surface (FSS) with their design considerations for 6G wireless communications. Several antenna models/designs are presented with their simulation results which are as follows. (1) A unit-cell planar rectangular microstrip antenna radiator with a resonance frequency of 112.42 GHz, and the achieved gain and impedance bandwidth are 6.78 dB and 6.04 GHz. (2) A unit-cell crown-shaped planar microstrip antenna radiator for a resonance frequency of 127.6 GHz; the achieved gain and impedance bandwidth were 5.67 dB and 5 GHz. (3) A unit-cell bandstop double circular ring-shaped FSS for a resonance frequency of 120 GHz. The achieved impedance bandwidth, S_{11}, and S_{21} were 25.54 GHz, −44.04 dB, and −0.05 dB, respectively. (4) a unit-cell bandstop discontinuous circular ring-shaped FSS for a resonance frequency of 119.68 GHz. The achieved impedance bandwidth, S_{11}, and S_{21} were 10.22 GHz, −42.48 dB, and −15.83 dB, respectively. (5) A unit-cell bandstop double rectangular ring-shaped FSS for a resonance frequency of 111.35 GHz. The achieved impedance bandwidth, S_{11}, and S_{21} were 12.81 GHz, −19.77 dB, and −44 dB, respectively. (6) A unit-cell bandstop double discontinuous rectangular ring-shaped and rectangular dipole FSS for a resonance frequency of 109.46 GHz. The achieved impedance bandwidth, S_{11}, and S_{21} were 6.38 GHz, −48.4 dB, and −13 dB, respectively. (7) A unit-cell bandstop and bandpass rectangular steering wheel-shaped FSSs for a resonance frequency of 129.95 GHz. The achieved impedance bandwidth, S_{11}, and S_{21} were 49.35 GHz, −50 dB, and −34.23 dB, respectively. (8) A unit-cell bandstop cross-shaped FSS for a resonance frequency of 104.64 GHz. The achieved impedance bandwidth, S_{11}, and S_{21} were 14.95 GHz, −42.07 dB, and −17 dB, respectively. (9) A unit-cell bandstop Jerusalem cross-shaped FSS for a resonance frequency of 105.9 GHz. The achieved impedance bandwidth, S_{11}, and S_{21} were 21.51 GHz, −41.13 dB, and −15.73 dB, respectively. (10) Finally, a bandstop and bandpass tilted cross-shaped FSSs for a resonance frequency of 153.83 GHz. The achieved impedance bandwidth, S_{11}, and S_{21} were 31.72 GHz, −47.12 dB, and −49.88 dB, respectively. The suggested THz models are designed and simulated with the FIT solver within the CST MWS. In contrast, the first and second proposed antenna models' simulation results are compared with the FEM solver within the CST MWS simulator, which shows excellent agreements.

Chapter 7 presents gain and bandwidth enhancement techniques for THz planar microstrip antennas. Four models were designed, and the simulation results are presented in this chapter. The developed THz antenna models are as follows: (1) low SLLs microstrip array antenna for a resonance frequency of 118.2 GHz, and the achieved gain and impedance bandwidth were 17.54 dB and >7.86 GHz. Further, the sub-model of this design was a THz low SLLs microstrip array antenna with bandstop FSSs for a resonance frequency of 118.2 GHz, and the achieved gain and impedance bandwidth is 21.16 dB and >6.15 GHz. (2) Slotted circular patch microstrip array antenna with silicon-integrated waveguide (SIW) for a resonance frequency of 121.18 GHz, and the achieved gain and impedance bandwidth were 23.73 dB and 12.86 GHz. The sub-model of this design is a THz slotting circular patch microstrip array antenna with SIW and bandstop FSSs for a resonance frequency of 121.3 GHz, and the achieved gain and impedance bandwidth is 26.78 dB and 12.47 GHz. (3) Parasitic patch microstrip array antenna for a resonance frequency of 114.07 GHz, and the achieved gain and impedance bandwidth were 20.4 dB and 18.01 GHz. The sub-model of this design was a THz parasitic patch microstrip array antenna with bandstop FSSs for a resonance frequency of 118.09 GHz, and the achieved gain and impedance bandwidth is 21.8 dB and 18.46 GHz. and (4) Log-periodic 24 × 16 microstrip array antenna for a resonance frequency of 96.66 GHz, and the achieved gain and impedance bandwidth are 18.78 dB and 52.6 GHz. The first sub-model of this antenna was a THz log-periodic 24 × 32 microstrip array antenna for a resonance frequency of 137.34 GHz, and the achieved gain and impedance bandwidth is 19.74 dB and 54.3 GHz. In comparison, the second sub-model of this antenna is a THz log-periodic 24 × 32 microstrip array antenna with bandstop FSSs for a resonance frequency of 128.46 GHz, and the achieved gain and impedance bandwidth is 21.42 dB and 53.9 GHz. From error analysis, the proposed THz log-periodic microstrip array antenna gains and the bandwidth because of the predictable fabrication etching tolerance accuracy ±10 μm, the microstrip laminates frequency-pendent dielectric properties (ε_r) and the frequency-pendent losses (tanδ) may change by about 1.81 dB, 3.2 GHz, because of the antennas' high frequency working beyond 100 GHz. The simulation results of the primary models were validated with another kind of antenna simulator, and a good agreement was achieved.

Chapter 8 discusses the multi-user, multiple-input, multiple-output (UM-MIMO) technology with multi-beam and beamforming THz array antenna for 6G communication. There are four designed THz antenna models, including three models of THz beamforming microstrip array antennas and one THz ultrawide band MIMO antenna model with their simulation results presented. The presented THz antenna models are (1) used for THz beamforming using a 30 × 16 E-shaped radiator microstrip array antenna for the resonance frequency of 146.55 GHz. The achieved gain, impedance bandwidth, and steering angles are 23.8 dB, >41.88 GHz, and ±13.7 degrees, respectively. (2) Used for THz beamforming 30 × 16 E-shaped radiators microstrip array and bandstop and passband FSSs antenna for the resonance frequency of 136.45 GHz, and the achieved gain, impedance bandwidth, and steering angles were 26.04 dB, >33.3 GHz, and ±13.5 degrees, respectively. (3) Used for THz beamforming Rotman lens microstrip array antenna for the resonance frequency of 118–143.2 GHz, and the achieved realized gain, impedance bandwidth, and steering angles were 14.09 dB, 31.02 GHz, and ±23.8 degrees, respectively. From error analysis, the proposed THz beamforming microstrip array antenna, gain, and bandwidth because of the predictable fabrication etching tolerance accuracy ±10 μm, the microstrip laminates frequency-pendent dielectric properties (ε_r) and the frequency-pendent losses (tan δ) may change by a maximum of about 2.1 dB and 6.31 GHz because of the antennas' high frequency working beyond 100 GHz. The simulation results of the primary models are verified with another kind of antenna simulator, and a good agreement is achieved. (4) Used a 4-ports MIMO microstrip antenna, including four antipodal Vivaldi microstrip radiators. Furthermore, the simulated results of the proposed THz MIMO antenna showed an impedance bandwidth, maximum gain at a solid angle, minimum envelope correlation coefficient (ECC), and minimum diversity gain (DG) of >1846.7 GHz, 8.66 dB, 23.48 dB, 0.000916, and 9.9965 dB, respectively. From state-of-the-art simulation results with future fabrication etching tolerance accuracy, microstrip substrate ε_r, and microstrip substrate tan δ deviations, which can be when an experimental verification with a prototype fabricated THz MIMO antenna, the impedance bandwidth and maximum gain can be altered by a maximum of about 1.25–2 GHz, 0.5–0.9 dB from the simulated results.

In Chap. 9, the current trends in technological advancement have escalated the demands of multifunctional components across the spectrum. In the THz regime of the spectrum, wireless communications demand efficient, reconfigurable, tunable, inexpensive, and electrically small antennas, which could be implemented in increasingly space-limited devices. This chapter presents a THz reconfigurable microstrip antenna to fulfill the demand for the 6G wireless communication system. Further, the reconfigurable antennas need intelligence and capacity to sustain the appropriate communications tactics established upon the signal feedback estimation and feedback from sensing the channel performances. The THz reconfigurable antennas for 6G will allow supplying a remedy, which permits the integration of numerous communication components into the system-on-chip (SoC) design. The idea of a THz ultrawide band reconfigurable planar microstrip antenna within a new-fangled gold radiator patch within a couple of PIN diodes mounted on a Benzocyclobutene (BCB) polymer is exploited, and different kinds of reconfigurable antenna, including beams, bandwidth, and frequency reconfigurations, are designed using CST MSW based on finite integration technique. It has been illustrated that the reconfigurable microstrip antenna's working frequency band can be changed up to 8 GHz when this antenna is made due to etching accuracy, which is ±5 μm. Following the simulation within the FIT solver at the CST MWS simulator, the offered THz ultra-wideband reconfigurable microstrip antenna has six operation bands. Moreover, the simulation results of the maximal bandwidth and total efficiency apperceived by the offered antenna are >201.77 GHz (>100.44%), 93.4%, appropriately.

In Chap. 10, the potential role of machine learning (ML)/deep learning (DL)/ artificial intelligence (AI) for 6G wireless communication systems is discussed. Developing an efficacious 6G communication network at THz frequency is more complicated than those at lower frequency systems because channels at the THz band are noted to be further insecure than those of lower frequencies. Therefore, 6G wireless networks will own the intelligence and capacity to encourage the most appropriate communications tactics based on channel activities sensing and signal quality estimation feedback using ML/DL/AI technologies. In addition, to increase the bandwidth and boost the data rate to new dimensions, we need to use THz communication to induce 6G wireless applications such as holographic communication and digital twinning. Moreover, the THz frequencies supply access to broader bandwidth, which allows interaction with communication devices to be changed by improving aspects like gesture recognition to assist XR-based applications, for example, the Metaverse.

In summary, the book provides a unified view of the state-of-the-art terahertz communication technology, which should be accessible to a readership with basic knowledge of electromagnetic and RF microwave antenna theory and wireless communication systems. The readership may find each chapter's rich set of references handy. We strongly recommend the book to graduate students, researchers, and engineers who intend to work in wireless communication, particularly 5G/B5G and 6G. Although numerous journal/conference publications, tutorials, and books on terahertz communication have been published in the last few years, most focus on the various terahertz source and detector technology. This book distinguishes itself from the existing prosperous literature on terahertz communication and next-generation antenna technologies. Such books emphasize the theoretical design and performance evaluation of terahertz antenna technologies for future-generation communication, such as B5G and 6G.

The authors are indebted to numerous colleagues for their valuable suggestions during the entire period of manuscript preparation. The first author conveys the most incredible gratitude to his parents, whose tremendous blessings and endless sacrifice always encourage the manuscript's writing and carrier growth. They have shown immense patience and support in every possible way, even though words alone can never express our gratitude to them. The first author also extends warm thanks to his wife, sons, and daughters; their innocent smiles and chats always bring cheerfulness, even at odd times. We would also like to thank publishers at Springer, in particular, Charles B. Glaser, Brian P. Halm, and Cynthya Pushparaj, for their helpful guidance and encouragement during the creation of this book.

Furthermore, the authors would not justify their work without showing gratitude to their family members, who have always been the source of strength to work tirelessly to accomplish the assignment. Last but not least, we thank the ultimate source of energy of every particle in the universe, the Almighty, for giving us enough energy and strength to complete the work. All praise and gratitude belong to him.

Johannesburg, South Africa Uri Nissanov
Ghanshyam Singh

Contents

List of Figures

List of Tables

About the Authors

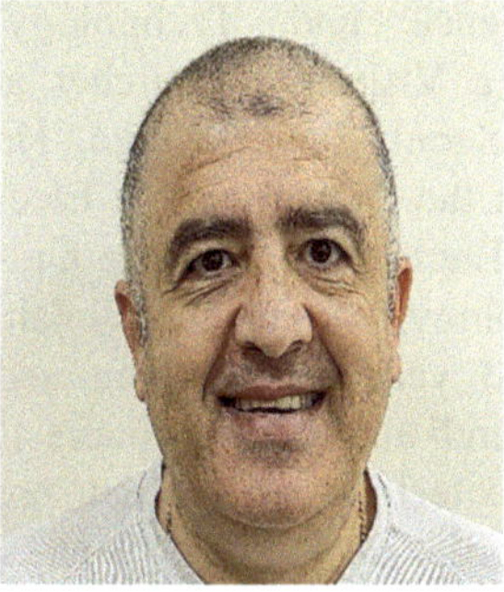

Uri Nissanov (Nissan) received a Ph.D. in Electrical and Electronic Engineering with Excellence from the University of Johannesburg, South Africa, in 2022, where his thesis was to analyze and design a Terahertz (THz) microstrip antenna for next-generation communication systems (B5G/6G). He is pursuing a post-doctoral research fellowship on antenna technology for THz wireless communication systems (B5G/6G) at the Centre for Smart Information and Communication Systems, Department of Electrical and Electronic Engineering Sciences, Auckland Park Kingsway Campus, University of Johannesburg, South Africa. He has authored over 30 publications in various journals and conferences. His research interests are in the design and simulation of THz band/sub-mmWave/mmWave, and sub-6GHz high-gain, broadband microstrip array antennas, reconfigurable antennas, beam steering microstrip antennas, multi-input-multi-output (MIMO) microstrip antennas, and components such as Wilkinson power dividers, T-power dividers, frequency selective surfaces (FSSs), substrate integrated waveguides (SIWs), and beam-steering Rotman Lens for 5G/B5G/6G wireless communication systems.

Ghanshyam Singh (Member, IEEE) received a Ph.D. in Electronics Engineering from the Indian Institute of Technology (Banaras Hindu University) Varanasi, Varanasi, India, in 2000. Currently, he is working as a Full Professor and Director of the Centre for Smart Information and Communication Systems, Department of Electrical and Electronic Engineering Science, Auckland Park Kingsway Campus, University of Johannesburg, South Africa. He was associated with the Central Electronics Engineering Research Institute, Pilani, and Institute for Plasma Research, Gandhinagar, India, respectively, where he was a Research Scientist. He also worked as an Assistant Professor with the Electronics and Communication Engineering Department, Nirma University of Science and Technology, Ahmedabad, India. He was a Visiting Researcher at Seoul National University, Seoul, South Korea. He also worked as a Professor at the Department of Electronics and Communication Engineering, Jaypee University of Information Technology, Wakanaghat, Solan, India. He has more than 23 years of teaching and research experience. His research and teaching interests include millimeter/THz wave technologies which are applications in communication and imaging, next-generation communication systems (5G/6G)/cognitive radio/NOMA, resource allocation and management, interference management, applications of 5G/6G in sustainable smart city, industry 4.0/5.0, healthcare 4.0, intelligent transport systems, energy management (IoE), and digital farming. He is the author/coauthor of more than 290 scientific papers in refereed journals and international conferences and several books and book chapters published by Springer, Wiley, IET, CRC, and Academic Press. He has supervised various M.Tech. and Ph.D. theses. He has worked as an editor/associate editor and a reviewer for several reputed journals and conferences.

Chapter 1
Introduction

1.1 Evolution of Wireless Communication

Wireless communication technology has permanently evolved to reach increasing demands and higher specification requirements. Since deploying first-generation (1G) mobile wireless networks, the telecommunication business has faced many new demands in technology, effective spectrum utilization, and, ultimately, security for end users. Future wireless communication technologies will provide ultra-fast, feature-rich, and highly secure mobile networks. The wireless communication system has undergone several evolutions in the past few decades after introducing the 1G mobile wireless network in the early 1980s. Due to the enormous demand for higher data rates and more connections worldwide, mobile communication standards advanced rapidly to support this demand.

The evolution began with Marconi, an Italian inventor, transmitting Morse-code signals wirelessly using radio-frequency (RF) waves to about 3 km in 1895. It was the first wireless transmission in the history of science. Since then, engineers and scientists have been working on efficient communication using RF waves. Telephones became popular during the mid of nineteenth century. However, due to wired connection and restricted mobility, engineers started developing a device that does not require a wired connection and transmits voice using RF waves [1, 2].

The 1G mobile wireless network was deployed in Japan by Nippon Telephone and Telegraph (NTT) company in Tokyo in 1979. At the beginning of the 1980s, it was deployed in the UK, Europe, Finland, and the USA. This system used analog signals, and the main features were transmitted and received frequencies were 800 megahertz (MHz) and 900 MHz, and the bandwidth (BW) was 10 MHz with 666 duplex channels with BW of 30 kilohertz (kHz). The technology was analog frequency modulation (FM) with frequency division multiple access (FDMA) techniques with only voice mode service. However, the technical limitations had many disadvantages, such as poor battery life and sound quality due to electromagnetic

U. Nissanov, G. Singh, *Antenna Technology for Terahertz Wireless Communication*,
https://doi.org/10.1007/978-3-031-35900-2_1

(EM) interference. In addition, with a limited number of users and cell coverage, large-sized mobile phones were not convenient to carry, roaming was not possible between similar systems, and less security, i.e., calls could be decoded by using an FM demodulator [3, 4].

The second-generation (2G) mobile wireless communication system introduced a state-of-the-art digital wireless transmission technology with a global mobile communication (GSM) system. GSM technology became the base standard for further development in wireless standards later. This standard could support up to 64 kilobits per second (kb/s) of data, sufficient for short message service (SMS) texts and email services. Qualcomm developed and implemented the code division multiple access (CDMA) systems in the mid-1990s. The CDMA has more features than GSM regarding data rate, user number, and spectral efficiency. The primary feathers of 2G were a digital cellular network with enhanced security and encrypted voice transmission and enabled SMS texts and internet at a lower data rate. In contrast, the main disadvantages were limited mobility, a limited number of users and hardware capability, and a lower data rate [4, 5].

The 2.5G with general packet radio service (GPRS) technology was developed to support a higher data rate up to a maximum of 171 kbps. At the same time, the CDMA2000 supported a data rate up to a maximum of 384 kb/s, and the enhanced data GSM evolution (EDGE) technology supported a data rate up to a maximum of 473.6 kb/s [4].

The third-generation (3G) wireless communication was initiated with universal mobile terrestrial/telecommunication systems (UMTS) technology. It was planned to implement a global frequency band in the 2000 MHz range by the International Telecommunication Union (ITU). UMTS has a data rate of up to 384 kbps; it supports video calling on mobile devices, cellular Internet of Things (IoT) applications, and internet protocol (IP)-based services.Subsequently introducing the 3G mobile communication system, smartphones evolved widely across the globe. As a result, specific applications were developed for smartphones that handle email, multimedia chat, games, video calling, social media, and healthcare. The advantages of the 3G were video calling, a higher data rate, enhanced security with more coverage and users, location maps and tracking, mobile app support, television (TV) streaming, and high-quality third-dimensions (3D) games. In contrast, the significant disadvantages were expensive spectrum licenses, higher BW requirements to support higher data rates, costly mobile devices, and costly infrastructure equipment and implementation [6, 7].

In order to enhance the data rate in existing 3G networks, the 3.5G wireless network was developed with other technology improvements that were embedded in the 3G network, including high-speed downlink packet access (HSDPA) and high-speed UL packet access (HSUPA) and long-term evolution (LTE). As a result, the 3.5G network can support higher data transmission rates of about 14 Mb/s [7].

Fourth-generation (4G) mobile wireless systems are an enhanced version of 3G networks that offer download data rates up to 100 Mb/s of either the data receiver or the server set is moving at a speed of 60 km/h and are capable of handling more advanced multimedia services with using the LTE and LTE advanced wireless technology. Furthermore, it has compatibility with the previous version. Thus, easier deployment and upgrade of LTE and LTE advanced networks are possible. In addition, it can transmit data and voice simultaneously over the LTE (VoLTE) network. Furthermore, all services, voice, and data can be transmitted over IP packets [7, 8].

Furthermore, wireless transmission technologies like worldwide interoperability for microwave access (WiMax) are involved in 4G systems to enhance data rate and network performance. The main advantages of the 4G are that it can transmit up to 1 Gb/s, reduce latency, and allow transmission of high-definition (HD) video streaming and HD gaming. In contrast, the main disadvantages are expensive infrastructure, hardware, and a highly costly spectrum [7, 8].

The fifth-generation (5G) wireless network, deployed globally, uses advanced technologies to deliver ultrafast multimedia and internet experience for users. The main features of 5G are low latency in milliseconds (ms), ultra-fast mobile internet uplink (UL) of up to 10 Gb/s, and downlink (DL) of up to 20 Gb/s, higher security, and reliable network, and uses technologies like small cells and beamforming to improve efficiency. In addition, cloud-based infrastructure that offers power efficiency, easy maintenance, and hardware upgrade will extend the analytical capabilities and functionalities for industries, security applications, healthcare, and autonomous driving. The 5G network will function in standalone and non-standalone modes in earlier deployments. In non-standalone mode, the LTE and 5G-new radio (5G-NR) spectrum will be used. Control signaling will be connected to the LTE core network in non-standalone mode. In addition, there will be a dedicated 5G core wireless network higher bandwidth 5G-NR spectrum for standalone mode. The sub-6-GHz spectrum of 4.1–7.125 GHz (FR1 ranges) is used in the headmost deployments of 5G wireless networks. In order to achieve a higher data rate, 5G technology will use millimeter waves (mmWaves) and an unlicensed spectrum for data transmission. A complex modulation technique has been developed to support a massive data rate for IoT [7–10].

1.2 Sixth-Generation Wireless Communication Systems

1.2.1 Key Performance Indicators

It is assumed that 6G will primarily consist of the same key performance indicators (KPIs) as prior generations; however, with significantly upper aspiration, the energy efficiency KPIs are outstanding, initially offered in 5G, although without specific goals. Nevertheless, while the KPIs were mainly self-reliant in 5G, a cross-relationship is acceptable in 6G via group definitions. Complete indicators in a group should be completed simultaneously; still, distinctive groups can have varied demands. Thus, 6G will require real-time scalable to overlap these diverse groups. Therefore, the following KIPs are needed in the future 6G wireless communication era.

(a) *Peak Data Rate and Experienced Data Rate*

To implement state-of-the-art multimedia servicing, high-fidelity 3D mobile holograms, digital duplicates, and genuinely enveloping extended realities (XRs), 6G requires an extreme data rate than 5G. Therefore, 5G was designed to accomplish a DL peak data rate of 20 Gb/s, whereas 6G desires to supply a peak data rate beyond 1 Tb/s and a user-experienced data rate of 1 Gb/s [11].

(b) *Latency and Jitter*

The latency-related operation demands are reduced significantly to supply the superior experience of delay-sensitive and real-time applications like latency-related rendition and interactive tactile internet (TI) performance. Performance goals include end-to-end (E2E) latency of less than 1 ms, air latency of fewer than 100 microseconds (μs), and enormously low jitter delay as μs. Along these demands fulfilled, the user-experienced latency can be below 10 ms, which is the latency requirement for XR services. The user-experienced latency demand is assigned to the accumulated latency components in wireline links, wireless links, and the calculation on the couple server and client sides [12].

(c) *Enhanced Energy Efficiency*

6G wireless technologies should enable better energy efficiency than the preceding generations. However, the energy efficiency endeavor must be accomplished with transmission equipment and user efficiency. The future 6G wireless technology is anticipated to produce an output that should attain Tb/s per Joule (Tb/s/J). So, it is vital to intensify that designing an extremely energy-efficient communication scheme is one of the significant issues of upcoming 6G wireless technology, which needs to be two times spectral efficiency higher than 5G [13].

(d) *Reliability*

The error rate of 6G needs to be 10^{-7}, so 6G wireless reliability needs to be 100 times lower than 5G to facilitate latency-sensitive services that need ultra-reliability for remote surgery, emergency response, and industrial automation [14].

(e) *Enhanced Spectral Efficiency*

The gross 6G wireless network performance needs to be enhanced, so 6G wireless has to have a higher spectral efficiency of 100 bits per second per hertz (b/s/Hz), about three times higher than 5G, to supply progressive multimedia services to many people [15].

(f) *Connection Density and Mobility*

Over past generations, network coverage has been essential and will stay highly significant in 6G. 6G needs to facilitate over ample coverage than 5G. The mobile device's maximum supported speed in 5G, i.e., mobility, is 500 km/h. It may need additional improvements in 6G, beyond 1000 km/h, pending on the development of hauling systems. The vast increase in connected machines will demand 6G to

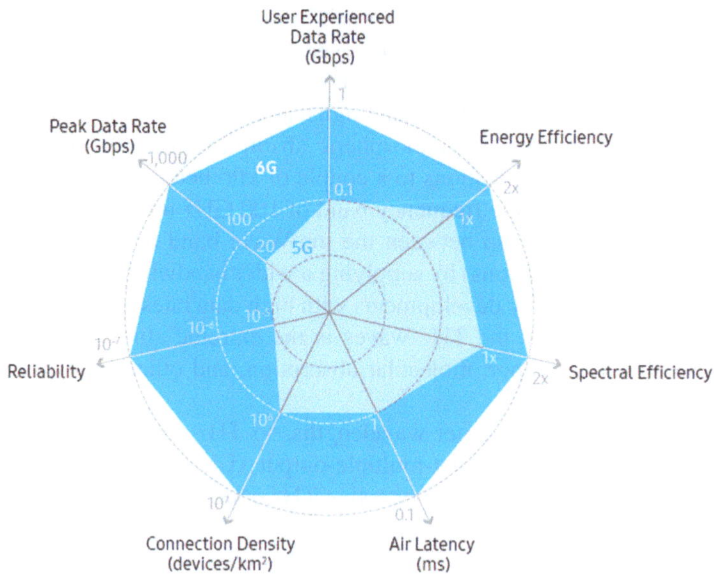

Fig. 1.1 Comparison of KIPs requirements between 5G and 6G [11]

support about 10^7 devices per square kilometer (km^2). This connection density is 10 times larger than the connection density necessity of 5G [16]. Figure 1.1 is attached, which describes and summarizes the comparison of KIPs requirements between 5G and 6G.

1.2.2 Enabling Technologies

Future sixth-generation (6G) systems will demand the upholding of state-of-the-art technologies to entitle unmatched functionalities and enabling technologies in the network. These technologies are realized to introduce new applications based on compelling necessities for reliability, latency, capacity, efficiency, and energy analogs to their 5G correlate. The state-of-the-art 6G enabling technologies are described in the following paragraphs.

(a) *Millimeter – and THz – Wave for 6G Wireless Communication*

One of the crucial challenges for implementing 6G wireless communication is the lack of available spectrum below 100 GHz for bandwidth requirements to support data rates up to a couple of terabits per second (Tb/s). The mmWave is the frequency band between 30 and 300 GHz; the corresponding wavelength is between 10 millimeters (mm) and 1 mm. Due to the short wavelengths, mmWave communications permit diminutive-sized antenna arrays with many small radiators. Therefore, a thin directional beam can produce multipath reflection and immunity against eavesdropping and jamming attacks. However, the mmWave has some drawbacks. For example, small component fabrications are a significant challenge due to the increased fabrication cost. In addition, the signal attenuation of more than 10 decibels per kilometer (dB/km) from the atmospheric absorption restricts the transmission range of mmWave communications to a couple of kilometers (km) [17].

The Terahertz (THz) is the frequency band of 100 GHz up to 10,000 GHz and is predicted to overpass the gap between the mmWave band and infrared (IR) light waves (optical communications) by supplying a wider bandwidth and facilitating the elaboration of novel employ developments with high data rates needs. However, one of the main drawbacks of the THz waves is the extreme atmospheric absorption losses caused by water vapor, molecular absorption, and other atmospheric conditions, such as rain.

Moreover, the THz has shorter wavelengths, so THz communication will integrate ultra-massive multiple-input-multiple-output (UM-MIMO) microstrip array antenna radiators with other passive or active THz components in a small integrated circuit (IC) and get a highly directional beam radiation pattern to mitigate the atmospheric absorption losses. However, THz-based communications need to rethink existing remedies and investigate state-of-the-art approaches, which suggest a smooth operation over the whole THz band [18].

(b) *Free-Space Optical Wireless Communication*

Free-space optical wireless communications (OWC) can supply incredibly high data rate communications via short-to-medium ranges, as line-of-sight (LoS) can be accomplished. At the same time, some applications are also achievable in non-LoS (NLoS) footings. In contrast, the achievable data rates needed for signal processing and modulation can differ. To this end, more research is required to study the architectures that supply the proper intricacy-cost-performance bargain. For example, we commonly differentiate between light-emitting-diode (LED) and light amplification using stimulated radiation (Laser) techniques. The first, also known as visible light communication (VLC), is mainly based on LEDs, which exist as lighting sources and transmission of information. Additionally, the optical transmission is aimed at the DL, although the UL needs to be supplied by conventional radio links, which will bring challenges in integrating 6G wireless fidelity (6G Wi-Fi) and 6G wireless that needs more investigation. Also, the accommodation to mobility establishes a robust challenge [19].

VLC can use a vast unlicensed band in indoor scenarios without cross-interference and cheap hardware. On the other hand, Laser-based systems can supply higher data rates; finally, fine beam widths are appropriate for fixed wireless situations. Further, they are susceptible to blockage of the LoS paths because no multipath diversity is attainable, while the detection and modulation methods appropriate in ambient with fast variations of channel conditions additionally require more research [20].

(c) *Programmable Metasurfaces*

Metasurfaces have lately developed as a state-of-the-art technology predicted to reform 6G wireless communications by making wireless system engineering designers manipulate the propagation of EM waves in the 6G wireless link. The programmable metasurfaces will facilitate unexampled abilities interacting with EM waves, including absorption, focusing, scattering, imaging, and polarization [21].

Furthermore, the programmable metasurface will allow beam steering, beam splitting, wave absorption, wave polarizing, and phase control. These programmable metasurfaces can mitigate the Doppler effect in the mmWave and THz communications and deal with the stringent multipath propagation in different environments, such as indoor and outdoor environments [21].

(d) *Integrated Terrestrial, Airborne, and Satellite Networks*

The next-generation wireless communication systems, i.e., 6G, are expected to supply all-over services in substitutive remote rural not already supplied across oceans and outer space. Moreover, these communication services will establish a flawlessly integrated connectivity framework, including marine and terrestrial networks as land-based, airborne networks such as aircraft, pseudo satellites, drones, balloons, and space-based networks such as low-earth orbit (LEO) satellites, geostationary earth orbiting (GEO) satellites, non-GEO satellite constellations base [22].

Combining terrestrials and non-terrestrials (NTNs) systems is anticipated that 6G will accomplish on-demand capacity and universal multi-connectivity coverage without cell architecture coverage as needed in previous 2G up to 5G. In the 6G era, the satellite network's shape will be viewed as a state-of-the-art network junction because of the solid integration of NTNs and terrestrial systems. Furthermore, merging the designs of the couple systems will make multi-connection joint operations extra functional and extra efficient joining further flexible and fast traverse-connection switching [23].

(e) *Holographic Communication*

6G will include holographic communication, while the holographic displays are the subsequent transformation in multimedia experience carrying 3D images from one or numerous sources to various destinations, supplying an enveloping 3D experience for the end user. The holographic ability will demand extremely high data rates beyond 1 Tb/s and ultralow latency up to 0.1 ms [24].

(f) *Tactile Internet*

The TI is an area of research, including 6G wireless communications to entitle human beings to real-time interaction based on touch-related data to sense over the internet via touchable ends, with audio-visual data as a response. This new TI technology is examined as the next developmental footstep for the IoT and is anticipated to fetch a significant change in autonomous vehicles, Industry 4.0, and Healthcare 4.0 to solve complex issues in a novel society [25, 26].

(g) *Backscatter Communications and Energy Harvesting*

Different fascinating technologies for self-dependent systems in future 6G networks are suggested, engaging in backscatter communications (BackCom) systems and energy harvesting. The RF signals received at an apparatus can be engineered as an energy source, so the basic idea of BackCom is to use the incident RF signals instead of using an internally generating new carrier by reflecting it after modulating the current information over the air modulation [27].

BackCom's conception can increase communication networks' spectral efficiency and efficiency, influencing ultra-massively connected wireless networks' spectrum demands and total energy. Furthermore, the conception of the number of devices forecasted in the paradigm of massive IoT networks obligates studies to design energy-efficient wireless networks without battery or wireless-powered devices. Furthermore, in the appearing paradigm of green communications with the need for highly dense connectivity in 6G wireless networks, BackCom can supply appearing services, like massive IoT, further increased broadband, and other novel technologies. Transceiver design with extreme-high power efficiency for next-generation IoT apparatus is necessary, where energy accumulation from RF sources is a potential solution [27].

(h) *Drone-Based Communications and Autonomous Systems*

A fundamental motivity behind the perception of 6G is the implementation of connected and autonomous vehicles (CAVs) systems and drone-based communication, known as unmanned aerial vehicles (UAVs). The drone-based communication categories are mobile-enabled drones and wireless architecture drones. The application of this feather at 6G wireless communication can be in monitoring, military, rescue, agriculture, mining, and logistics. This feather in 6G wireless communication will have low cost, improved 6G network capacity, enhanced 6G coverage, flexibility, and less affected 6G wireless communications by environments [28].

1.2.3 Terahertz Wave Challenges for Sixth-Generation

Comparable to previous generations, 6G will promise to essentially alter how businesses and consumers communicate and facilitate the future generation of use cases capable of capitalizing on the capacity, latency, speed, and changeability it suggests, in addition to the artificial intelligence (AI), visualization, and parallel computing evolutions. It is anticipated that merely incorporating different technologies into a system can sufficiently address the following challenge's requirements.

(a) *Ultra Higher Data Rates*

BWs of 10 GHz and beyond are needed to accomplish the imagined wireless links data rates of a few Tb/s. However, these BWs can only be fulfilled at frequencies above 100 GHz, referred to as THz. The transmission channel for frequencies above

100 GHz, mainly for mobile medium-range (50–200 m) outdoor and small-range (up to 10 m) indoor LoS communications, is not adequately researched, and state-of-the-art channel models must be designed [29].

Additionally, the high frequencies enhance the Doppler rate when the broad BW causes hardware deteriorations such as nonlinearities and phase noise that must be offset [30, 31].

(b) *Support a Vast Number of Devices*

The current RF band has recently become inadequate to carry the demands of future 6G wireless networks. As a result, the vast number of devices and their needed data rate will reach a "spectrum crunch" when no new apparatus can connect to the network because of a lack of available BW. In addition, 6G wireless networks will connect cellular smartphones, wearables, personal devices, and IoT sensors in smart city deployments, UAVs, vehicles, and robots. So, this load increase will not sustain any user's connectivity requests [20]. As a result, 6G will be specified by a notably increased number of devices and applications with more diverse applications and needs to support many devices to supply the needful quality-of-service (QoS) quality-of-experience (QoE) levels among users and one possible solution to mitigate this problem is by using distributed UM-MIMO or cell-free networks. As a result, the ITU has predicted that connected cellular and IoT devices will climb to about 10^{11} by 2030 [32, 33].

(c) *Smart High Gain Antennas*

One of the crucial drawbacks of the THz waves is the extreme atmospheric absorption losses caused by water vapor molecular and water absorption and other atmospheric conditions, such as rain. So, to compensate for these atmospheric absorption losses, we need to design 6G wireless communication networks in a frequency band in which atmospheric absorption losses are less than 0.1 dB/m. And to allow medium-range, 100–300 m, LoS outdoor 6G wireless communication networks, high gain antennas need to design due to the lack of THz power emitter sources beyond 100 milliwatt (mW) to compensate these atmospheric absorption losses [34, 35].

(d) *Ubiquitous Availability*

Other crucial aspects of 6G will be ubiquitous availability, i.e., everywhere availability, and prolonging the coverage to achieve global connectivity. This direction will extend by interconnecting and including multiple distinct satellite orbits and network architectures. These architectures will be acquired coming up 6G, where diverse airborne, space-borne platforms, UAVs, and planes will be attached to the terrestrial 6G network on a 3D multilayer demanding a flexible radio access network (RAN), core network, and dynamic architecture. This scope of modern small networks spans established spreadings like those presently predictable for enterprise shop floors, provisional networks with different base stations, enhancing the local capacity, nomadic formations, supplying coverage for a placed margin with a place such as a construction site or music festival, up to mobile deployments as needed by public transport, maritime, logistics, and public protection and disaster relief (PPDR) [36].

(e) *Ultra-Precise Positioning and Sensing*

Three technology enablers primarily push ultra-precise positioning and sensing in 6G wireless communication networks: ultra-dense networks, collectively processed correlated multipoint transmission, and distributed UM-MIMO that generate the foundation of cell-free UM-MIMO systems. Cell-free UM-MIMO does not require a constant network, only dynamically changeable arrays of access points that ensure the user. With this alteration change, new challenges and opportunities originate. The linked integration of 6G wireless communication networks additionally arrives with supreme signaling upward to gauge positioning-tied signals. Therefore, state-of-the-art ways or a standard reference signal for positioning and communication are needed. Furthermore, THz signals with ultra-dense networks and enormous bandwidth supply the capability of ultra-accurate positioning with millimeter (mm) precision. So, there is a need to research suitable THz UM-MIMO antenna arrays, their characteristics, and their influence on positioning and sensing achievement [37].

(f) *Channel Modeling for THz Communications*

Available low-frequency channel models cannot precisely predict high-frequency THz channel links that experience extreme attenuation because of the molecular absorption and the free-space path loss. In addition, the multipath channel of THz communications contains NLoS and LoS propagations. LoS propagation attenuation, expressed by path loss, is measured by expanding the molecular absorption and spreading losses due to wave expansion. Furthermore, because of the deficiency of the LoS elements in a few schemes, the THz channel link can be defined as the NLoS scheme that can be sorted into diffracted EM waves, diffusely scattered waves, and specularly reflected waves. Then, for a proper THz channel assuming, it is needed to accurately footprint the diffraction coefficients, scattering, and reflection of the incident EM beam in the THz communication system, depending on the surface material, geometry, surface material, and incident angle. Consequently, developing an accurate and realistic THz channel model link is still a research dilemma requiring investigation to implement an effective THz 6G wireless communication system [38].

(g) *Sustainability*

6G will perform massively in sustainability, diminishing its resources, energy, and emissions footstep and enhancing sustainability in different industries and society. Using sustainability for the future 6G will stop the energy curve, i.e., dampen the magnification in energy utilization with increasing traffic, which requires the carbon footstep of mobile networks by power optimization and user equipment (UE) energy diminution. Sustainability will permit extreme-adaptable 6G wireless communication networks by lowering overhead through fast adaptation and resource allocation decisions [39].

(h) *Artificial Intelligence and Reconfigurable Capabilities*

Machine learning (ML), with deep neural networks (DNNs) and AI, is a change technology challenge that drives new research occasions in diverse areas, as well as 6G wireless communications and IoTs [40]. The growing demand for 6G communications has motivated recent research on state-of-the-art transceiver hardware architectures and relevant communication algorithms. Such hardware architectures contain vast numbers of EM radiators, motivating UM-MIMO communications. The 6G wireless communications must be reconfigurable using AI, DNNs, and ML features to mitigate this issue. They will own the intelligence and capacity to encourage the most appropriate communications tactics based on feedback from sensing and signal quality estimation [41, 42].

(i) *Cell-Free Design*

Drastic network revolutions are predictable in 6G wireless systems to move out through the traditional operation techniques to achieve improved network achievement. One feasible revolution is reformulating the approach of network cells to establish distributed UM-MIMO or cell-free networks. This kind of network's cell-free design scheme challenge combines several access points (APs) provided within specific antennas uniformly distributed in the coverage region. In addition, all the APs are joined to a centralized processing unit (CPU) to synchronize the serving of users. Therefore, it will provide a consistent service quality standard among all network users, specifically under UM-MIMO systems. Another profit is fighting the adverse effects of signal propagation, such as shadowing and signal-fading [43].

(j) *Terahertz Transceiver and Power Amplifier Designs*

One of the most significant issues regarding using the THz regime is designing a novel transceiver module because the present designs do not fit THz frequencies [44]. Therefore, designing a state-of-the-art THz band transceiver architecture has become a crucial demand. The main problems of incomplete THz transceiver hardware components are limited modulation index, phase noise, limited power amplifier (PA) source beyond about 20 decibel-milliwatts (dBm) [35], nonlinear amplifier, low-noise amplifier (LNA), limited signal detector sensitivity, band-limited convertor, limited antenna gain, and energy-efficient signal processing may seriously damage the property of transmitted signals. Additionally, digitalizing signals with significant bandwidth introduces stringent limitations on analog-to-digital converters (ADCs) and digital-to-analog converters (DACs). Hence, state-of-the-art THz transceiver technologies have evolved to fabricate THz transceivers and PA sources, particularly beyond 300 GHz. There are some promising technologies to fabricate THz transceivers and signal sources on IC, which are complementary metal-oxide-semiconductor (CMOS), bipolar CMOS (BiCMOS), silicon-germanium (SiGe), gallium-arsenide (GaAs), gallium-nitride (GaN), indium phosphide (InP), and III–V band-gap semiconductors-based heterojunction bipolar transistor (HBT), and high electron mobility transistor (HEMT) as GaN HEMT, InP HEMT, and InP HBT in electronics technologies, and photo-conductive or photomixers antennas, quantum cascade lasers (QCLs), and uni-traveling carrier photodiodes (UTCs) in photonic technologies [44].

1.2.4 Features of Terahertz Wave for Wireless Communication

(a) *Secure and Stealth Communications*

Due to extreme atmospheric absorption losses caused by water vapor absorption, we must design a high-gain antenna that leads to narrow and directional beam signals. So the THz signal resists leakage in the defense forces' short-range strategic wireless communication. Meanwhile, THz transmission via the atmosphere will have significant excess fading losses, and the THz wave is difficult to trace in the long-range, so it can be used to secure and stealth military wireless communication [45].

(b) *Signal Propagation via Reflection and Penetration*

The transmit THz beam signals can propagate through both penetration and reflection, where the penetration may be through a lens, passive optics, and mirrors like the "photonic" aspect of the THz transmit beam. However, on the other hand, reflection and diffraction are like metallic sheets, walls, buildings, and concrete, like the "electronic" aspects of the THz beam. Therefore, we can simultaneously use the NLoS diffracted EM waves for THz wireless communication as LoS propagates signals [46].

(c) *Material Transparency*

The THz radiation beam can propagate via nonpolar and nonmetallic materials, facilitating THz systems to "see-through," eliminating obstacles such as clothing, walls, packaging, layers, and interfaces obscured in dielectric materials. They thus can enable short-distance indoor 6G wireless communication, despite these obstacles [47].

(d) *In Vivo Nano-Network, Wireless Communications and Energy Harvesting*

An encouraging utilization of THz wireless communication enables active communication links through nano-sensors with EM communications at THz frequencies recognized as one of the highest applicable techniques to facilitate wireless communication between nano-sensors due to biological tissues keeping non-ionized at THz frequencies. Additionally, we get resistant to the scattering phenomenon by using THz frequencies. Furthermore, using THz frequencies for in vivo nanoscale sensor communication networks uses molecular resonance. Finally, we consider these network nodes to harvest energy power wirelessly [48].

(e) *Non-Ionizing Wireless Communications*

The 6G wireless communication, which will use the THz waves, is low-energy, non-ionizing, and thus innocuous to humans, as previous 1G up to 5G wireless communications [49, 50].

(f) *Joint THz Wireless Communication and THz Sensing Systems and THz-Tailored Architectures*

THz systems can gain an advantage from their quasi-opticality feature and apply any communication challenge to have a sensing feature, achieving numerous functions with novel THz 6G wireless communication with THz sensing applications. On the other hand, setting up THz wireless communication systems demands upper congestion of small base stations (SBSs) in a shorter communication distance to deliver multiple communication, imaging, and sensing functions and unique channel settings. These aspects cause the adoption of additional adaptable THz-tailored architectures networks that can utilize THz wireless communication systems [51].

(g) *Synergy with Lower Frequency Bands*

Incorporating THz wireless communications with sub-6 GHz and mmWave bands supplies numerous opportunities. This synergy between the THz and lower frequency bands can make future 6G wireless communication systems realize universal coverage and bring other adaptable network remedies by improving blockage prediction and sharing control information data using the links with a more reliable range and extending over the lower frequency bands [52–54].

1.2.5 Pillars of Sixth-Generation

(a) *Hyper-Connectivity*

6G will supply worldwide high-performance wireless hyper-connectivity connections and finite attempts within optical fibers speed equivalent. A peak data rate of Tb/s, experienced data rate of 20–100 Gb/s, latency up to 100 μs, cm-level imaging, mm-level localization, E2E system fidelity, and a decimal enhancement in the congestion of 5G links, established on manageable error spreading. Hyper-connectivity will facilitate human-centric enveloping ultimate services and speed up comprehensive digital revolution and productivity improvement [55, 56].

(b) *Integrated Terrestrial and Non-Terrestrials*

6G will combine terrestrial and NTNs, defined as base stations, to supply ubiquitous users with continuous, high-fidelity utilizations. Many LEO satellites or very-LEO (VLEO) satellites and high-altitude-pseudo satellites (HAPS) will be set up to compose an enormous satellite arrangement in the NTNs networks. These aerial wireless networks will enlarge the covering of the terrestrial cellular networks and allow low-latency remedies for the ultra-extended transmission range [57, 58].

(c) *Native Artificial Intelligence*

6G will magnify oneself its native AI ability. Its network and air interface designs will use AI, E2E, and ML to realize tailor-made automated operations and maintenance (O&M) and optimization. Besides that, any 6G network component will naturally incorporate communication, sensing, and computing abilities, mitigating the development of concentrated intelligence everywhere in cloud intelligence. A distributed ML engineering based on deep-edge intelligence will be vital to reaching manufacturing and future society [59, 60].

(d) *Sustainability*

A sustainable and green upgrade is the extreme goal and foundation demand of terminal and network designs in the 6G era. By incorporating a green design term and native AI ability, 6G goals to enhance the total energy efficiency by a 100 times over the network to block the total energy dissipation of information and communications technology (ICT) terminals and infrastructure from exceeding 5G, when either assures optimal experience and service performance. 6G will make distinctive contributions to humankind's sustainable development [61, 62].

(e) *Trustworthiness*

The 6G wireless network will incorporate numerous sensing, communication, intelligence, and computing abilities, making it essential to reevaluate the network architecture. The state-of-the-art network architecture should uphold trustworthiness natively and be resilient and adjusted for assignments, including distributed learning and cooperative sensing, to multiply large-scale AI implementations. Intelligence, data, and knowledge drive the 6G network architecture redesign, in which new properties will be designed to entitle native trustworthiness E2E. These involve novel data administration architectures supporting invention and compliance, quantum defense, and sophisticated individuality protection technologies [63, 64].

(f) *Networked Sensing*

6G will have a network-sensing ability. 6G wireless communications systems will incorporate wireless sensing abilities to investigate the substantial world via RF signal transmission, reflection, scattering, and echolocation. They will further supply high-resolution sensing, environment reorganization, imaging, and localization capabilities to enhance communication abilities and assist a wide range of network service schemes, laying an infrastructure for an innovative digital world [65, 66].

(g) *Improved Antenna Technology*

UM-MIMO and beam-steering antenna technologies must be improved to facilitate the use of THz frequencies, which may include metamaterial (MtM)-based antenna and orbital angular momentum and RF front-end technologies [67, 68].

(h) *Evolution in Duplex Technology*

5G introduced dynamic time-division-duplex (TDD) to duplex resilient, but there is still work to be done to eliminate the constraints that UL and downlink (DL) must use jointly, especially time-frequency implements [69, 70].

1.2.6 *Sixth-Generation Roadmap Timeline*

Since 2018, various actions have been started for 6G wireless communication research. Academic and industry meetings, such as the ITU World Radio-Communication Conference 2019 (ITU WRC-19) [71], the third-generation partnership project 5G (3GPP- 5G) R16 and R17 standards [72, 73] in the USA, South Korea, Japan, China, and Europe, have discovered the typical implementation scenarios, possibilities technologies, and primary critical abilities for the future 6G wireless network [74]. ITU-R is starting novel activities for 2030 and beyond as the dominant international organization by vision capabilities study and KPIs [75] in mid-2023. In addition, ITU-R working party 5D (WP 5D) has studied the following technology vision and trends vision for next-generation IMT-2030 standards [76]. ITU-R will fulfill the vision study following the present program by discussing the 6G frequencies spectrum in mid-2023, before the World Radio Congress 2023 (WRC-23). In addition, it will supply overall framework objectives for 6G fundamental ability needs and usage sketches. As more industry study thoroughly evaluates how these demands will influence the design of next-generation wireless communication systems, 3GPP may begin an all-inclusive study of 6G through the end of 2025. We hope the 6G specification standardization's first version will be released around 2030 and begin target distribution in about 2030 further [74]. The future 6G roadmap timeline is presented in the Fig. 1.2.

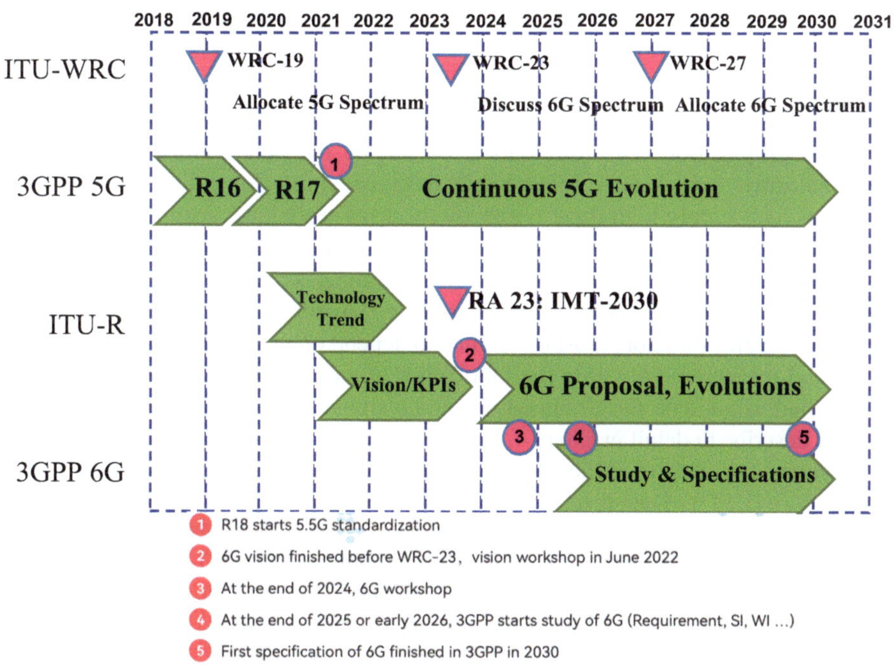

Fig. 1.2 The 6G roadmap timeline [74]

1.3 Organization of the Book

The remainder of this book is organized as follows. THz waves are the least researched and used in the EM spectrum. However, the THz waves are located between the optical frequency and mmW bands, allowing electronics-based silicon-electronics, GaAs/InP electronics, and photonics-based technologies to generate the transmit EM waves. Furthermore, the photonics-based techniques are based on Lasers diodes, modulators, and photodiodes, which are already available; these technologies will be summarized in Chap. 2. Researchers have not studied the THz waves in profundity because of the absence of developments in electronics power and transmitter sources with power over 20 dBm, high-directivity antennas, and power detectors at the indicated regime, so Chap.2 will summarize the design considerations of THz transceivers, i.e., THz transmitter and receivers to supply high-rate physical layer up to 1000 Gb/s.

It is known that the propagation channel determines the fundamental limits of wireless communication and the actual performance of any practical wireless communication system. Therefore, the best channel propagation models are essential for developing and accessing new communication systems. Furthermore, the channel models are essential to compare different system proposals by international standards. For these reasons, developing a realistic, easy-to-use model for wireless propagation channels has been an important and active research area for several decades. So Chap. 3 will discuss the channel properties due to weather and atmospheric effect, the Friss free-space path loss, blockages, delay spread, and Doppler spread. These feathers influence the link budget needed for future 6G wireless communication.

Compared to the mmWave band, the THz signals will tolerate high path losses due to the humid atmospheric conditions, which results in very short communication distance propagation. The indicated issue can be reduced using crowded Nano-antenna arrays, which provide high-directivity antennas. Some THz antenna technologies may supply high directivities by the array technique, such as horn, dielectric, Nano, Plasmonic antennas, and microstrip antennas, described in Chap. 4. Besides that, this chapter will describe the advantages and disadvantages of each antenna technology. In addition, the results apperceived with the prognosis with antenna design simulators must be verified to depend on the simulation results. So an empirical measurement within the fabrication of an archetype THz antenna should be applied to validate the designed antenna's simulation results, which is the best way to describe in detail in Chap. 4. However, solvers comparisons can be chosen at THz antenna models to reduce this validation's high cost, reaching beyond 10,000 US dollars (USD). Therefore, it is the most acceptable way to utilize a couple of mercantile EM simulation solvers that verb variously to prepare a reasonable simulation equalization where the simulation results will be similar. For example, a full-wave 3D commercial finite integration technique (FIT) solver in the Computer Simulation Technology Microwave Studio (CST MWS) simulator can be used for designing THz antennas. In contrast, other full-wave 3D commercial finite element method (FEM) solvers in the Ansys high-frequency structure simulator (HFSS) can validate the THz antennas, which are good simulators for designing THz antennas, and other THz components, which will be described in detail in Chap. 4.

The potential demanding selection of the suitable microstrip laminate is very noticeable at mmWave and THz waves frequencies due to the conductor and substrate losses, and the substrate with low relative permittivity (ε_r) and low loss tangent (tan δ) needs to be used to reduce these losses, which will be described in Chap. 5. Furthermore, an antenna with a tremendous gain is essential for 6G wireless communication to mitigate the atmosphere's tremendous path loss due to resonance with water vapor, which is one of the main concerns in the THz band. Thus, an array of antennas will enhance the microstrip antenna single radiator's gain. So, it is required to utilize a feeding network that demands appropriate minimum losses power dividers, for example, the T-junction power splitter and Wilkinson power splitter. However, at a frequency above 110 GHz, a "subminiature version a" (SMA) connector is unsuitable because of its small size, and the substrate-integrated waveguide (SIW) technologies have to be used, which will also be described in detail in Chap. 5.

Each THz antenna design starts with a unit-cell antenna design, where the unit-cell design considerations will be described in detail in Chap. 6. In addition, Chap. 6 discusses the design consideration of unit-cell frequency selective surfaces (FSSs), and the bandpass and bandstop FSSs need to be classified into two categories to find analytical formulations of the surface impedance. The first is called a grid-type configuration, while the second is called a patch-type configuration.

The array technique must enhance the unit-cell (Unit-cell) microstrip antenna gain, which cannot exceed 8 dB. The array technology can be divided into the following methods series feed, corporate feed, and hybrid feed, described in Chap. 7, to allow signal propagation and LoS medium-distance 6G wireless communication for a couple of 100 m. In addition, the FSSs techniques can be used to increase the gain of the microstrip array antenna. The FSSs are artificial structures that do not exist in nature that set up a bandstop filter and bandpass filter features to absorb, transmit, or reflect EM waves based on the EM transmit waves' frequency, which will be described in detail in Chap. 7.

Another main problem of a microstrip antenna is its limited impedance BW, which is about 2–5%, and it allows transmission of a data rate of up to 1000 Gb/s, which is also needed for 6G wireless communications. Therefore, some helpful techniques to enhance THz microstrip antenna impedance BW include parasitic patch microstrip antenna, microstrip antenna with slots, log-periodic microstrip antenna, and proximity coupled microstrip antenna feed, which is also described in detail in Chap. 7.

Another possible solution to mitigate the high propagation losses of the THz band's atmosphere is beamforming/beam-steering antennas and UM-MIMO antennas. A beamforming antenna is an electronically scanned phased array antenna controlled by a computer through phase shifters. At the same time, there are a couple of helpful beamforming technologies, such as microelectromechanical switches (MEMS) beamforming and phased array beamforming. Two sections can split up phase shifters. Analog phase shifters supply a tuneable continuous phase through a maximum and minimum phase shift angle using the Rotman lens or Butler matrix. However, digital phase shifters require different control voltages to straighten the control circuit, which causes less precise beam-steering, pendent on the number of phase bits. Furthermore, the advantages of the UM-MIMO antennas can be expanded by enhancing the system gain diversity by using an efficacious diversity antenna or transmitting numerous corresponding data streams to enhance data rates and spectral efficiency, which will be described in detail in Chap. 8.

Starting in the contemporary period of wireless communications, the spreading psychic has been noticeable, like a casual behaving unity through the receiver and the transmitter that usually reduces the received signal quality because of the not controllable reciprocal actions of the transmitted RF within the environment substances. Still, this common perception has been altered by the recent revelation of reconfigurable intelligent surfaces (RISs) and hypersurfaces (HSFs), which enable proactive control of the RF wave's refraction, scattering, and reflection characteristics conquering the harmful effects of the natural environment wireless spreading. The growing demand for 6G wireless communications has motivated recent research on state-of-the-art transceiver hardware architectures and relevant communication algorithms. Such hardware architectures contain vast numbers of EM radiators, motivating the way for UM-MIMO communications. The 6G wireless communications need reconfiguring to mitigate this issue described in Chap. 9.

Finally, designing efficacious communication networks in THz 6G wireless communication systems is more complicated than those at lower frequency systems because channels at the THz band are noted to be further insecure than those of lower frequencies. To treat this complication while promising near-real-time function, machine learning (ML) and deep learning (DL), which are pillars of artificial intelligence (AI), can be used as scalable and supple tools to foresighted improve resources for accomplishing network-level and user-level performance goals. 6G wireless networks will own the intelligence and capacity to encourage the most appropriate communications tactics based on channel activities sensing and signal quality estimation feedback by ML, DL, and AI technologies, described in Chap. 10.

In addition, to increase BW and boost data rate to new dimensions, we need to use THz communication to induce 6G wireless applications such as holographic communication and digital twinning. Moreover, THz frequencies supply access to wider BWs. These wider BWs allow us to change how we interact with our apparatus by improving aspects like gesture recognition to assist extended reality (XR)-based applications, for example, the Metaverse. XR is a term including augmented reality (AR), virtual reality (VR), and mixed reality (MR), which will also be described in Chap. 10.

1.4 Summary

Every decade, wireless communication technology has permanently evolved to meet the increasing demands of higher specifications, such as higher data rates and connections worldwide. At the same time, the 2G allowed a data rate of 64 kb/s and 5G allowed a UL data rate of up to 10 Gb/s, and a downlink data rate of up to 20 Gb/s, whereas the future 6G will allow a peak data rate beyond 1 Tb/s. Furthermore, the 1G was analog FM modulation with BW of 10 MHz, while the next 2G up to the future 6G are digital modulation, where the 6G will demand at least 10 GHz. To support the 6G data rate and connections, 6G will use sub-6 GHz, mmWave, and THz frequencies. As a result, the future 6G KIPs will supply a peak data rate beyond 1 Tb/s and a user-experienced data rate of 1 Gb/s, E2E latency of less than 1 ms, air latency of fewer than 100 μs, and enormously low jitter delay as μs. The future 6G wireless technology is anticipated to produce an output that should attain Tb/s/J, where the error rate reliability of 6G needs to be $10^{-7,}$ and the spectral efficiency needs to be 100 b/s/Hz with a mobility beyond 1000 km/h, while it will support about 107 devices per km^2.

The 6G enabling technologies will be mmWave and THz technologies because of the lack of available spectrum below 100 GHz for the BW requirements to support data rates beyond 1 Tb/s. The 6G will use free-space OWC to supply incredibly high data rate communications via short-to-medium ranges, as LoS and NLoS propagations can be accomplished. In addition, the 6G will use programmable metasurfaces technologies to manipulate the propagation of EM waves in the 6G wireless link. 6G will use integrated terrestrial, NTNs, airborne, and satellite networks to supply all-over services in substitutive remote rural not already supplied across oceans and outer space. 6G will include 3D holographic communication technologies, TI technologies, BackCom technologies, energy harvesting technologies, drone-based communications, and autonomous systems technologies.

The THz challenges will be that the BW of at least 10 GHz is needed to supply the needed data rate beyond 1 Tb/s, so mmWave, and THz frequencies should be used, so the high frequencies enhance the Doppler rate when the broad BW causes hardware deteriorations such as nonlinearities and phase noise that must be offset. In addition, 6G wireless networks will connect cellular smartphones, wearables, personal devices, and IoT sensors in smart city deployments, UAVs, vehicles, and robots. So, this load increase will not sustain any user's connectivity requests. As a result, the ITU has predicted that connected cellular and IoT devices will climb to about 10^{11} by 2030. One of the crucial drawbacks of the THz waves is the extreme atmospheric absorption losses caused by water vapor molecular absorption and other atmospheric conditions, such as rain. So to compensate for these atmospheric absorption losses, we must design 6G wireless communication networks in a frequency band in which atmospheric absorption losses are less than 0.1 dB/m.

Furthermore, to allow medium-range (50–200 m) LoS outdoor 6G wireless communication networks, high gain antennas need to be designed due to the lack of THz emitter sources beyond 100 mW to compensate for these atmospheric absorption losses. Other crucial aspects of 6G will be ubiquitous availability, i.e., availability everywhere, and prolonging the coverage to achieve global connectivity. 6G will allow ultra-precise positioning and sensing besides wireless communication, so there is a need to research suitable THz UM-MIMO antenna arrays, their characteristics, and their influence on positioning and sense achievement. Channel modeling for THz communication is needed to predict high-frequency THz channel link attenuation precisely. 6G will perform massively in sustainability, diminishing its resources, energy, and emissions footstep and enhancing sustainability in different industries and society. 6G will have ML, AI, and DNNs. 6G needs to be a cell-free or distributed UM-MIMO network design. One of the most significant issues regarding using the THz regime is designing a novel transceiver module because the present designs do not fit THz frequencies. By using THz waves, 6G will have stealth and secure communication. The THz beam signals can propagate through penetration and reflection; therefore, we can simultaneously use the NLoS diffracted EM waves for THz wireless communication as LoS propagates signals. 6G will have material transparency by using THz waves, thus can enable short-distance indoor 6G wireless communication, despite these obstacles. The 6G wireless communication, which will use the THz waves, is of low energy, non-ionizing, and thus innocuous to humans. The 6G pillars should have hyper-connectivity; 6 G will combine terrestrial and NTNs, native AI ability. 6G will make distinctive contributions to humankind's sustainable development with trustworthiness and network sensing.

References

1. E. H. Armstrong, "The spirit of discovery an appreciation of the work of Marconi," *Electrical Engineering,* vol. 72, no. 8, pp. 670–676, Aug. 1953.
2. K. P. Bandyopadhyay, "Guglielmo Marconi – The father of long-distance radio communication – An engineer's tribute," *Proceedings of IEEE 25th European Microwave Conference*, Bologna, Italy, 4 April 1995, pp. 879–885.
3. J. Rodriguez, *Fundamentals of 5G Mobile Networks.* John Wiley and Sons, pp. 1–336, 2015.
4. A. Agarwal, K. Agarwal, S. Agarwal, and G. Misra, Evolution of mobile communication technology towards 5G networks and challenges," *American Journal of Electrical and Electronic Engineering,* vol. 7, no. 2, pp. 34–37, 2019.
5. J. L. Burbank, J. Andrusenko, J.S. Everett, and W. T. M. Kasch, *Wireless Networking: Understanding Internetworking Challenges.* John Wiley and Sons, chapter 6, pp. 250–365, July 2013.
6. A. Kumar, "3G networks: opportunities and challenges," *Bulletin of Mathematical Sciences and Applications*, vol. 3, pp. 28–36, 2013.
7. M. Meraj and S. Kumar, Evolution of mobile wireless technology from 0G to 5G," *International Journal of Computer Science and Information Technologies*, vol. 6, no. 3, pp. 2545–2551, 2015.
8. R. Mishra, *Fundamentals of Network Planning and Optimization 2G/3G/4G: Evolution to 5G.* John Wiley and Sons, 2nd edition, pp. 235–294, 2018.
9. M. Liu, "Research on the development of intelligent logistics established on 5G technology," *Proceedings of IEEE 2nd International Conference on Urban Engineering and Management Science (ICUEMS)*, Sanya, China, 29–31 Jan. 2021, pp. 1–3.

10. C. Oestges and F. Quitin, *Inclusive Radio Communications for 5G and Beyond.* Academic Press, pp. 1–394, May 2021.
11. 6G-the Next Hyper-Connected Experience for All, *Samsung Research (White Paper)*, pp. 1–46, Jul. 2020.
12. Z. Chen, C. Han; Y. Wu, L. Li, C. Huang, Z. Zhang, G. Wang, and W. Tong, "Terahertz wireless communications for 2030 and beyond-A cutting-edge frontier," *IEEE Communications Magazine*, vol. 59, no. 11, pp. 66 – 72, Nov. 2021.
13. S. Elmeadawy, R. M. Shubair, "6G wireless communications: future technologies and research challenges," *Proceedings of IEEE International Conference on Electrical and Computing Technologies and Applications (ICECTA)*, Ras Al Khaimah, United Arab Emirates, 19–21 Nov. 2019, pp. 1–5.
14. 6G Vision, *Mediatek (White Paper)*, pp. 1–38, Jan. 2022.
15. A. Mourad, R. Yang, P. H. Lehne, and A. de la Oliva, "Towards 6G: evolution of key performance indicators and technology trends," *Proceedings of IEEE 2nd 6G Wireless Summit (6G SUMMIT),* Levi, Finland 17–20, Mar. 2020, pp. 1–5.
16. S. Alraih, I. Shayea, M. Behjati, R. Nordin, N. F. Abdullah, A. Abu-Samah, and D. Nandi, "Revolution or evolution? Technical requirements and considerations towards 6G mobile communications," *Sensors*, vol. 22, no. 3, pp. 762/1–19, Jan. 2022.
17. D. Soldani, "6G fundamentals: Vision and enabling technologies," *Journal of Telecommunications and the Digital Economy*, vol. *9, no.* 3, pp. 58–86, 2021.
18. L. Bariah, L. Mohjazi, S. Muhaidat, P. C. Sofotasios, G. K. Kurt, H. Yanikomeroglu, and O. A. Dobre, "A prospective look: key enabling technologies, applications, and open research topics in 6G networks," *IEEE Access*, vol. 8, pp. 174792–174820, Oct. 2021.
19. H. Tataria, M. Shafi, A. F. Molisch, M. Dohler, and H. Sjöland, F. Tufvesson, "6G wireless systems: vision, requirements, challenges, insights, and opportunities," *Proceeding of the IEEE*, vol. 109, no. 7, pp. 1166–1199, July 2021.
20. Y. Wu, S. Singh, T. Taleb, A. Roy, H. S. Dhillon, M. R. Kanagarathinam, and A. De, *6G Mobile Wireless Networks*. Springer Nature, Switzerland, 2021.
21. C. Liaskos, S. Nie, A. Tsioliaridou, A. Pitsillides, S. Ioannidis, and I. Akyildiz, "A new wireless communication paradigm through software-controlled metasurfaces," *IEEE Communication. Magazine*, vol. 56, no. 9, pp. 162–169, Sep. 2018.
22. X. Huang, J. A. Zhang, R. P. Liu, Y. J. Guo, and L. Hanzo, "Airplane-aided integrated networking for 6G wireless: Will it work?," *IEEE Vehicular Technology Magazine*, vol. 14, no. 3, pp. 84–91, Sep. 2019.
23. H. Tataria, K. Haneda, A. F. Molisch, Mansoor Shafi, and F. Tufvesson "Standardization of propagation models for terrestrial cellular systems: A historical perspective," *International Journal of Wireless Information Networks*, vol. 28, no. 1, pp. 20–44, 2021.
24. K. B. Letaief, W. Chen, Y. Shi, J. Zhang, and Y. J. A. Zhang, "The roadmap to 6G: AI-empowered wireless networks," *IEEE Communication Magazine*, vol. 57, no. 8, pp. 84–90, Aug. 2019.
25. V. Fanibhare, N. I. Sarkar, and A. Al-Anbuky, "A survey of the tactile internet: design issues and challenges, applications, and future directions," *Electronics,* vol. 10, no. 17, pp. 2171/1–35, Aug. 2021.
26. A. Aijaz, M. Dohler, A. H. Aghvami, V. Friderikos, and M. Frodigh, "Realizing the tactile internet: haptic communications over next-generation 5G cellular networks," *IEEE Wireless Communication*, vol. 24, no. 2, pp. 82–89, Apr. 2017.
27. S. J. Nawaz, S. K. Sharma, B. Mansoor, M. N. Patwary, and N. M. Khan, "Non-coherent and backscatter communications: enabling ultra-massive connectivity in 6G wireless networks," *IEEE Access*, vol. 9, pp. 38144–38186, Mar. 2021.
28. G. Amponis, T. Lagkas, M. Zevgara, G. Katsikas, T. Xirofotos, I. Moscholios, and P. Sarigiannidis, "Drones in B5G/6G networks as flying base stations," *Drones*, vol. 6, no. 2, pp. 39/1–18, Jan. 2022.

29. H. Sarieddeen, N. Saeed, T. Y. Al-Naffouri, and M. S. Alouini, "Next-generation terahertz communications: a rendezvous of sensing, imaging, and localization," *IEEE Communications Magazine*, vol. 58, no. 5, pp. 69–75, May 2020.
30. C. Castro, R. Elschner, T. Merkle, C. Schubert, and R. Freund, "Experimental demonstrations of high-capacity THz-wireless transmission systems for beyond 5G," *IEEE Communications Magazine*, vol. 58, no. 11, pp. 41–47, Nov. 2020.
31. A. A. A. Boulogeorgos T. Merkle, C. Schubert, R. Elschner, A. Katsiotis, P. Stavrianos, D. Kritharidis, P. K. Chartsias, J. Kokkoniemi, M. Juntti, J. Lehtomki, A. Teixeira, and F. Rodrigues, "Terahertz technologies to deliver optical network quality of experience in wireless systems beyond 5G," *IEEE Communications Magazine*, vol. 56, no. 6, pp. 144–151, June 2018.
32. S. Tripathi, N. V. Sabu, A. K. Gupta, and H. S. Dhillon, *Millimeter-Wave and Terahertz Spectrum for 6G Wireless*. In: Y. Wu et al. *6G Mobile Wireless Networks*. Springer, Cham Switzerland, Feb. 2021.
33. International Telecommunication Union, "IMT traffic estimates for the years 2020 to 2030," Report ITU-R M.2370-0, pp. 1–49, July 2015.
34. D. Pavlidis, *Fundamentals of Terahertz Devices and Applications*. John Wiley & Sons, Chapter 11, pp. 447–477, July 2021.
35. M. J. W. Rodwell, "Beyond 5G: 100–340GHz transistor, IC, and system design," *Short Course: Device Research Conference*, University of Michigan, USA, 23 June 2019, pp. 1–68.
36. On the Road to 6G: Drivers, Challenges and Enabling Technologies, *A Fraunhofer 6G (White Paper)*, pp. 1–15, Nov. 2021. https://www.iis.fraunhofer.de/en/ff/kom/mobile-kom/6g-sentinel/6g-sentinel-white-paper.html
37. C. Yeh, G. D. Jo, Y. J. Ko, and H. K. Chung, "Perspectives on 6G wireless communications," *ICT Express.*, vol. 9, no. 1, pp. 82–91, Feb. 2023.
38. L. Bariah, L. Mohjazi, S. Muradian, P. C. Sulfatases, and G. K. Kurt, "A prospective look: key enabling technologies, applications, and open research topics in 6G networks," *IEEE Access*, vol. 8, pp. 174792–174820, Aug. 2020.
39. European vision for the 6G network ecosystem, *The 5G IA (White Paper)*, pp. 1–51, June 2021. https://5g-ppp.eu/european-vision-for-the-6g-network-ecosystem/
40. M. H. Afsharid, A. H. Kelechi, M. A. Aubree, S. A. Chaudhry, M. S. Zia, and S. Kim, "Sixth generation (6G) wireless networks: vision, research activities, challenges, and potential solutions," *Symmetry*, vol. 12, no. 4, pp. 676/1–21, Apr. 2020.
41. N. Schlesinger, G. C. Alexandropoulos, M. F. Imani, Y. C. Elda, and D. R. Smith, "Dynamic metasurface antennas for 6G extreme massive MIMO communications," *IEEE Wireless Communications*, vol. 28, no. 2, pp. 106–113, April 2021.
42. N. O. Parchin, H. J. Basherlou, Yasir I. A. Al-Yasir, A. M. Abdulkhaleq, and R. A. Abd-Alhameed, "Reconfigurable antennas: switching techniques-a survey," *Electronics*, vol. 9, no. 2, pp. 36/1–14, May 2020.
43. A. I. Salameh and M. E. Tarhuni, "From 5G to 6G–challenges, technologies, and applications," *Future Internet*, vol. 14, no. 4, pp. 117/1–35, Apr. 2022.
44. A. A. A. Solyman and I. A. Elhaty, "Potential key challenges for terahertz communication systems," *International Journal of Electrical and Computer Engineering*, vol. 11, no. 4, pp. 3403–3409, Aug. 2021.
45. M. C. Kemp, P. F. Taday, B. E. Cole, J. A. Cluff, A. J. Fitzgerald, and W. R. Tribe, "Security applications of terahertz imaging in terahertz for military and security applications," *Proceedings of SPIE*, vol. 5070, pp. 44–52, 2003.
46. M. H. Rahaman, A. Bandyopadhyay, S. Pal, and K. P. Ray, "Reviewing the scope of THz communication and a technology roadmap for implementation," *IETE Technical Review,* vol. 38, no. 5, pp. 465–478, June 2020.
47. A. J. Seeds, H. Shams, M. J. Fice, and C. C. Renaud, "Terahertz photonics for wireless communications," *IEEE Journal of Lightwave Technology*, vol. 33, no. 3, pp. 579–587, 2015.

48. M. Usman, S. Ansari, A. Taha, A. Zahid, Q. H. Abbasi, and M. A. Imran, "Terahertz-based joint communication and sensing for precision agriculture: a 6G use-case," *Frontiers in Communications and Networks*, vol. 3, pp. 836506/1–7, Mar. 2022.
49. V. A. Revkova, I. V. Ilina, S. A. Gurova, R. O. Shatalova, M. A. Konoplyannikov, V. A. Kalsin, V. P. Baklaushev, and D. S. Sitnikov, "Effects of high-intensity non-ionizing pulses of terahertz radiation on human skin fibroblasts," *Proceedings of SPIE- Advances in Terahertz Biomedical Imaging and Spectroscopy*, vol. 11975, pp. 1–16, Mar. 2022.
50. D. S. Sitnikov, I. V. Ilina, V. A. Revkova, S. A. Rodionov, S. A. Gurova, R. O. Shatalova, A. V. Kovalev, A. V. Ovchinnikov, O. V. Chefonov, M. A. Konoplyannikov, V. A. Kalsin, and V. P. Baklaushev, "Effects of high-intensity non-ionizing terahertz radiation on human skin fibroblasts," *OPTICA: Biomedical Optics Express*, vol. 12, no. 11, pp. 7122–7138, 2021.
51. C. Chaccour, M. N. Soorki, W. Saad, M. Bennis, P. Popovski and M. Debbah, "Seven defining features of terahertz (THz) wireless systems: A fellowship of communication and sensing," *IEEE Communications Surveys & Tutorials*, vol. 24, no. 2, pp. 967–993, Second quarter 2022.
52. W. Saad, M. Bennis, and M. Chen, "A vision of 6G wireless systems: applications, trends, technologies, and open research problems," *IEEE Network*, vol. 34, no. 3, pp. 134–142, May. 2020.
53. H. Elayan, O. Amin, B. Shihada, R. M. Shubair, and M.-S. Alouini, "Terahertz band: the last piece of RF spectrum puzzle for communication systems," *IEEE Open Journal of the Communications Society*, vol. 1, pp. 1–32, Nov. 2019.
54. C. Oestges and F. Quitin, *Inclusive Radio Communications for 5G and Beyond.* Academic Press, 1st edition, Belgium, 2021.
55. W. Tong and P. Zhu, *6G The Next Horizon: From Connected People and Things to Connected Intelligence.* Cambridge University Press, pp. 1–462, Apr. 2021.
56. H. Lee, B. Lee, H. Yang, J. Kim, S. Kim, W. Shin, B. Shim, and H. V. Poor, "Towards 6G hyper-connectivity: Vision, challenges, and key enabling technologies," [arXiv], pp 1–16, Jan. 2023, https://arxiv.org/abs/2301.11111.
57. G. Araniti, A. Lera, S. Pizzi, and F. Rinaldi, "Toward 6G non-terrestrial networks," *IEEE Network*, vol. 36, no. 1, pp. 113–120, Nov. 2021.
58. Xiangming Zhu and Chunxiao Jiang, "Integrated satellite-terrestrial networks toward 6G: architectures, applications, and challenges," *IEEE Internet of Things Journal*, vol. 9, no. 1, pp. 437–461, Jan. 2022.
59. H. Yang, A. Alphones, Z. Xiong, D. Niyato, J. Zhao, and K. Wu, "Artificial-intelligence-enabled intelligent 6G networks," *IEEE Network*, vol. 34, no. 6, pp. 272–280, Nov. 2020.
60. S. Zhang and Dali Zhu, "Towards artificial intelligence-enabled 6G: State of the art, challenges, and opportunities," *Computer Networks*, vol. 183, no. 107556, pp. 1–28, Sep. 2020.
61. S. Chen, J. Zhang, Y. Jin, and B. Ai, "Wireless powered IoE for 6G: massive access meets scalable cell-free massive MIMO," *China Communications*, vol. 17, no. 12, pp. 92–109, Dec. 2020.
62. M. Matinmikko-Blue, S. Yrjölä, P. Ahokangas K. Ojutkangas, and E. Rossi, "6G and the UN SDGs: where is the connection," *Wireless Personal Communications*, vol. 121, pp. 1339–1360, Aug. 2021.
63. V. Ziegler, P. Schneider, H. Viswanathan, M. Montag, S Kanugovi, and A. Rezaki, Security and trust in the 6G era," *IEEE Access*, vol. 9, pp. 142314–142327, Oct. 2021.
64. E. R. Griffor, C. Greer, D. A. Wollman, and M. J. Burns, "Framework for cyber-physical systems: volume 1, overview, version 1," *NIST Special Publication* 1500–201, pp. 1–79, June. 2017.
65. T. Wild, V. Braun, and H. Viswanathan, "Joint design of communication and sensing for beyond 5G and 6G systems," *IEEE Access*, vol. 9, pp. 30845–30857, Feb. 2021.
66. K. B. Cooper, R. J. Dengler, N. Llombart, B. Thomas, G. Chattopadhyay, and P. H. Siegel, "THz imaging radar for standoff personnel screening," *IEEE Transactions on Terahertz Science and Technology*, vol. 1, no. 1, pp. 169–182, Sep. 2011.

67. M. H. Loukil, H. Sarieddeen, M. S. Alouini, and T. Y. Al-Naffouri, "Terahertz-band MIMO systems: Adaptive transmission and blind parameter estimation," *IEEE Communications Letters*, vol. 25, no. 2, pp. 641–645, Feb. 2021.
68. P. Lu, T. Haddad, B. Sievert, B. Khani, S. Makhlouf, S. Dülme, J. F. Estévez, A. Rennings, D. Erni, U. Pfeiffer, and A. Stöhr, "InP-based THz beam steering leaky-wave antenna," *IEEE Transactions on Terahertz Science and Technology*, vol. 11, no. 2, pp. 218–230, Mar. 2021.
69. J. Lee, M. Rim, and C. G. Kang, "Decentralized slot-ordered cross-link interference control scheme for dynamic time division duplexing (TDD) in 5G cellular system," *IEEE Access*, vol. 9, pp. 63567–63579, Apr. 2021.
70. H. Ji, Y. Kim, T. Kim, K. Muhammad, C. Tarver, M. Tonnemacher, J. Oh, B. Yu, and Gary Xu, "Enabling advanced duplex in 6G," *Proceedings of IEEE International Conference on Communications Workshops (ICC Workshops)*, Montreal, Canada, 14–23 June 2021, pp. 1–3.
71. World Radio-communication Conference 2019 (WRC-19) Final Acts, *International Telecommunication Union (ITU)*, pp. 1–567, 2019.
72. 3GPP-5G R16, [online], https://www.3gpp.org/release-16.
73. 3GPP-5G R17, [online], https://www.3gpp.org/release-17.
74. 6G: The Next Horizon Fom Connected People Things to Connected Intelligence, *Huawei Technology (White Paper)*, pp. 1–35, Nov. 2021.
75. E. C. Strinati, S. Barbarossa, J. L. G. Jimenez, D. Ktnas, N. Cassiau, L. Maret, and C. Dehos, "6G: the next frontier from holographic messaging to artificial intelligence using subterahertz and visible light communication," *IEEE Vehicular Technology Magazine*, vol. 14, no. 3, pp. 42–50, Sep. 2019.
76. ITU-R Radio-communication Study Groups 2020, *ITU Publications*, pp. 1–62, 2020.

Chapter 2
Terahertz Wireless Communication Systems

2.1 Introduction

THz wireless communication transmission is a complementary wireless technology for communication networks, allowing accelerated wireless evolution of the optical fibers beyond 5G (B5G), i.e., 6G [1].

The generation of wireless apparatus and the enlarged number of superior emerging wireless employments have raised the requirements for high data rate transmission and spectral BW transmission concerns. In addition, the wireless communication world is altering, facing the 5G era with much technological leading, such as good facilitators, full-duplexing, mmWave, and MIMO communications. Nevertheless, on the other hand, there appear to be severe restrictions on flexibly and efficiently controlling the enormous number of QoE/QoS-oriented data substituted in an inevitable big-data-driven community with an extremely high data rate and approximately zero-latency needs. Accordingly, wireless Tb/s communications and the encouraging backhaul network foundation will become the primary technology aim in the following 10 years [2, 3].

6G wireless networks are forecast to supply unmatched high-quality data rates targeting the Tb/s data rate and constitutionally facilitating an extensive dynamic range of state-of-the-art applications and service scenarios that merge these maximum data rates with reliability, agility, AI, and zero-time response. 3D printing, virtual presence, cyber-physical systems for intelligent transportation, and Industry 4.0 are highly challenging, predictable use cases [4].

Although 5G looks over prepared to facilitate some game-changing design fundamentals, such as software, virtualization, and merchandise of resources, to increase scalability, flexibility, and efficient resource uses, it can be inferred that significant rendition confines correlated to accessible BW, processing delay, transmission, and energy and cost utilization yet determine the cloak of 5G abilities [5].

U. Nissanov, G. Singh, *Antenna Technology for Terahertz Wireless Communication*,
https://doi.org/10.1007/978-3-031-35900-2_2

To tackle these obstacles in 6G wireless networks, one must bring little-investigated technologies and resources to verification and utilization. Furthermore, where the THz wireless communication is particular of those technologies, in the offing, the users in remote regions or rural that are hard to access, e.g., islands and mountains, should be linked with extreme data rates of up to 10 Gb/s per consumer. This high-speed connection is expensive or impracticable when exclusively using optical fiber remedies. Therefore, THz transmission as a wireless backhaul prolonging the optical fibers will be essential in mitigating this challenge and responsibility for 6G wireless high-speed internet access [6].

Furthermore, the growing quantity of fixed and mobile users in the industry, the free enterprise, and the customer services will demand hundreds of Gb/s in communication between backhaul, e.g., cell towers, or through fronthaul, e.g., cell towers' remote radio wireless communication. Therefore, THz wireless communication can perform in complementary schemes [7].

Additionally, THz communications are predictable to implement flawless interconnection via extreme high-speed wired networks, e.g., fiber-optic links, and private wireless apparatus, such as tablet-like apparatus and laptops, accomplishing adequate lucidity and rate merging via wired and wireless links. This transparency will moderate BW-intensive implementations beyond mobile and static users, primarily local access and indoor schemes. Some individual utilizations are HD holographic video conferencing, e.g., virtual reality (VR) or extreme-high-speed wireless data sharing in data centers [8].

Finally, ultimately maintaining digital networking in the community services, commerce, and industry, along with autonomous driving and traffic control, far health monitoring services, supply chain, safety, and security forms, production lines, and extensive production sites automation, place demanding needs for Tb/s access class subject to quick response disciplines. These cyber-physical systems schemes image the "true colors" of what is frequently known as TI, primarily defying the ability of the 6G wireless systems. So, THz wireless is a fair technology to the inferior supple and additional expensive optical-fiber connections in every scenario above [9, 10].

2.2 Wireless Communication Systems

Figure 2.1 shows the significant modules of a wireless communication transceiver system, i.e., transmitter (Tx) and receiver (Rx) modules. The input message can be a human voice, a video signal from a charged-coupled device (CCD) camera, and an email message. These messages need to convert to an electric signal, referred to as the baseband signal, by the input transducer via physical apparatus such as a microphone, CCD camera, or computer keyboard. The transmitter may include an analog-to-digital converter (ADC), modulator, upconverter mixer, high-power amplifier, and transmit antenna to convert the electric signal to EM waves to transmit through the channel. The channel is a physical medium that acts like a filter that attenuates and distorts the transmitted EM signal. The signal attenuation enhances

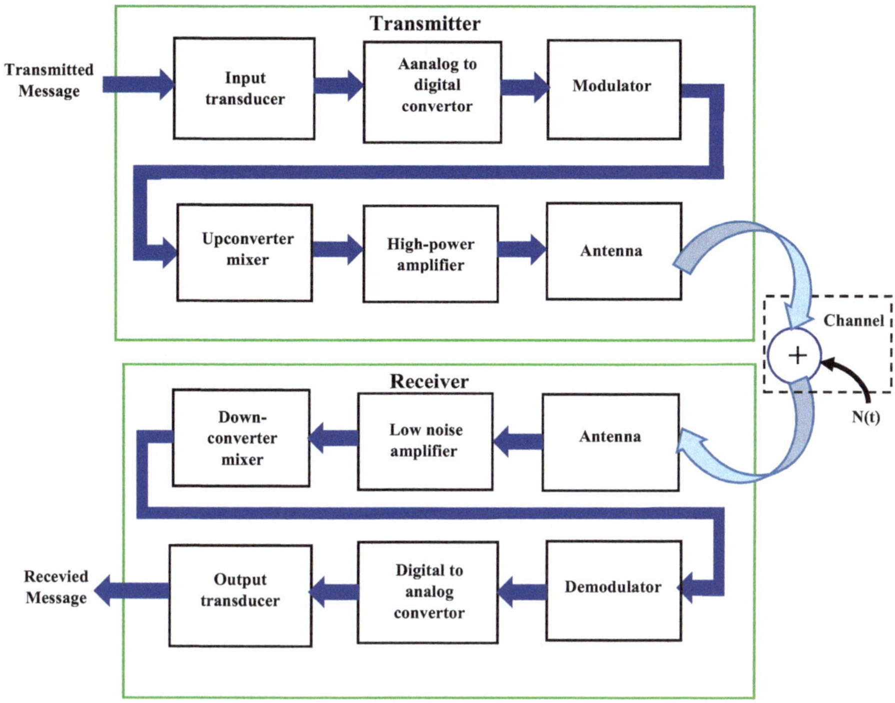

Fig. 2.1 The typical wireless communication transceiver [12]

with the length of the channel, where the signals are distorted because of path loss effects, multipath effects, shadowing effects, Doppler shift effects, and frequency-dependent gains. As a result, the receiver will receive the signal with added random and unpredictable channel interference noises N(t). The receiver includes a receiving antenna, low noise amplifier (LNA), downconverter mixer, demodulator, digital-to-analog converter (DAC), and output transducer. The receiver reprocesses the signal received from the channel by reversing the signal alteration made by the transmitter and mitigating the channel medium's distortion. The final module at the receiver is fed to the output transducer that converts the electric signal to its original message form [11, 12].

2.3 Ways to Generate Terahertz Radiation

Extensive assignment of THz communication technology applications is not widespread due to the essential restricted accessibility of THz powerful PA and THz emitter sources. THz technologies have vast possibilities in various demands, but the scarcity of concise and efficient detectors and sources piloted the THz band being named the "THz gap." Various powerful THz emitter sources, such as vacuum tube sources (VTSs), cyclotrons, or gyrotron, can supply up to 1 kW output power. Still, they are large and cannot incorporate in IC with other THz components, while they demand extremely high EM fields. So, they cannot, in any circumstances, be used for practical THz communication applications. However, due to the THz regime being located betwixt mmWave and lightwaves, several electronic and optical devices are being researched for THz communication sources [13].

On the other hand, THz emitter sources can be fabricated on three major technologies: photonic solid-state sources, VTSs, and solid-state electronic sources. Although most mobile photonic THz peak output power sources cannot be beyond 1 mW, mobile electronic sources can be more than 10 mW. However, solid-state devices' output power reduces as the signal frequency increases, so the performance of solid-state electronic sources drops dramatically above 0.5 THz. In contrast, the photonic sources endeavor to supply high powers equivalent to the all-solid-state electronic-based sources [14, 15].

Terahertz solid-state emitter sources are generally designed in two ways. The first is by THz high-frequency oscillator. The second uses a coupled microwave oscillator circuit and frequency multiplier conversion circuits, such as diode frequency multipliers or a frequency multiplier with a high-frequency transistor. In the multiplier of such conversion, the level of microwave power of the output signal on the Nth harmonic Pout is ~1/N. Diode frequency multipliers exploit the resistive and or reactive nonlinearity of the diode to achieve harmonics of an input pumping signal by supplying suitable impedances matching at any integer multiplication of the input frequency. This way, it is likely to design frequency multipliers, triplers, or more. In comparison, when using the frequency multiplier with a high-frequency transistor, an advantage is equalized with passive techniques based on diodes because they can supply additional gain besides losses. On the other hand, the significant restriction in transistor-based multipliers is the maximum operation frequency of THz emitters, which needs up to two-thirds of the cutoff frequency of the high-frequency transistors [16, 17].

III–V electronic solid-state semiconductors, such as gallium arsenide (GaAs), indium phosphide (InP), gallium nitride (GaN), and indium arsenide (InAs), can be used with other solid-state technology to design and fabricate THz emitter sources. Because these semiconductors provided a high-performance substitute for THz mobile communication and an optimum solution of higher transmit power and linearity with low power consumption, higher cutoff frequencies, and better integration with other THz components [18].

In addition, the THz emitter sources can be divided into three different types, depending on the type of radiation, i.e., continuous wave (CW) radiation, additionally named as narrowband THz source, pulsed radiation, additionally named as broadband THz source and quasi-CW radiation. As a result, relatively high peak power can be achieved when radiation in pulses of about 5 kW [19, 20].

2.3.1 Electronic Solid-State Sources

(a) *High-Frequency Gunn Effect Diodes.*

Gunn effect diodes, or transferred electron devices (TEDs), are two-terminal negative differential resistance (NDR) devices that generate microwave RF power sources. A Gunn diode uses the electron transfer effect by supplying an equaling electrical field threshold to generate electron domains transmitted through the diode-like waves. By subsequent emission generation of EM waves compatible with the time-domain traveling, Gunn diode's planar architectures supply substantial advantages from Gunn diode vertical structures as they can be incorporated in microwave monolithic integrated circuits (MMICs). In addition, these devices can work at a frequency of 100–400 GHz approximately with power beyond 100 mW (20 dBm) at 100 GHz when the power drops at higher frequencies [21–24].

(b) *High-Frequency Resonant Tunneling Diodes.*

A resonant tunneling diode (RTD) comprises dual barriers and a quantum well of a sandwich of fine semiconductor multilayers containing emitter, RTD, and collector layers. By supplying a direct current (DC) bias voltage at a specific level where the conduction edge band of the emitter is adjusted over the quantum well resonance level, the current via the RTD decreases while the applied DC bias voltage increases, so current-voltage (I–V) curvature present an NDR region. This region generates THz radiation for room-temperature environments without needing a cooling device. In addition, these devices can work between 200 GHz and 1100 GHz, where the peak power is 0 dBm (1 mW) at 200–1000 GHz when the power drops at higher frequencies [25–28].

(c) *High-Frequency Impact Ionization Avalanche Transit-Time Diodes.*

Different solid-state emitter sources are applicable for generating RF power at mmWave and microwave frequency bands. However, the highest research activities for mmWave systems using impact ionization avalanche transit-time (IMPATT) diode emitter sources are focused on window frequencies, i.e., 94, 140, and 220 GHz, where atmospheric attenuation is comparatively tiny. IMPATT diode devices are reverse DC-biased p-n junction diodes in an integrated resonant cavity activated at avalanche breakdown voltage, where they present oscillation at mmWave and microwave frequencies to generate power. As a result, silicon (Si) IMPATT diodes have appeared as the first solid-state RF power emitter sources in pulsed and CW operation modes at mmWave frequency bands. The IMPATT diodes can be fabricated with GaAs technology besides the Si technologies. In addition, these devices can work between 100 GHz and 500 GHz, where the maximum CW power is 110 mW at 140 GHz, while the maximum CW is 50 mW at 220 GHz. The maximum power is more than 1 W at 140 GHz in pulsed mode, while the maximum pulsed power is about 500 mW at 220 GHz [29–32].

(d) *High-Frequency Tunneling Transit-Time Diodes.*

Tunneling transit-time (TUNNETT) diodes are low-noise signal oscillators formed on transit-time delay and tunneling injection of electrons through a P-type semiconductor anode to an undoped drift sheet with an external metal waveguide resonator. Under reversed bias DC voltage, an appropriate phase-shift angle among the displacement current and tunneling electron current in the tunneling region of the device, the delay accomplished by the electrons in the drift region presents the NDR region, so the current via the TUNNETT diode decreases, where the applied DC bias voltage increases, so the TUNNETT diode is transformed into a signal generator. In addition, these devices can work between 100 GHz and 500 GHz, where the pulsed peak power is 1 W (30 dBm) at 100 GHz when the power drops at higher frequencies. For example, the peak power is 100 mW (20 dBm) at about 250 GHz, and the CW peak power is 140 μw at 355 GHz. Therefore, the TUNNETT diodes are appropriate as a THz emitter component source of THz communication [33–36].

(e) *High-Frequency Schottky Diodes.*

The Schottky diode is a couple-end device containing a metal-semiconductor junction, establishing a barrier to the electrons' flow. As a result, Schottky diodes are a susceptible, resilient, and trustable nonlinear device with solid achievement via the THz frequency range. In addition, Schottky diodes can operate at room temperatures, forming them appropriate for THz communication emitter sources needing rugged, compact, and low-power systems. Schottky diodes are generally used for frequency multiplication because they are commonly combined with active electronic devices such as IMPATT, Gunn oscillators, or transistor amplifiers. These devices can work between 100 GHz and 2500 GHz, where the peak output power can be 110 mW at 220 GHz, while at 1640 GHz, the peak power is 0.7 mW, whereas, at 2500 GHz, the peak power is 50 μW [37–40].

(f) *High-Frequency Heterojunction Bipolar Transistors.*

The heterojunction bipolar transistor (HBT) is a bipolar junction transistor (BJT) that uses different III–V electronic semiconductors to conceive a heterostructure-junction for the base and emitter regions. The HBT technology upgrades the BJT technology to operate between 200 GHz and 1000 GHz. The peak power signal stage InP HBTs is ten mW at 200 GHz, while the peak power is one mW at 600 GHz [41, 42].

InP HBTs with MMICs at a frequency of 210 GHz achieved maximum power with 16 PAs, and suitable Wilkinson power splitters are more than 200 mW. Therefore, InP HBTs with MMICs are ideal for THz power emitter sources with PAs [43, 44].

(g) *High-Frequency Complementary Metal-Oxide-Semiconductor.*

The silicon-complementary metal-oxide-semiconductor (CMOS) fabrication technology with coupled oscillators and voltage control oscillator (VCO) can be used to design and fabricate THz solid-state emitter between the frequency 200 GHz and 1000 GHz [45]. For example, Steyaert et al. [46] designed and fabricated a

CMOS signal generator for 540 GHz using third harmonics and based VCO with an inductor and capacitor (LC) filter and two cross-coupled transistors to generate a sinusoidal with a fundamental frequency of 180 GHz. Next, the VCO is connected to a differential amplifier, generating the third harmonic, 539.6–561.5 GHz, with a peak output power of −31 dBm (0.8 μW) at 540–543 GHz. Khamaisi et al. [47] presented THz emitter source on CMOS with Colpitts VCO. The oscillation is based on two field-effect transistors (FETs) with common gate transistors and positive feedback through the capacitor. Ikamas et al. [48] have designed and fabricated a Colpitts VCO oscillator with two FETs based on the CMOS process, where the maximum third harmonic radiated power was 43 μW, and the operating frequency was 252–266 GHz.

(h) *High-Frequency High-Electron-Mobility Transistor.*

It is feasible to accomplish high-frequency operation with considerable output power with high-electron-mobility transistors (HEMTs) technology owing to their high carrier mobility and wide bandgap [49]. HEMTs can be used as a THz emitter using frequency multipliers or oscillators. The operating frequency of THz HEMTs emitter sources can be between 200 GHz and 1000 GHz with a peak power of about 30 μW at 1000 GHz [25 and 50–51].

Deal et al. [52] demonstrated an InP HEMT frequency tripler with an input frequency source of 100 GHz to the output frequency of 300 GHz with a peak power of 0.6 mW at 300 GHz when the input power was 63 mW at 100 GHz. Zamora et al. [53] designed and fabricated a twice-frequency tripler chain (x9) in InP HEMT technology with a peak power of 6 mW at 400 GHz, while peak output power at 395 GHz was 6.9 mW when the input power was 15 mW at 44.4 GHz. Radisic et al. [54] demonstrated an MMIC common gate InP HEMT oscillator, where the oscillator was applied in coplanar waveguide (CPW) technology with a maximum output power of 0.27 mW at 330 GHz. The frequency range of this InP HEMT oscillator was 300–350 GHz.

2.3.2 Photonics Solid-State Sources

(a) *Photomixers.*

Most THz photomixers include a couple of independent diode laser emitter sources. Therefore, the frequency range of the couple lasers must permit tuning the lasers to produce the frequency variation, i.e., $f_2 - f_1$, in the THz range. These laser radiations are "mixed" in a photoconductive antenna (PCA) or an ultrafast semiconductor photodetector (PD). In addition, the THz photomixer needs careful tuning with a focusing lens to focus the CW infrared (IR) radiation of the laser diodes at the PCA surface. The second difficulty is correctly posting the PCA at the silicon (Si) lens surface [55]. Applicable photomixer devices contain PCAs, uni-traveling-carrier photodiodes (UTC-PDs), or positive-intrinsic-negative photodiodes (PIN-PDs), where the peak output power of photomixer with UTC-PDs is 0.1 mW at 250 GHz, about 11 μW at 1040 GHz, while photomixer with PIN-PD peak power is 30 μW at 250 GHz [56–58].

(b) *Quantum Cascade Lasers.*

THz quantum cascade lasers (QCLs) are unipolar semiconductor lasers dependent on interband transitions in AlGaAs/GaAs quantum wells that can be manipulated to generate the desired transmitted radiation wavelengths. These QCLs are solid, but tuning their broadband frequency is demanding, and QCLs produce THz waves above 1000 GHz. In addition, they mainly operate at low temperatures and demand a cryogenic chilling system [59]. The advantage of QCLs working as optical frequency scanners is that they can operate as emitter sources of light with a diapason that comprises center modes and simultaneously as photomixers, equipped that their gain changes are decently fast, generating microwaves of high purity spectral straight through the laser cavity [60]. For example, Scalari et al. [61] presented a QCL with a peak power of 0.1 mW at 1300 GHz at −260 °C, while Acharyya et al. [62] presented a QCL with a pulsed peak power of 2 mW at 4400 GHz at −220 °C.

2.3.3 Vacuum Tubes Sources

Because of their efficacious synergy mechanism through an EM wave and an electron beam in vacuum tubes, THz VTSs are potential applicants as high-power THz emitter sources. The VTSs were extensively used as a high-power radiation emitter source of up to 1 kW for radar systems, diverse heating systems, analysis, and communication systems. There are essential aspects to developing VTSs beyond 100 GHz, such as a high-current-density electron beam is ordinarily essential to generate high-power THz wave transmitting [63].

Some vacuum tubes can generate THz radiation, such as free-electron lasers (FELs) [64], backward wave oscillators (BWOs) [65], traveling wave tubes (TWTs) [66], Gyrotrons [67], Klystrons [68], and Magnetrons [69]. These VTSs can work between 100 GHz and 10,000 GHz and supply peak power between 10 and 100 W at 100–500 GHz [64–69].

To summarize this paragraph, Fig. 2.2 shows the current (Jan. 2021) THz emitter sources' output power vs. the frequency [25]. While the attached Table 2.1 shows the THz power sources technology comparison with maximum output power sources and working frequency range. From Table 2.1, it may have been shown that the peak power of solid-state devices is about 110 mW (20.4 dBm) at 220 GHz when the power drops at higher frequencies.

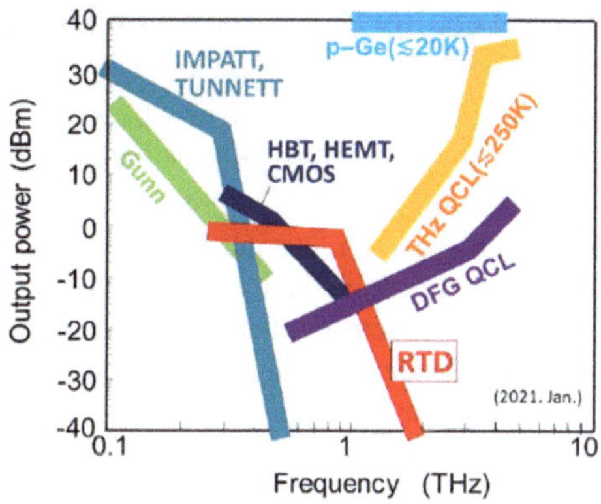

Fig. 2.2 THz emitter sources' current status (January 2021) graph [25]

Table 2.1 THz power sources technology comparison

Ref.	Technology	Maximum CW output power	Possible frequency range [GHz]
[21–24]	Gunn diodes	>100 mW @ 100 GHz	100–400
[25–28]	RTDs	1 mW @ 1000 GHz	200–1100
[29–32]	IMPATT diodes	110 mW @ 140 GHz	100–500
[33–36]	TUNNETT diodes	100 mW @ 250 GHz	100–500
[37–40]	Schottky diodes	110 mW @ 220 GHz	100–2500
[41, 42]	HBT	10 mW @ 200 GHz 1 mW @ 600 GHz	200–1000
[43, 44]	InP HBTs MMIC with 16 PAs	>200 mW @ 220 GHz	
[45, 46]	CMOS with coupled FETs VCO	0.8 μW @ 543 GHz	200–1000
[47, 48]	CMOS with Colpitts VCO	43 μW @ 260 GHz	
[49–51]	HEMTs	~30 μW @ 1000 GHz	200–1000
[52]	InP HEMT tripler	0.6 mW @ 300 GHz	
[53]	InP HEMT (x9 multiplier chain)	6.9 mW @ 395 GHz	
[54]	MMIC InP HEMT oscillator	0.27 mW @ 330 GHz	
[56–58]	Photomixer with UTC-PD	0.1 mW @ 250 GHz	100–1040
		~11 μW @ 1040 GHz	
	Photomixer with PIN-PD	30 μW @ 250 GHz	
[61]	QCLs ($\simeq -240$ °C)	0.1 mW @ 1300 GHz	1000–10,000
[62]		2 mW (pulsed) @ 4400 GHz	
[64–69]	VTSs	10–100 W @ (100–500 GHz)	100–10,000

2.4 Terahertz Transceiver Design Concepts

Next-generation wireless networks will supply higher communication data rates for further users in different places. Wide frequency bands are considered the leading solution to fulfill this objective. Still, the frequency bands in the sub-6 GHz spectrum are expensive and minimal. Therefore, the 5G mobile networks have yet planned to operate in mmWave bands to facilitate larger BWs [70].

The THz spectrum, therefore, defines a critical opportunity to set up high data rate wireless services. Furthermore, due to the development of photonic and electronic semiconductor technologies, it is presently perspective to anticipate radio-frequency (RF) communications in the THz regime. Therefore, wireless communications in the THz bands are expected to be based on B5G wireless networks [71].

Nevertheless, standard transmission schemes cannot be used because they do not contemplate certain communications features of THz frequencies. Furthermore, using a new spectrum necessitates different technological challenges and constraints involving the substantial phase disorders caused by high-frequency oscillators, the upper sampling rates needed by the ADCs, the severe signal propagation conditions losses, and the demanding design of high-gain antennas [72–74].

Further research is needed to design practical physical layer innovations to accomplish high-rate THz communications. It has been highlighted that several design criteria originate from the physical layer of THz transceiver systems: low complexity vs. spectral-efficient. Current THz communication systems primarily center on spectral-efficient systems joining coherent THz transceivers with channel bonding. In this case, the fundamental goal is to actualize a significant communication bit rate without constraints on the THz transceiver's complexity or power dissipating. High-capacity backhauling is a predicted application for spectral-efficient THz communication systems [75].

Nevertheless, designing cost-efficient THz communication systems with energy or complexity restrictions with impulse radio transceivers or spatial multiplexing is regarded as implementing high-rate communications with limited transceiver complexity, so electronic and photonic technologies are upgraded to facilitate the communication transceivers operating at THz frequencies [76].

Sarmah et al. [77] presented a 240 GHz in a 0.13 nm SiGe BiCMOS integrated quadrature transceiver scheme with a transmit peak power of 0.36 mW and noise fig. (NF) of 15 dB, while Wang et al. [78] presented a 210 GHz CMOS transceiver scheme with on-off-keying (OOK) modulation, with an effective isotropic radiated power (EIRP) of 3.3 mW and an NF of 11 dB.

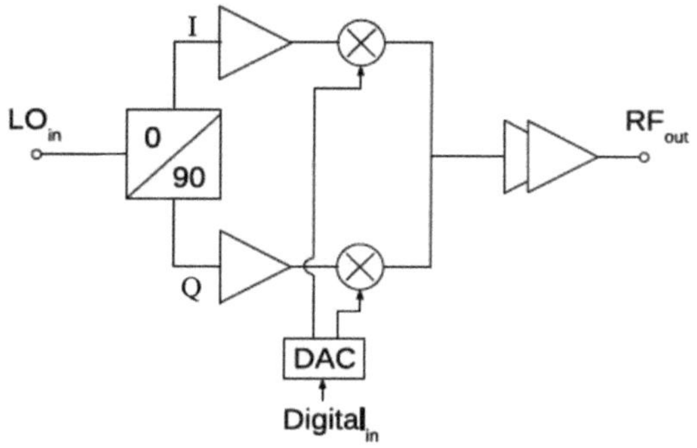

Fig. 2.3 The traditional RF analog-based IQ transmitter [79]

2.5 Terahertz Transmitter Design Concepts

Developing high-performance-integrated THz transmitters to process data rates of a couple hundred Gb/s for 6G wireless communication networks is necessary. So, to stand the specifications of high-speed data 6G wireless communication, it needs to reduce the power consumption in the THz transmitters as feasible. The traditional RF analog-based in-phase/quadrature (IQ) transmitter is attached in Fig. 2.3.

Due to the needed data rate for 6G wireless communication networks, a traditional RF analog-based IQ transmitter with a single-step upconversion mixer needs a high-performance DAC sample rate, raising the DAC's energy consumption in a traditional RF analog-based IQ transmitter. In addition, upconversion mixers should be highly linear to block distortion of the analog signal baseband after this upconversion [79].

A direct digital THz carrier modulation concept must be used to compete for the problems introduced in a conventional RF architecture. The most straightforward and also generally used in high-speed transmitter technique is amplitude shift keying (ASK) modulation [80–81], or more advanced high-order digital modulation such as quadrature amplitude modulation (QAM), 16-QAM, 32-QAM, and 64-QAM, which have high spectral efficiencies [82–85]. For example, Wang et al. [86] demonstrated a total spectral efficiency of more than 2.8 bit/s/Hz at a 340 GHz wireless transmitter supporting 16-QAM. In comparison, König et al. [87] demonstrated a 237.5 GHz wireless transmitter with spectral efficiency of 4 bit/s/Hz when the transmitter was photonic and the receiver was electronic. While Hermelo et al. [88] demonstrated a 325 GHz coherent-radio-over-fiber (CRoF) wireless transmitter using a 64-QAM and orthogonal-frequency-division-multiplexing (OFDM) signals and total spectral efficiency of 5.9 bit/s/Hz for BW of 10 GHz is achieved, which yield a peak data rate of 59 Gb/s.

2.5.1 High-Speed Analog to Digital Converter Design Concepts

The demand for a higher sample rate at the THz wireless communication requires extreme-high-speed and high BW ADCs, one of the critical units in the direct digital THz wireless communication transmitter. This ADC converts the analog signal to a digital signal needed for a digital-based IQ transmitter [89]. Many characteristics determine the function and quality of ADCs, such as BW, sampling rate, the effective number of bits (ENOBs), and resolution [90–91]. On the other hand, the current electronic ADCs maintain a few technical and fundamental limitations that may relate to different kinds of noise, nonlinearities, provisional jitter, and the moderately low ineffectualness of archetypal architectures, which are growing antiquated and not matching the current needs of THz telecommunications electronics. These limitations prevent the elevation of analog-to-digital conversion rate, and correspondingly, there is no capability to implement reliable methods to the electronic design for THz wireless communications [92–95].

Khatib et al. [96] demonstrated an antenna-coupled FET detector with one-bit ADC at 325–375 GHz for the THz imaging detectors, while Aghadjani et al. [97] demonstrated a THz Mach-Zehnder interferometer (MZI) ADC utilizing single-sided spoof-surface Plasmon polariton (SSPP) waveguide, which can support sampling rate above 45 Giga-sampling per second (GS/s). In contrast, Fang et al. [98] demonstrated a 320 GHz ADC using photonic-electronic at the same integrated circuit (IC). As a result, photonic and electronic combinations for high-speed ADC at the same IC can facilitate efficient extremely high BW and a high sampling rate for 6G wireless communication systems in the THz frequencies [98].

2.5.2 Modulator Design Concepts

The THz modulator is an essential device in the THz wireless communication system. The intelligent use of THz modulators can dramatically lower THz wireless communication's cost, complexity, and configuration. Signal modulation uses the modulation signal, the transmitted message, after converting to digital via ADC, to manage one or more parameters (phase, amplitude, and frequency) of the THz carrier signal [99]. In comparison, the THz modulators can be categorized by their physical parameter, e.g., phase, amplitude, orbital angular momentum, and spectrum, or by the various excitations used, e.g., electronic modulation, all-photonic modulation, magnetic modulation, and thermal modulation. Nevertheless, the fundamentals of the THz dynamic devices and modulator pendants in the semiconductor substance determine the device's performance and functionality. Various semiconductor substances can change diverse functionalities in manipulating the THz spatial characteristics and waves to shape different modulators, such as phase, amplitude, frequency, programmable, and polarization modulators [100].

The most straightforward to direct digital THz wireless modulation is by the ASK modulator or the OOK modulator, in which the amplitude of the THz carrier modulates the digital message. However, this modulator has a maximum total spectral efficiency below one bit/s/Hz. This total spectral efficiency means that for a data rate of 100 Gb/s, an impedance BW of more than 100 GHz is needed. Causing other demanding challenges in designing the front-end THz components, such as THz mixers, PAs, and antennas [101]. A high-order advanced digital modulator has higher spectral efficiency, reducing the needed impedance BW. QPSK modulator offers a couple of the spectral efficiency of the ASK modulator scheme, and the ability of frequency multiplexing with mild complexity would be an improved alternative for THz wireless communications [102]. QAM, 16-QAM, 32-QAM, and 64-QAM modulators offer beyond two bit/s/Hz spectral efficiencies, facilitating more than 100 Gb/s data rates with impedance BW below 40 GHz needed by future 6G wireless communication systems [103].

High-order modulators with multicarrier OFDM signals are the highest effective modulation scheme in 4G and 5G wireless communication because they allow maximum communication capacity across frequency-selective fading channels. OFDM incorporates multiplexing and modulation approaches to enhance spectral efficiency. A transmission channel is split into a lot of lesser sub-carriers or sub-channels. The sub-carrier frequencies and spacings are chosen so they are perpendicular. As a result, their spectra will not interrupt one another. Therefore, thus no guard bands are needed, where multicarrier OFDM signals mitigate the inter-symbol interference (ISI) [104].

However, the OFDM has significant challenges: sensitivity to frequency dispersion, limitation of spectral efficiency because of the cyclic prefix that mitigates delay dispersion issues, and high peak-to-average power ratio (PAPR) for highly linear PAs. These challenges outcomes are becoming more demanding at mmWave and THz frequencies since frequency dispersion escalations because of the higher phase noise and higher Doppler shifts [105]. So, high-order modulators with multicarrier OFDM signals can be used in the future 6G wireless communication because of the backward compatibility with 4G and 5G wireless communications [106].

Rout et al. [107] demonstrated multilevel ASK using THz spatial light modulators (SLM) with active MtM for 447–455 GHz. This modulator may be used at THz high-speed wireless communication. Ducournau et al. [108] demonstrated a QPSK modulator for 385 GHz using a photonics-based THz source with double heterodyne THz electronic detection. This modulator can provide a data rate of 32 Gb/s with an impedance BW of 30 GHz for THz wireless communication, while Song et al. [109] demonstrated a direct QPSK for 300 GHz. This modulator can supply a data rate of 50 Gb/s with a bit error rate (BER) of $6 \cdot 10^{-8}$ and an impedance BW of 30 GHz. Wang et al. [110] demonstrated a THz radio-over-fiber (RoF) system operating at 350 GHz with a 16-QAM modulator and OFDM signals. This THz RoF can supply a data rate of beyond 100 Gb/s over single-mode fiber (SMF) at a distance of 10 km. In contrast, Jia et al. [111] demonstrated a THz photonic wireless communication with a 64-QAM modulator with probabilistic shaping (PS) and OFDM signals at 320–380 GHz. This THz photonic wireless communication can supply a total system data rate of 610 Gb/s (2x250 Gb/s net rate) with spectral efficiency of about 4.5 bit/s/Hz per antenna.

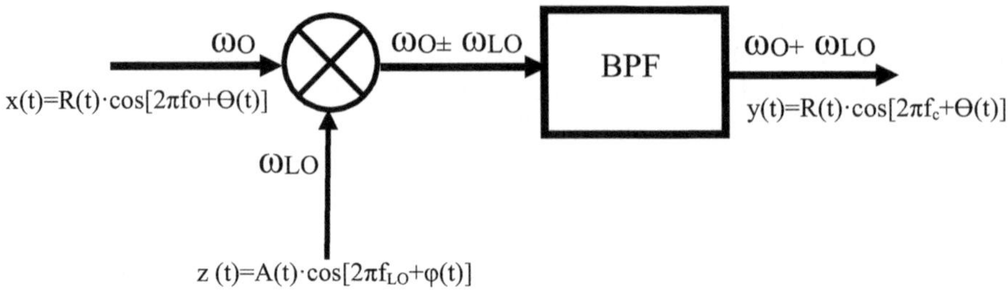

Fig. 2.4 The basic block diagram of a THz upconverter mixer [112]

2.5.3 Upconverter Mixer Design Concepts

The THz upconverter mixer is a frequency convertor used to change the input signal frequency from the THz modulator to higher output signal frequency using a mixer and bandpass filter (BPF), which is a nonlinear device such as a diode or a transistor acting as a signal multiplication; the input modulated signal is from the output THz modulator and local oscillator (LO) signal. So, a THz upconverter mixer alters the carrier frequency f_O of a modulated signal $x(t) = R(t) \cdot \cos[2\pi f_o + \Theta(t)]$ from the THz modulator to a higher frequency $f_c >> f_o$ for transmission in the designed THz spectrum without influencing the phase $\Theta(t)$ and the envelope shape $R(t)$ of the modulated signal from the THz modulator. In other words, an upconverting mixer combines a couple of frequencies or harmonics of a combination of frequencies. For instance, a higher frequency ($\omega_O + \omega_{LO}$) can be attained utilizing a couple of frequencies (ω_O, ω_{LO}). The motive is that it is less overpriced to design a THz modulator at the low intermediate frequency (IF) f_O than at the high THz frequency fc [112]. Figure 2.4 presents the basic block diagram of a THz upconverter mixer.

Shan et al. [113] demonstrated a THz upconverter-based sub-harmonic mixer (SHM) with a Schottky diode. This upconverter changed the IF input signal from 63.2–67.2 GHz to a 129–144 GHz output signal. In contrast, Erickson [114] demonstrated a THz upconverter based on the Schottky diode balanced mixer to 1500 GHz using a LO source based on an optically pumped Laser at 1.562 THz when the input IF signal was 10–40 GHz. The upconverter mixer obtained a conversion loss of about 12 dB. Finally, Deal et al. [115] demonstrated a THz upconverter using InP HEMT and GaAs Schottky diode SHM to 660 GHz when the IF input signal was about 18 GHz with low conversion loss by adding an InP HEMT amplifier to the SHM.

2.5.4 High Power Amplifier Design Concepts

One of the main components of THz transmitter wireless communication is the power amplifier (PA) used to amplify the power level of a modulated THz carrier signal to drive the transmitter antenna [79] because the PA's power level determines the signal propagation distance of the THz wireless communication channel [116] and due to the extreme path loss of the THz signal [117]. In addition, these PAs also are driving components for frequency multiplier and mixer stages. Nevertheless, small breakdown voltage and low transistor power gain restrict the PA performance for THz wireless communication. This limitation typically results in amplifiers with low output power and efficiency. Accepted techniques to enhance the solid-state power amplifier (SSPA) output power based on transistors at lower frequency bands are the amplifier cascode connection scheme, increasing the output transistor size, and power combining [118].

With transistor cascode, because there are a couple or more transistors connected on top of everyone, there is a potential to increase the power supply rail. However, the major disadvantage of this scheme at THz is the lack of certainty of the bulk influence, which enhances the uncertainty of total design. Furthermore, stability derived from the cascode transistor nodes may be developed [119].

Increasing the width of the output transistor enhances output power. Nevertheless, the interconnect parasitic effects become considerable after a specific dimension, lowering performance. The most significant detrimental effects of the transistors cascode and increasing the output transistor size are the gate interconnect resistance enhancement, increasing the transistor nodes' parasitic capacitances, which become prominent at the THz regime, and lowering the efficiency and power gain [120].

However, the power combiner scheme adds more insertion loss than the other techniques. This loss also lowers amplifier gain, efficiency, and output power. There is confinement at which power combining becomes further efficient than enhancing transistor size. That restriction is achieved when the further interconnect loss becomes more significant than the extra insertion loss of the power combining devices, such as the T-power divider and combiner. A further deficiency of the power combining scheme is increased circuit size and complexity [44 and 121].

With the cascode scheme, Yishay et al. [122] demonstrated a D-band SSPA at 110–137 GHz in SiGe BiCMOS technology. This SSPA achieved a maximum power of 60 mW at 115 GHz, while Hamada et al. [123] demonstrated a THz SSPA at 278–302 GHz with cascode and power combiner schemes at InP HEMT technology. This SSPA achieved maximum power of 16 mW at 295 GHz. Finally, Reed et al. [124] demonstrated a THz SSPA in InP HBT and MMIC technology at 210–230 GHz with a power combiner scheme, while a peak power of 180 mW at 214 GHz was achieved.

2.6 Terahertz Antenna Design Concepts

THz wireless communication's main components are the transmitter and the receiver antennas. The reasonably slow evolution of the THz band for imaging, communication, and other applications can be assigned to a duo of significant challenges. First, the attenuation level of THz EM waves in the Earth's atmosphere is extreme and originates mainly from the high absorption by water and water vapor molecules.

The extreme attenuation will confine the long-distance propagation of the THz EM waves in the air – second insufficient devices to exploit THz EM waves, including detection and generation. For example, there is a lack of solid-state THz electronic or photonic sources with sufficient output power beyond ten dBm [125], which surpasses the −20 dBm level regarded to be the minimum for serviceable THz communication applications [126]. Due to these reasons and free-space path loss (FSPL) [117], to facilitate 6G wireless outdoor communication to a line-of-sight (LoS) medium distance of 50–200 m [127], high gain transmitter and receiver antennas are needed to compensate for the high absorption by water and water vapor and FSPLs.

Furthermore, 6G wireless communication should supply data rates between 20–1000 Gb/s [128], so these antennas should be broadband to support these data rates besides the high gain antenna requirement. Two antenna technologies, such as a planar and horn antenna, can supply high gain as needed in THz wireless communication and can be used for THz wireless communication [129]. Another method scheme that can enhance capacity and moderate the high free-space path loss and high absorption loss by water vapor at the THz wireless communication is by using a THz ultra-massive multi-input-multi-output (THz UM-MIMO) planar antennas [130] or by THz beam steering planar antennas [131].

Nissanov et al. [132] demonstrated THz antennas with substrate-integrated waveguide (SIW) at 110 GHz with the planar microstrip array antenna. The peak gain and impedance BW were 26.8 dB and 12.47 GHz, while Nissanov [133] demonstrated THz beam steering with Rotman lens planar microstrip array antenna at 110–145 GHz. The peak obtained realized gain, impedance BW, and steering angles was 14 dB at 115 GHz, more than 35 GHz, and − 23.35° up to 23.76°, respectively. In contrast, Aqlan et al. [134] demonstrated a THz circularly polarized horn antenna at 270–330 GHz. The peak obtained realized gain and impedance BW was 18.4 dB at 312 GHz and more than 60 GHz, while Saxena et al. [135] demonstrated a THz MIMO planar microstrip antenna at 0.33–10 THz. The peak obtained realized gain and impedance BW was 19 dB at 6.5 THz and 9.5 THz, respectively. These antennas [132–135] can be a good candidate for future 6G wireless communication.

2.7 Terahertz Receiver Design Concepts

Digital communication modulation enciphers a bit flow of final symbol duration within a single of numerous possible transmitted symbols. In the same way, the digital communication receiver reduces the possibility of detection mistakes, despite the noises and equipment imperfections, by deciphering the received signal as the signal in the value of likely transmitted symbols that are "closest" to the one received. In addition, the receiver must maintain symbol timing with sufficient accuracy to mitigate the ISI and internal distortions, not to increase the bit-error rate (BER) [136].

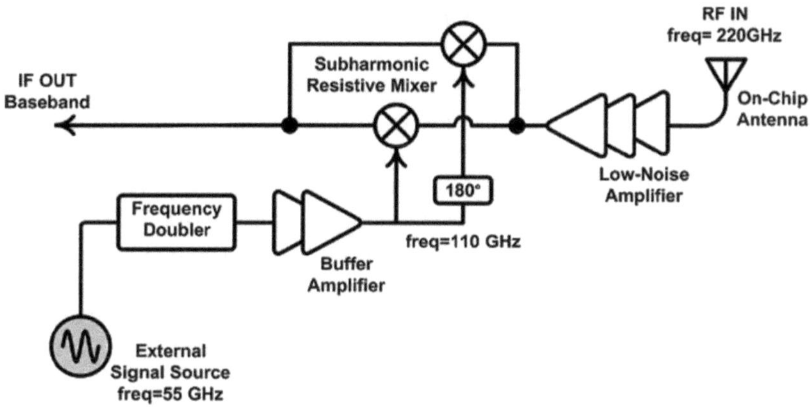

Fig. 2.5 Block diagram of the THz receiver within a low-IF heterodyne scheme [139]

6G wireless communication systems rely on efficient QAM schemes, which modulate data on a THz carrier's phase and amplitude. THz receiver can be implemented by a direct conversion (zero-IF) receiver scheme or low-IF heterodyne downconversion receiver. However, coherent downconversion receiver of analogous signals usually demands complicated THz receivers, which include a continuous-wave (CW) LO as a phase reference and a THz high-speed mixer for spectral downconversion, where a significant parameter of the receiver is the receiver sensitivity [137]. On the contrary, the dilemma with the direct-conversion architecture correlated to the standard two-step conversion scheme is the demand for a higher frequency LO signal. In addition, a direct conversion receiver frequently endures the dc offset generated by the nonlinearity of the receiver [138]. The block diagram of the THz receiver within a low-IF heterodyne scheme and improved receiver sensitivity in MMIC technology are presented in Fig. 2.5.

The THz low-IF heterodyne receiver scheme drastically enhances the receiver sensitivity and, accordingly, its response time equalizes to the direct THz receiver scheme. It is, moreover, feasible to digitize the low-IF signal instantaneously later on the RF front-end for latterly digital signal processing [139].

Dan et al. [140] demonstrated THz photonic transmitter with an active electronic receiver at 300 GHz. This photonic transmitter is based on a 64-QAM transmitter, where the receiver was with a low-IF signal heterodyne based on MMIC technology. The LO signal at the receiver was at 100 GHz after using a multiplier chain with an input signal of 8.33 GHz. The maximum data rate achieved was 100 Gb/s with a signal BW of 37.5 GHz. On the other end, Yi et al. [141] demonstrated a 300 GHz receiver including a carrier lock-in Costas loop with two mixers and a voltage control oscillator (VCO). This receiver was based on a direct conversion of a QPSK-modulated signal.

2.7.1 Low Noise Amplifier Design Concepts

A significant component of the THz receiver is the low-noise amplifier (LNA), which sets the THz receiver's noise fig. (NF). LNA is an electronic amplifier that enhances a weak power signal without substantially lowering its signal-to-noise ratio (SNR). LNAs are designed to diminish their added noise [76]. The guideline for designing THz LNAs is that the LNAs are setting the noise performance of the entire THz receiver, especially at the first stage cascode LNAs. The first stage of cascode LNAs gain should be high gain with low noise as much as can be to reduce the THz receiver's NF. If the gain is too low, it may lead to inadequate inhibition of the noise from the following stages or the loss of the gain at the first stage [125]. The following equation sets the total NF of the THz receiver [79]:

$$NF_{RX} = NF_{LNA} + (NF_{RX\ \text{without}\ LNA})/(Gain_{LNA}) \quad (2.1)$$

NF_{RX} is the receiver's NF, NF_{LNA} is the NF of the LNA, $NF_{RX\ \text{without}\ LNA}$ is the NF of the receiver without the LNA, and $Gain_{LNA}$ is the gain of the LNA. The above equation shows that high gain LNA with low NF will reduce the total NF of the THz receivers.

Gadallah et al. [142] demonstrated THz LNA in SiGe BiCMOS technology with a gain of 10.8 dB at 325 GHz while the NF was below 12.7 dB, while Liu et al. [143] proposed a D-band LNA in CMOS technology for 123–138 GHz. The LNA contained three stage common-source (CS) amplifier that achieved simulation results of a peak gain of 19.7 dB with NF 6.1 dB at 130 GHz.

2.7.2 Downconverter Mixer Design Concepts

The THz downconverter mixer is a frequency convertor used to change the input signal frequency from the LNA at the THz receiver to lower output signal frequency using a mixer. So, a THz downconverter alters the carrier frequency f_c of a modulated received signal $y(t) = R(t) \cdot \cos\ [2\pi f_o + \Theta(t)]$ from the THz LNA to a lower frequency $f_0 < <f_c$ for THz demodulation without influencing the phase $\Theta(t)$ and the envelope shape of $R(t)$. In other words, a downconverting mixer is accessible to subtract a couple of frequencies. For example, a lower frequency ($\omega_O - \omega_{LO}$) can be attained utilizing a couple of frequencies (ω_O, ω_{LO}). The motive is that designing a THz demodulator at the low IF (f_0) is less expensive than at the high THz frequency f_c. In addition, the THz downconverter mixers' performance sets the sensitivity of the THz receiver [144]. Figure 2.6 presents the basic block diagram of a THz downconverter mixer.

Grötsch et al. [145] demonstrated a THz-integrated active downconverter mixer in MMIC and InGaAs HEMT technology at 281–319 GHz for THz superheterodyne receiver. The Lo signal can be operated between 70–80 GHz, during the IF output signal can be between 65–95 GHz. The measured conversion loss without the LNA

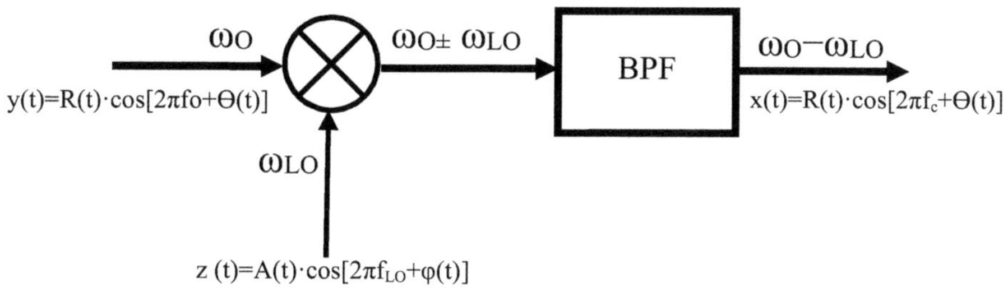

Fig. 2.6 The basic block diagram of a THz downconverter mixer [112]

was about 7.5–11 dB at 285–315 GHz, while the LO was 72–75.5 GHz, with the LNA, the conversion gain was 10–15 dB at 285–315 GHz. In addition, in CMOS technology, Lee et al. [146] demonstrated a D-band active downconversion mixer. The LO signal was 120 GHz, while the output IF signal was 2–7 GHz. The conversion gain was $-11-(-13$ dB) at 113–127 GHz.

2.7.3 Demodulator Design Concepts

The digital modulate transmitter signal in the THz transmitter should be done through an inverse process of demodulation in the THz receiver to recover precisely the phase, amplitude, or frequency of the transmit signals in the THz transmitter wireless communication, i.e., the demodulator is used to decrease the transmitter high-frequency modulated signal to a low-frequency baseband signal [147]. The design considerations of the THz demodulator are simple to design and mild complexity, low power consumption, and high noise immunity because of the channel delay, channel distortions, and phase errors. Therefore, the digital demodulator in the THz transmitter wireless communication should be with the same scheme as the modulator of the THz receiver wireless communication and can be with the ASK, OOK, or high order demodulators such as QPSK, QAM, 16-QAM, 32-QAM, 64-QAM, or high order demodulators with multicarrier OFDM signals [148–149].

Park et al. [150] demonstrated a THz CMOS transceiver for 260 GHz. This transceiver includes an OOK demodulator based on an envelope detector to derive the transmitted digital data when the IF input signal to the OOK demodulator is 65 GHz. The achieved data rate was 6 Gb/s with a BER of $4.3\cdot10^{-12}$ at an impedance BW of 12 GHz. While to achieve a high data rate, Song et al. [151] demonstrated a 300 GHz direct conversion QPSK modulator and demodulator in MMIC technology, where the data rate was 50 Gb/s with a BER of $1\cdot10^{-8}$ and an impedance BW of 30 GHz.

2.7.4 *High-Speed Digital to Analog Converter Design Concepts*

As the THz demodulator, the digitally transmitted signal via the THz transmitter should be done through an inverse process of converting the digital signal to an analog signal via the DAC in the THz receiver module. Integrating CMOS technology with high-speed InP devices permits to design high-speed and high-frequency DACs needed for THz wireless communication [152]. High-speed DACs with impedance BW of dozens of GHz with high resolutions beyond to first Nyquist criterion zone [153] facilitate a data rate beyond 100 Gb/s needed for 6G wireless communication [128] and high selectivity BPF confines the maximum synthesizable frequency, which causes distortion and demands linear post-DAC amplification and return-to-zero (RZ) DACs. Due to their rigorous timing synchronization, these demands cause difficulty in designing high-speed and high-frequency DACs [154]. Besides that, a noise shaping scheme to merge RF-DACs with an RF mixer to enhance DAC's signal spectral quality to synthesize high frequencies has been made for direct modulation transmitter, where this scheme demands fewer input bits and lowers clock jitter sensitivity for a high sampling rate and at direct demodulator receiver there is a need for ADCs [155–156]. Another approach to designing high-speed and high-frequency DACs was using a data-and-hold-interleaving scheme for beyond-Nyquist criteria, which facilitates image replica elimination and is shown to mitigate timing synchronization and gain mismatches [157]. Another helpful way to design high-speed DACs with bettered efficiency is using a power DAC that can straightly drive an optical modulator, excluding requiring PA. The sampling rate can be doubled by using this power DAC as a multiplexer because a time-inter-leaved power DAC is attained so that this DAC can be used at high-speed 6G wireless communication requirements [158].

Chen et al. [159] demonstrated a high-speed and high sample-rate DAC with clock leakage offset based on an optical discrete-multi-tone (DMT) scheme to mitigate clock leakage at a high sample-rate DAC. Using 16-QAM optical DMT, a data rate up to 121 Gb/s was achieved with a BER of $3.8 \cdot 10^{-3}$ and clock leakage compensation at 20 GHz, while Li et al. [160] demonstrated a high-speed DAC by hybrid optical access and a spectral null filling at 28 GHz and nonlinear equalizer to mitigate crosstalk. Using hybrid optical access with 4-level digital signal pulse amplitude modulation (PAM-4), spectral null filling at 28 GHz, and a nonlinear equalizer, a data rate of 56 Gb/s were achieved with a BER of $1 \cdot 10^{-3}$. Finally, to achieve a higher data rate, Kim et al. [161] demonstrated a high-speed 7-bit digital signal processor (DSP) DAC-based PAM-4 transmitters and feed-forward equalizer (FFE) in 10 nm FinFET with low-jitter CMOS clocks up to 29 GHz where the data rate was 224 Gb/s with a BER of $1 \cdot 10^{-4}$, so this DAC has acceptable resolution and dynamic range for 6G wireless communication.

2.8 Terahertz High-Rate Physical-Layer Design Concepts

6G wireless networks will be established by incorporating 5G and LTE with a variant of specific technologies, which are not based adequately to be incorporated in 5G. In addition, 6G wireless networks will be established on variant future state-of-art schemes. The physical (PHY) layer is fundamental to accomplishing unquestionable high-speed data transmission across wireless channels. At the transmitter, the bit-streams are handled over equivalent modules like channel, channel coding, signal modulation, pre-coding UM-MIMO, and high-order modulators with multicarrier OFDM signals; the reverse approach is attained at the receiver to restore the wanted digital bits. The critical parameters at the algorithmic/protocol cover emphasized coding, waveforms, and modulation to attain low latency, state-of-the-art trustworthiness, and complete multiplicity. Various compromises will be required to flawlessly assist feasible use cases such as massive machine-type communications (MMTC) and ultra-reliable low latency communication (URLLC). The source capacity can be improved using a discrete blend of full-duplex radios, interference leading established on rate decentralization, distribution, coded caching, and machine-learning-established development [76].

Accurately implementing THz bands demands designing a set of THz PHY layer schemes. Accordingly, concerning the deficiency of compatibility through classic PHY layer schemes employed and the demands of the THz wireless communications, the THz PHY layer feathers should be adequately studied concerning THz relaying, THz coding, and THz modulation as an example [7].

The THz high-rate PHY layer design will include the following design challenges to supply a high data rate of up to 1000 Gb/s as needed for future 6G wireless communication networks (a) Design 6G transmitter and receiver PHY layer at THz and mmWave bands (b) Design cell-free UM-MIMO, (C) Signal processing (SP) with accurate channel estimation and powerful selective and filtering (c) Design intelligence reflective surfaces (IRSs), which can modulate the input EM signal by reflection to desired wavefront (d) Design AI and ML with AI radio hardware architectures and security issues, which will be enablers of clandestineness in 6G wireless networks connectivity [162–163].

2.9 Open Research Challenges

This chapter has several open research challenges: (1) Design and fabrication of THz high-power solid-state sources with power beyond 100 mW (20 dBm) for at least 100–500 GHz. (2) Design and fabricate high-speed and high-sample-rate ADCs with a sample rate of beyond 46 GS/s and a data rate of 50 Gb/s for at least 100–500 GHz. (3) Design and fabricate high-order modulators with OFDM multicarrier signals with spectral efficiency beyond 4.5 bits/s/Hz for at least 100–500 GHz. (4) Design and fabricate THz upconverters with conversion loss below 12 dB for at least 100–500 GHz. (5) Design and fabricate high PAs with power beyond 180 mW (22 dBm) with large operating BW for at least 100–500 GHz. (6) THz PAs operate within high power rails, reducing the THz transceivers' total efficiency, so we need to design and fabricate high-sample rate ADCs and DACs with low energy consumption. In addition, the ADC/DAC resolution increase with BW, making their design and fabrication extremely expensive and challenging. (7) Design and fabricate THz planar microstrip antennas with gain beyond 30 dB, impedance BW beyond 35 GHz, and steering angle above ±25° for at least 100–500 GHz. (h) Design and fabricate THz LNAs with gain beyond 19.7 dB and NF below 6.1 dB for at least 100–500 GHz. (8) Design and fabricate THz downconverters with conversion gain beyond 15 dB for at least 100–500 GHz and IF for 10–100 GHz. (9) Design and fabricate THz high-order demodulators with OFDM multicarrier signals with data rate beyond 50 Gb/s and spectral efficiency beyond 4.5 bits/s/Hz for at least 100–500 GHz. (10) Design and fabricate high-speed and high sample rate DACs with data rates beyond 224 Gb/s for at least 20–500 GHz.

2.10 Summary

This chapter discusses the ways to generate THz EM radiation, which can be with THz solid-state sources, THz photonic solid-state sources, and THz VTSs. In addition, this chapter presents the major design concepts of the THz wireless communication modules in transmitter and receiver. In the end, the design concepts of the THz high-rate PHY layer are presented. To fulfill the THz high-rate PHY layer in 6G wireless networks design challenges, all of the described THz transmitter and receiver wireless communication modules should work in synergy and optimally because the 6G wireless communication network will not provide the demands when there is a failure in a specific module.

References

1. C. Oestges and F. Quitin, *Inclusive Radio Communications for 5G and Beyond*. Academic Press, first-edition, May 2021.
2. S. Tripathi, N. V. Sabu, A. K. Gupta, and H. S. Dhillon, *Millimeter-Wave and Terahertz Spectrum for 6G Wireless*. Y. Wu et al., *6G Mobile Wireless Networks*. Springer, Cham. 2021.
3. H. Sarieddeen, M. S. Alouini, and T. Y. Al-Naffouri, "Terahertz-band ultra-massive spatial modulation MIMO," *IEEE Journal on Selected Areas in Communication*, vol. 37, no. 9, pp. 1–19, June 2019.
4. E. C. Strinati, S. Barbarossa, J. L. Gonzalez-Jimenez, D. Kténas, Nicolas Cassiau, L. Maret, and C. Dehos, "6G: the next frontier: from holographic messaging to artificial intelligence using subterahertz and visible light communication," *IEEE Vehicular Technology Magazine*, vol. 14, no. 3, pp. 42–50, Sept. 2019.

5. M. Liu, "Research on the development of intelligent logistics established on 5G technology," *Proceeding of the IEEE 2nd International Conference on Urban Engineering and Management Science* (ICUEMS), Sanya, China, 29–31 Jan. 2021, pp. 1–3.
6. X. Huang, J. A. Zhang, R. P. Liu, Y. J. Guo, and L. Hanzo, "Airplane-aided integrated networking for 6G wireless: Will it work?" *IEEE Vehicular Technology Magazine*, vol. 14, no. 3, pp. 84–91, Sep. 2019.
7. K. Tekbiyik, A. R. Ekti, G. K. Kurt and A. Gorcin, "Terahertz band communication systems: challenges, novelties, and standardization efforts," *Elsevier Physical Communication*, vol. 35, pp. 100700/1–28, May 2019.
8. Y. L. Lee, D. Qin, L. C. Wang, and G. H. Sim, "6G massive radio access networks: key applications, requirements, and challenges," *IEEE Open Journal of Vehicular Technology*, vol. 2, pp. 54 – 66, Jan. 2021.
9. A. Slalmi, H. Chaibi, A. Chehri, R. Saadane, and G. Jeon, "Toward 6G: understanding network requirements and key performance indicators," *Transaction on Emerging Telecommunication Technology*, vol. 32, no. 3, pp. e4201/ 1–15, Dec. 2020.
10. H. Tataria, M. Shafi, A. F. Molisch, M. Dohler, and H. Sjöland, F. Tufvesson, "6G wireless systems: vision, requirements, challenges, insights, and opportunities," *Proceeding of the IEEE*, vol. 109, no. 7, pp. 1166–1199, July 2021.
11. B. P. Lathi and Z. Ding, *Modern Digital and Analog Communication Systems*. Oxford University Press, 4th Edition, New York, 2009.
12. Tri. T. Ha, *Theory and Design of Digital Communication*. Cambridge University Press, UK, 2011.
13. M. H. Rahaman, A. Bandyopadhyay, S. Pal, and K. P. Ray, "Reviewing the scope of THz communication and a technology roadmap for implementation," *IETE Technical Review,* vol. 38, no. 5, pp. 465–478, June 2020.
14. K. Sengupta, T. Nagatsuma, and D. M. Mittleman, "Terahertz integrated electronic and hybrid electronic–photonic systems," *Nature Electronics*, vol. 1, pp. 622–635, 2018.
15. M. A. B. Othman, "Sub-Terahertz: Generation and Detection," Ph.D. Electrical and Electronic Engineering Thesis, University of Nottingham, UK, June 2013.
16. Y. Zhang, C. Wu, X. Liu, L. Wang, C. Dai, J. Cui, Y. Li, and N. Kinar, "The development of frequency multipliers for Terahertz remote sensing system," *Remote Sensing*, vol. 14, no. 10, pep. 2486/1–26, May 2022.
17. N. Karushkin, E. Smirnova, and L. Chelyshev, "400 GHz electromagnetic radiation sources based on IMPATT diodes," *Proceeding of the ITM Web of Conferences* (CriMiCo'2019), 2019, pp. 1–3.
18. N. M. Burforda and M. O. El-Shenawe, "Review of terahertz photoconductive antenna technology," *Optical Engineering*, vol. 56, no. 1, pp. 0109011–01090120, Jan. 2017.
19. G. P. Gallerano and S. Biedron, "Overview of terahertz radiation sources," *Proceedings of the 26th International Free Electron Laser Conference and 11th FEL Users Workshop,* Washington DC USA, 12–14 Feb. 2004, pp. 216–221.
20. I. Malhotra and G. Singh, *Terahertz Antenna Technology for Imaging and Sensing Applications*. Springer Nature, Switzerland, May 2021.
21. A. Mindil, G. Dunn, A. Khalid, and C. Oxley, "Investigation of THz frequency shaped anode planar Gunn diodes operating in delayed mode," *IEEE Journal of the Electron Devices Society*, vol. 8, pp. 1072–1075, Oct. 2020.
22. A. S. Hajo, O. Yilmazoglu, A. Dadgar; F. Küppers, and T. Kusserow, "Reliable GaN-based THz Gunn diodes with side-contact and field-plate technologies," *IEEE Access*, vol. 8, pp. 84116 – 84122, Apr. 2020.
23. E. Alekseev and D. Pavlidis, " GaN Gunn diodes for THz signal generation," *Proceedings of IEEE MTT-S International Microwave Symposium Digest*, Boston, USA 11–16 June 2000, pp. 1– 5.

24. A. Khalid, G. M. Dunn, R. F. Macpherson, S. Thoms, D. Macintyre, C. Li, M. J. Steer, V. Papageorgiou, I. G.Thayne, M. Kuball, C. H. Oxley, M. Montes Bajo, A. Stephen, J. Glover, and D. R. S. Cumming, "Terahertz oscillations in an $In_{0.53}Ga_{0.47}As$ submicron planar Gunn diode," *AIP Publishing Journal of Applied Physics*, vol. 115, no. 11, pp. 1145021–1145026, 2014.
25. D. Pavlidis, *Fundamentals of Terahertz Devices and Applications*. John Wiley & Sons, Chapter 11, pp. 447–477, July 2021.
26. M. Asada and S. Suzuki, "Terahertz emitter using resonant-tunneling diode and applications," *Sensors*, vol. 21, no. 4, pp. 1384/1–19, Feb. 2021.
27. M. Villani, S. Clochiatti, W. Prost, N. Weimann, and X. Oriols, "There is plenty of room for THz tunneling electron devices beyond the transit time limit," *IEEE Electron Device Lettres*, vol. 42, no. 2, pp. 224–227, Feb. 2021.
28. A. Al-Khalidi, K. H. Alharbi, J. Wang, R. Morariu, L. Wang, A. Khalid, J. M. L. Figueiredo, and E. Wasige, "Resonant Tunneling diode terahertz sources with up to 1 mW output power in the J-band," *IEEE Transactions on Terahertz Science and Technology*, vol. 10, no. 2, pp. 150–157, March 2020.
29. A. Banerjee, B. Chakraborty, H. Inokawa, and J. Roy, *Terahertz Biomedical and Healthcare Technologies: Materials and Devices*. Elsevier, Aug. 2020.
30. T.A. Midford and R. L. Bernick, "Millimeter-wave CW IMPATT diodes and oscillators," *IEEE Transactions on Microwave Theory and Techniques*, vol. 27, no. 5, pp. 483–492, May 1979.
31. T. Ishibashi and M. Ohmori, "200 GHz 50 mW CW oscillation with Si DDR IMPATT diode," *IEEE Transactions on Microwave Theory and Techniques*, vol. 24, no. 11, pp. 858–859, Nov. 1976.
32. S. Chakraborty; A. Acharyya; A. Biswas, and A. K. Kundu, "Multi-stage-multi-iterative optimization algorithm for design optimization of multi-quantum well Terahertz IMPATT sources," *Proceeding of IEEE VLSI Device Circuit and System (VLSI DCS)*, Kolkata, India, 18–19 July 2020, pp. 129–133.
33. J. Nishizawa, P. Płotka, T. Kurabayashi, and H. Makabe, "706-GHz GaAs CW fundamental-mode TUNNETT diodes fabricated with molecular layer epitaxy," *Wiley InterScience Current Topics in Solid-State Physics,* vol. 5, no. 9, pp. 2802–2804, July 2008.
34. H. Eisele, "GaAs TUNNETT diodes and InP Gunn devices for efficient second-harmonic power generation above 200 GHz," *Proceeding of the 9th International of Space Terahertz Technology*, Pasadena California, 17–19 March 1998, pp. 587–596.
35. H. Eisele, M. Naftaly, and R. Kamoua, "Generation of submillimeter-wave radiation with GaAs TUNNETT diodes and InP Gunn devices in a second or higher harmonic mode," *International Journal of Infrared and Millimeter Waves,* vol. 26, no. 1, pp. 1–14, Jan. 2005.
36. V. V. Buniatyan, V. M. Aroutiounian, K. Zekentes, N. Camara, and P. Soukiassian, "Silicon carbide TUNNETT diodes," *Solid-State Electronics*, vol. 48, no.9, pp. 1569–1577, Sep. 2004.
37. J. V. Siles, K. B. Cooper, C. Lee, R. H. Lin, G. Chattopadhyay, and I. Mehdi, "A new generation of room-temperature frequency-multiplied sources with up to 10× higher output power in the 160-GHz–1.6-THz range," *IEEE Transactions on Terahertz Science and Technology*, vol. 8, no. 6, pp. 596–604, Nov. 2018.
38. G. Chattopadhyay, "Technology, capabilities, and performance of low power terahertz sources," *IEEE Transactions on Terahertz Science and Technology*, vol. 1, no. 1, pp. 33–53, Sep. 2011.
39. A. S. Hajo O. Yilmazoglu, F. Küppers, and T. Kusserow, "Integration and characterization of Schottky diodes with a pre-amplifier for THz applications," *Proceedings of IEEE 45th International Conference on Infrared, Millimeter, and Terahertz Waves (IRMMW-THz)*, pp. 1–2, Buffalo, NY, USA, 08–13 Nov. 2020.
40. Y. Tian, K. Huang, Y. He, L. Li, J. Cen, C. Lin, and J. Zhang, "A novel balanced frequency tripler with improved power capacity for submillimeter-wave application," *IEEE Microwave and Wireless Components Letters*, vol. 31, no. 8, pp. 925 – 928, Aug. 2021.

41. V. Radisic, D. Sawdai, D. Scott, W. R. Deal, L. Dang, D. Li, and R. Lai, "Demonstration of a 311-GHz fundamental oscillator using InP HBT technology," *IEEE Transactions of Microwave Theory and Techniques*, vol. 55, no.11, pp. 2329–2335, Nov. 2007.
42. S. Fregonese, M. Deng; M. Cabbia, C. Yadav, M. D. Matos, and T. Zimmer, "THz characterization and modeling of SiGe HBTs: Review," *IEEE Journal of the Electron Devices Society*, vol. 8, pp. 1363–1372, Nov. 2020.
43. Z. Griffith, M. Urteaga, P. Rowell, and R. Pierson, "A 23.2dBm at 210GHz to 21.0dBm at 235GHz 16-way PA-cell combined InP HBT SSPA MMIC," *Proceedings of IEEE Compound Semiconductor Integrated Circuit Symposium (CSICS)*, 19–22 October 2014, pp. 1–4.
44. M. Urteaga, Z. Griffith, M. Seo, J. Hacker, and M. J. W. Rodwell, "InP HBT technologies for THz integrated circuits," *Proceedings of the IEEE*, vol. 105, no. 6, pp. 1051–1067, June 2017.
45. Y. M. Tousi, O. Momeni, and E. Afshari, "A novel CMOS high-power terahertz VCO based on coupled oscillators: theory and implementation," *IEEE Journal of Solid-State Circuits*, vol. 47, no. 12, pp. 3032–3042, Dec. 2012.
46. W. Steyaert, and P. Reynaert, "A 0.54 THz signal generator in 40 nm bulk CMOS with 22 GHz tuning range and integrated planar antenna," *IEEE Journal of Solid-State Circuits,* vol. 49, no. 7, pp. 1617–1626, Jul. 2014.
47. B. Khamaisi, S. Jameson, E. Socher, "A 210–227 GHz transmitter with integrated on-chip antenna in 90 nm CMOS technology," *IEEE Transactions on Terahertz Science and Technology*, vol. 3, no. 2, pp. 141–150, Mar. 2013.
48. K. Ikamas, D. B. But, A. Cesiul, C. Kołacinski, T. Lisauskas, W. Knap, and A. Lisauskas, "All-electronic emitter-detector pairs for 250 GHz in silicon," *Sensors*, vol. 21, no. 17, pp. 5795/1–19, Aug. 2021.
49. G. Zhu, K. Zhang, Y. Kong, C. Li, H. Lu, X. Yu., Z. Li, and T. Chen, "Quaternary InAlGaN barrier high-electron-mobility transistors with $f_{max} > 400$GHz," *Applied Physics Express*, vol. 10, pp. 114101/1–13, Oct. 2017.
50. Y. Kwon, D. Pavlidis, P. Marsh, and T. Brock, "180 GHz InAlAs/InGaAs HEMT monolithic integrated frequency doubler," *Proceedings of IEEE 13th Annual Gallium Arsenide Integrated Circuit (GaAs IC) Symposium*, Monterey, CA, USA, 1991, pp. 165–168.
51. Y. Kwon, D. Pavlidis, and M. N. Tutt, "An evaluation of HEMT potential for millimeter-wave signal sources using interpolation and harmonic balance techniques," *IEEE Microwave and Guided Wave Letters*, vol. 1, no. 12, pp. 365–367, 1991.
52. W. Deal, X. B. Mei, K. M. K. H. Leong, V. Radisic, S. Sarkozy, and R. Lai, "THz monolithic integrated circuits using InP high electron mobility transistors," *IEEE Transactions on Terahertz Science and Technology*, vol. 1, no. 1, pp. 25–32, Sep. 2011.
53. A. Zamora, G. Mei, K. M. K. H. Leong, M. Lange, J. Lee, W. Yoshida, B. S. Gorospe, J. G. Padilla, and W. R. Deal, "A submillimeter-wave InP HEMT multiplier chain," *IEEE Microwave and Wireless Components Letters*, vol. 25, no. 9, pp. 591–593, Sep. 2015.
54. V. Radisic, L. Samoska, W. R. Deal, X.B. Mei, W. Yoshida, P. H. Liu, J. Uyeda, A. Fung, T. Gaier, and R. Lai "A 330-GHz MMIC oscillator module," *Proceedings of IEEE International Microwave Theory and Technique Symposium (MTT-S) Digest*, Atlanta, GA, USA, 15–20 June 2008, pp. 395–398.
55. E. F. Plinski, "Terahertz photomixer," *Bulletin of the Polish Academy of Sciences, Technical Sciences*, vol. 58, no. 4, pp. 463–470, 2010.
56. J. Xie, W. Ye, L. Zhou, X. Guo, X. Zang, L. Chen, and Y. Zhu, "A review on terahertz technologies accelerated by silicon photonics," *Nanomaterials*, vol. 11, no. 7, pp. 1646/1–23, June 2021.
57. H. Ito, F. Nakajima, T. Furuta, T. Ishibashi, "Continuous THz-wave generation using antenna-integrated uni-traveling-carrier photodiodes," *IOP Science*: *Semiconductor Science and Technology*, vol. 20, no. 7, pp. S191–S198, June 2005.
58. S. Nellen, T. Ishibashi, A. Deninger, R. B. Kohlhaas, L. Liebermeister, M. Schell, and B. Globish, "Experimental comparison of UTC and PIN-photodiodes for continuous-wave terahertz generation," *Journal of Infrared, Millimeter, and Terahertz Waves*, vol. 41, pp. 343–354, Oct. 2019.

59. J. Mikołajczyk, "Data link with a high-power pulsed quantum cascade laser operating at the wavelength of 4.5 μm," *Sensors*, vol. 21, no. 9, pp. 3231/1–13, May 2021.
60. M. Piccardo, M. Tamagnone, B. Schwarz, P. Chevalier, N. A. Rubin, Y. Wang, C. A. Wang, M. K. Connors, D. McNulty, A. Belyanin, and F. Capasso, "Radio frequency transmitter based on a laser frequency comb," *Proceedings of the National Academy of Sciences*, vol. 116, no. 19, pp. 9181–9185, May 2019.
61. G. Scalari, C. Walther, M. Fischer, R. Terazzi, H. Beere, D. Ritchie, and J. Faist, " THz and sub-THz quantum cascade laser," *Laser Photonics Reviews*, vol. 3, no. 1–2, pp. 45–66, Feb. 2009.
62. A. Acharyya and P. Das, *Advanced Materials for Future Terahertz Devices, Circuits, and Systems*. Springer Nature: Lecture Notes in Electrical Engineering, vol. 727, pp. 1–333, 2021.
63. H. J. Song and T. Nagatsuma, *Handbook of Terahertz Technologies Devices and Applications*. CRC Press: Taylor & Francis Group, Boca Raton, FL, pp. 1–585, 2015.
64. M. Nakajima, G. Isoyama, and T. Kurihara, "Magnetic domain control of ErFeO3 by intense terahertz free-electron laser pulses," *IEEE Transactions on Plasma Science*, vol: 49, no. 11, pp. 3344 – 3350, Nov. 2021.
65. L. Yang, W. Liu, K. Zhao, Zhi. Zhang, Zha. Zhang, Y. Ou, Z. Zhao, Z. Yang, and Y. Wei, "An efficiency-enhanced 0.5-THz BWO using double-slot embedded electron beams to drive a grating-loaded rectangular waveguide," *IEEE Transactions on Plasma Science*, vol. 50, no. 1, pp. 169–177, Jan. 2022.
66. A. E. Ploskih, N. M. Ryskin, A. A. Burtsev, A. V. Danilushkin, and I. A. Navrotsky, "Performance improvement of a sub-THz traveling-wave tube by using an electron optic system with a converging sheet electron beam," *Results in Physics*, vol. 19, pp. 793–803, Mar. 2019.
67. J. Huang, T. Song, P. Liang, X. Qi, D. Ran, C. Zhang, Z. Yan, Z. Wu, K. Zhang, T. Zhao, M. Hu, Y. Wei, Y. Gong, W. Wang, and D. Liu, "Detailed investigation on nonstationary behavior in a frequency-tunable gyrotron," *IEEE Transactions on Electron Devices*, vol. 69, no. 6, pp. 3400 – 3406, May 2022.
68. N. Guo, Q. Xue, Z. Qu, K. Liu, W. Song, X. Zhang, D. Zhao, and H. Ding, "Study of a 0.34-THz ladder-type extended interaction Klystron with narrow coupling cavities," *IEEE Transactions on Electron Devices*, vol. 68, no. 11, pp. 5851–5857, Nov. 2021.
69. A. Kuleshov, E. Khutoryan, S. Kishko, S. Ponomarenko, M. Glyavin, I. Bandurkin, V. Manuilov, A. Fedotov, T. Saito, Y. Ishikawa, Y. Tatematsu, S. Mitsudo, and T. Idehara, "Low-voltage operation of the double-beam gyrotron at 400 GHz," *IEEE Transactions on Electron Devices*, vol. 67, no. 2, pp. 673–676, Feb. 2020.
70. A. N. Uwaechia and N. M. Mahyuddin, "A comprehensive survey on millimeter-wave communications for fifth-generation wireless networks: feasibility and challenges," *IEEE Access*, vol. 8, pp. 62367–62414, Mar. 2020.
71. S. R. Moon, E. S. Kim, M. Sung, H. S. R. Moon, E. S. Kim, M. Sung, H. Y. Rha, E. S. Lee, I. M. Lee, K. H. Park, J. K. Lee, S. H. C. Y. Rha, E. S. Lee, I. M. Lee, K. H. Park, J. K. Lee, and S. H. Cho, "6G indoor network enabled by photonics- and electronics-based sub-THz technology," *IEEE Journal of Lightwave Technology*, vol. 40, no. 2, pp. 499–510, Sep. 2021.
72. Z. Sha and Z. Wang, "Channel estimation and equalization for terahertz receiver with RF impairments," *IEEE Journal on Selected Areas in Communications*, vol. 39, no. 6, pp. 1621–1635, Apr. 2021.
73. M. Khatib and M. Perenzoni, "A low-noise direct incremental A/D converter for FET-based THz imaging detectors," *Sensors*, vol. 18, no. 6, pp. 1867/1–19, June 2018.
74. U. Nissanov (Nissan), G. Singh, and A. Akinola, "Sixth-generation (6G) microstrip antenna with high-gain," *International Journal on Communications Antenna and Propagation*, vol. 11, no. 4, pp. 279–287, Aug. 2021.
75. S. Bicais, " Design of the physical layer for future sub-terahertz communication systems," Ph. D. Electrical and Electronic Engineering Thesis, University of Grenoble Alpes (UGA) France, Mar. 2021.

76. RF Enabling 6G – Opportunities and Challenges from Technology to Spectrum, *White paper 6G Research Vision*, no.13, pp. 1–68, University of Oulu, Apr. 2021.
77. N. Sarmah, J. Grzyb, K. Statnikov, S. Malz, P. R. Vazquez, W. Föerster, B. Heinemann, and U. R. Pfeiffer, "A fully integrated 240-GHz direct-conversion quadrature transmitter and receiver chipset in SiGe technology," *IEEE Transactions of Microwave Theory and Techniques*, vol. 64, no. 2, pp. 562–574, Feb. 2016.
78. Z. Wang, P. Chiang, P. Nazari, C. Wang, Z. Chen, and P. Heydari, "A CMOS 210-GHz fundamental transceiver with OOK modulation," *IEEE Journal of Solid-State Circuits*, vol. 49, no. 3, pp. 564–580, Mar. 2014.
79. N. Deferm and P. Reynaert, *CMOS Front Ends for Millimeter-Wave Wireless Communication Systems*. Springer International Publishing, Switzerland, 2015.
80. N. Deferm, J. F. Osorio, A. de Graauw, and P. Reynaert, "A 5Gb/s F-band ASK transmitter in 45nm low power CMOS," *Proceedings of the Microwave Integrated Circuits Conference (EuMIC)*, Jan. 2013, pp. 1–3.
81. G. Ducournau, F. Pavanello, A. Beck, L. Tohme, S. Blin, P. Nouvel, E. Peytavit, M. Zaknoune, P. Szriftgiser, and J. F. Lampin, "High-definition television transmission at 600 GHz combining THz photonics hotspot and high-sensitivity heterodyne receiver," *IET Electronics Letters*, vol. 50, no. 5, pp. 413–415, Feb. 2014.
82. N. Sarmah, J. Grzyb, K. Statnikov, S. Malz, P. R. Vazquez, W. Föerster, B. Heinemann, and U. R. Pfeiffer "A fully integrated 240-GHz direct-conversion quadrature transmitter and receiver chipset in SiGe technology," *IEEE Transactions on Microwave Theory and Techniques,* vol. 64, no. 2, pp. 562–574, Feb. 2016.
83. K. Katayama, K. Takano, S. Amakawa, S. Hara, A. Kasamatsu, K. Mizuno, K. Takahashi, T. Yoshida, and M. Fujishima, "A 300 GHz CMOS transmitter with 32-QAM 17.5 Gb/s/ch capability over six channels," *IEEE Journal of Solid-State Circuits*, vol. 51, no. 12, pp. 3037–3048, Sep. 2016.
84. K. Takano, S. Amakawa, K. Katayama, S. Hara, R. Dong, A. Kasamatsu, I. Hosako, K. Mizuno, K. Takahashi, T. Yoshida, and M. Fujishima, "A 105Gb/s 300GHz CMOS transmitter," *Proceedings of IEEE International Solid-State Circuits Conference (ISSCC)*, San Francisco, USA, 05–09 Feb. 2017, pp. 308–310.
85. M. Fujishima, "Future of 300 GHz band wireless communications and their enabler, CMOS transceiver technologies," *IOP Publishing: Japanese Journal of Applied Physics*, vol. 60, pp. SB08031–SB08039, Jan. 2021.
86. C. Wang, B. Lu, C. Lin, Q. Chen, X. Deng, and J. Zhang, "0.34-THz wireless link based on higher-order modulation for future wireless local area network applications," *IEEE Transactions on Terahertz Science and Technology*, vol. 4, no. 1, pp. 75–85, Jan. 2014.
87. S. Koenig, D. Lopez-Diaz, J. Antes, R. Henneberger, R. Schmogrow, D. Hillerkuss, R. Palmer, T. Zwick, C. Koos, W. Freude, O. Ambacher, I. Kallfass, and J. Leuthold, "100 Gbit/s wireless links with mm-wave photonics," *Proceedings of IEEE Optical Fiber Communication Conference and Exposition and the National Fiber Optic Engineers Conference (OFC/NFOEC)*, Anaheim USA, 17–21 March 2013, pp. 1–4.
88. M. F. Hermelo, P. T. B. Shih, M. Steeg, A. Ngoma, and A. Stöhr, "Spectral efficient 64-QAM-OFDM terahertz communication link," *Optics Express*, vol. 25, no. 16, pp. 19360–19370, Mar. 2017.
89. N. A. Borodulya, V. V. Florentsev, V. Y. Zhdamirov, R O. Rezaev, S. S. Lagunov, A. N. Tokarev, and A. P. Biryukov, "Method of analog-to-digital conversion of sub-terahertz signals by photonic time-stretched analog-to-digital conversion of continuous modulated optical waves," *Proceedings of IEEE XIV International Scientific-Technical Conference on Actual Problems of Electronic Instrument Engineering (APEIE)*, Novosibirsk, Russia, 02–06 October 2018, pp. 23–27.
90. L. Wang, C. Ren, R. Liu, and Z. Zheng, "Direction-of-arrival estimation for nested array using mixed-resolution ADCs, " *IEEE Communications Letters*, vol. 26, no. 8, pp. 1868 – 1872, Aug. 2022.

91. H. Sarieddeen, M. S. Alouini, and T. Y. Al-Naffouri, "An overview of signal processing techniques for terahertz communications, " *Proceedings of the IEEE*, vol. 109, no. 10, pp. 1628 – 1665, Aug. 2021.
92. B. Hui, "Terahertz communications," *ofinno.com*, pp. 1–9, Apr. 2021, https://1djhuc37mndh3l0ow71uejzf-wpengine.netdna-ssl.com/wp-content/uploads/2021/05/OFNO_White_Sheet_042221.pdf.
93. U. Jayamohan, "Not your grandfather's ADC: RF Sampling ADCs offer advantages in systems design," *Analog Devices technical article*, pp. 1–8, 2014, https://www.analog.com/media/en/technical-documentation/tech-articles/Not-Your-Grandfathers-ADC-RF-Sampling-ADCs-Offer-Advantages-in-Systems-Design.pdf.
94. D. Robertson. "High-speed converters: what are they, and how do they work?" *Electronic Design*, pp. 1–6, 2014, https://www.electronicdesign.com/technologies/analog/article/21800422/highspeed-converters-what-are-they-and-how-do-they-work.
95. A. Zazzi, J. Müller, S. Gudyriev, P. Marin-Palomo, D. Fang, J. C. Scheytt, C. Koos, and J. Witzens, "Fundamental limitations of spectrally-sliced optically enabled data converters arising from MLL timing jitter," *Optical Express*, vol. 28, no. 13, pp. 18790–18813, June 2020.
96. M. Khatib and M. Perenzoni, "A low-noise direct incremental A/D converter for FET-based THz imaging detectors," *Sensors*, vol. 18, no. 6, pp. 1867/1–18, June 2018.
97. M. Aghadjani and P. Mazumder, "THz analog to digital converter using single-sided spoof-surface plasmon polariton waveguide," *Proceedings of the 16th International Conference on Nanotechnology*, Sendai, Japan, 22–25 Aug. 2016, pp. 845–848.
98. D. Fang, D. Drayss, G. Lihachev, P. Marin-Palomo, H. Peng, C. Füllner, A. Kuzmin, J. Liu, R. Wang, V. Snigirev, A. Lukashchuk, M. Zhang, P. Kharel, J. Witzens, C. Scheytt, W. Freude, S. Randel, T. J. Kippenberg, and C. Koos, "320 GHz analog-to-digital converter exploiting Kerr soliton combs and photonic-electronic spectral stitching," *Proceedings of IEEE European Conference on Optical Communication (ECOC)*, Bordeaux, France, 13–16 Sept. 2021, pp. 1–4.
99. Z. T. Ma, Z. X. Geng, Z. Y. Fan, J. Liu, and H. D. Chen, "Modulators for terahertz communication: The current state of the art," *SPJ Research Journal*, vol. 2019, no. 6482975, pp. 1–22, Mar. 2019.
100. L. Wang, Y. Zhang, X. Guo, T. Chen, H. Liang, X. Hao, X. Hou, W. Kou, Y. Zhao, T. Zhou, S. Liang, and Z. Yang, "A review of THz modulators with dynamic tunable metasurfaces," *Nanomaterials*, vol. 9, no. 7, pp. 965/1–33, June 2019.
101. H. J. Song, K. Ajito, Y. Muramoto, A. Wakatsuki, T. Nagatsuma, and N. Kukutsu, "24 Gbit/s data transmission in 300 GHz band for future terahertz communications," *IET Electronics Letters*, vol. 48, no. 15, pp. 953–954, July 2012.
102. J. Treuttel, A. Maestrini, J. Sarrazin, and F. Joint, "330 GHz and 600 GHz Schottky heterodyne systems for QPSK terahertz telecommunication," *Proceedings of IEEE 25th International Conference on Telecommunications (ICT)*, Saint-Malo, France, 26–28 June 2018, pp. 291–294.
103. Z. Lai, C. Li, Y. Li, X. Lai, J. Li, and K. Guan, "A high-order modulation (64-QAM) broadband THz communication system over 100 Gbps," *Proceedings of IEEE 4th International Conference on Electronic Information and Communication Technology (ICEICT)*, Xi'an, China, 18–20
104. J. Y. Kaakinen, A. Loulou, T. Levanen, K. Pajukoski, A. Palin, M. Renfors, and M. Valkama, "Frequency-domain signal processing for spectrally-enhanced CP-OFDM waveforms in 5G new radio," *IEEE Transactions on Wireless Communications,* vol. 20, no. 10, pp. 6867–6883, May 2021.
105. N. Marchetti, M. I. Rahman, S. Kumar. and R. Prasad, *OFDM: Principles and Challenges*. V. Tarokh, (eds), *New Directions in Wireless Communications Research*. Springer, Boston, pp 29–62, Aug. 2009.

106. X. Chen, Z. Feng, Z. Wei, P. Zhang, and X. Yuan, "Code-division OFDM joint communication and sensing system for 6G machine-type communication," *IEEE Internet of Things Journal*, vol. 8, no. 15, pp. 12093–12105, Aug. 2021.
107. S. Rout and S. Sonkusale, "Wireless multi-level terahertz amplitude modulator using active metamaterial-based spatial light modulation," *Optics Express*, vol. 24, no. 13, pp. 14618-14631, Jun. 2016.
108. G. Ducournau, K. Engenhardt, P. Szriftgiser, D. Bacquet, M. Zaknoune, R. Kassi, E. Lecomte, and J. F. Lampin, "32 Gbit/s QPSK transmission at 385 GHz using coherent fiber-optic technologies and THz double heterodyne detection," *IET Electronics Letters*, vol. 51, no. 12, pp. 915–917, June 2015.
109. H. J. Song, J. Y. Kim; K. Ajito, N. Kukutsu, and M. Yaita, "50-Gb/s direct conversion QPSK modulator and demodulator MMICs for terahertz communications at 300 GHz," *IEEE Transactions on Microwave Theory and Techniques*, vol. 62, no.3, pp. 600 – 609, Mar. 2014.
110. S. Wang, Z. Lu, W. Li, S. Jia, L. Zhang, M. Qiao, X. Pang, N. Idrees, M. Saqlain, X. Gao, X. Cao, C. Lin, Q. Wu, and X. Zhang, and Xianbin Yu, "26.8-m THz wireless transmission of probabilistic shaping 16-QAM-OFDM signals," *APL Photonics*, vol. 5, pp. 0561051–0561059, Apr. 2020.
111. S. Jia, L. Zhang, S. Wang, W. Li, M. Qiao, Z. Lu, N. M. Idrees. X. Pang, H. Hu, X. Zhang, L.K. Oxenløwe, and X. Yu, "2×300 Gbit/s line rate PS-64QAM-OFDM THz photonic-wireless transmission," *IEEE Journal of Lightwave Technology*, vol. 38, no. 17, pp. 4715–4721, Sept. 2020.
112. B. P. Lathi and Z. Ding, *Modern Digital and Analog Communication Systems.* Oxford University Press, 4th Edition, pp. 184–189, 2009.
113. G. Shan, X. Zhao, H. Zhu, and C. H. Shek, "Critical components in 0.14 THz communication systems, " [Online] arXiv, pp. 1–4, 2012. https://arxiv.org/abs/1204.3014.
114. N. R. Erickson, "A Schottky-diode balanced mixer for 1.5 THz," *Proceedings of the 19th International Symposium on Space Terahertz Technology*, Groningen, 28–30 April 2008, pp. 221–223.
115. W. R. Deal, K. Leong, A. Zamora, B. Gorospe, K. Nguyen, and X. B. Mei, "A 660 GHz upconverter for THz communications," *Proceedings of IEEE Compound Semiconductor Integrated Circuit Symposium (CSICS)*, Miami, USA, 22–25 October 2017, pp. 1–4.
116. Y. Huang, Y. Shen, and J. Wang, "From terahertz imaging to terahertz wireless communications, " *Elsevier Engineering*, pp. 1– 19, Sep. 2022, https://doi.org/https://doi.org/10.1016/j.eng.2022.06.023.
117. H. T. Friss, "A note on a simple transmission formula," *Proceedings of the IRE and Waves and Electrons*, vol. 34, no. 5, pp. 254–256, May 1946.
118. D. Simic and P. Reynaert, "A 14.8 dBm 20.3 dB power amplifier for D-band applications in 40 nm CMOS," *Proceedings of IEEE Radio Frequency Integrated Circuits Symposium (RFIC)*, Philadelphia USA, 10–12 June 2018, pp. 232–235.
119. A. Ali, J. Yun, F. Giannini, H. J. Ng, D. Kissinger, and P. Colantonio, "168–195 GHz power amplifier with output power larger than 18 dBm in BiCMOS technology," *IEEE Access*, vol. 8, pp. 79299–79309, Apr. 2020.
120. L. A. Samoska, "An overview of solid-state integrated circuit amplifiers in the submillimeter-rave and THz regime," *IEEE Transactions on Terahertz Science and Technology*, vol. 1, no. 1, pp. 9–24, Sept. 2011.
121. Z. Griffith, M. Urteaga, P. Rowell, and R. Pierson, "A 23.2dBm at 210GHz to 21.0dBm at 235GHz 16-way PA-cell combined InP HBT SSPA MMIC," *Proceedings of IEEE Compound Semiconductor Integrated Circuit Symposium (CSICS)*, La Jolla, USA, 19–22 Oct. 2014, pp. 1–4.
122. R. B. Yishay and D. Elad, "A 17.8 dBm 110-130 GHz power amplifier and doubler chain in SiGe BiCMOS technology," *Proceedings of IEEE Radio Frequency Integrated Circuits Symposium (RFIC)*, Phoenix, USA, 17–19 May 2015, pp. 391–394.

123. H. Hamada, T. Tsutsumi, H. Matsuzaki, H. Sugiyama, and H. Nosaka, "Ultra-high-speed 300-GHz InP IC technology for beyond 5G," *NTT Technical Review*, vol. 19, no. 5, pp. 74–83, May 2021.
124. T. B. Reed, Z. Griffith, P. Rowell, M. Field, M. Rodwell, "A 180mW InP HBT power amplifier MMIC at 214 GHz," *Proceedings of IEEE Compound Semiconductor Integrated Circuit Symposium (CSICS)*, Monterey, USA, 13–16 Oct. 2013, pp. 1–4.
125. J. S. Rieh, *Introduction to Terahertz Electronics*. Springer Nature, Switzerland, 2021.
126. M. Piccardo, M. Tamagnone, B. Schwarz, P. Chevalier, N. A. Rubin, Y. Wang, C. A. Wang, M. K. Connors, D. McNulty, A. Belyanin, and F. Capasso, "Radio frequency transmitter based on a laser frequency comb," *Proceedings of the National Academic Science USA (PNAS)*, vol. 116, no. 19, pp. 9181–9185, May 2019.
127. H. Sarieddeen, N. Saeed, T. Y. Al-Naffouri, and M. S. Alouini, "Next-generation terahertz communications: a rendezvous of sensing, imaging, and localization," *IEEE Communications Magazine*, vol. 58, no. 5, pp. 69–75, May 2020.
128. S. Alraih, I. Shayea, M. Behjati, R. Nordin, N. F. Abdullah, A. Abu-Samah, and D. Nandi, "Revolution or evolution? Technical requirements and considerations towards 6G mobile communications," *Sensors*, vol. 22, no. 3, pp. 762/1–19, Jan. 2022.
129. K. R. Jha and G. Singh, *Terahertz Planar Antennas for Next Generation Communication*. Springer, Switzerland, 2014.
130. H. Sarieddeen, M. S. Alouini, and T. Y. Al-Naffouri, "Terahertz-band ultra-massive spatial modulation MIMO," *IEEE Journal On Areas In Communication*, vol. 37, no. 9, pp. 1–19, June 2019.
131. P. Lu, T. Haddad, B. Sievert, B. Khani, S. Makhlouf, S. Dülme, J. F. Estévez, A. Rennings, D. Erni, U. Pfeiffer, and A. Stöhr, "InP-based THz beam steering leaky-wave antenna," *IEEE Transactions on Terahertz Science and Technology*, vol. 11, no. 2, pp. 218–230, Mar. 2021.
132. U. Nissanov, G. Singh, and N. Kumar, "High gain microstrip array antenna with SIW and FSS for beyond 5G at THz band," *Optik: International Journal for Light and Electron Optics*, vol. 236c, pp. 166568/1–13, March 2021.
133. U. Nissanov, "6G Rotman lens D-band beam-steering microstrip antenna," *Journal of Computational Electronics*, pp. 1–14, Jan. 2022.
134. B. Aqlan, M. Himdi, L. L. Coq, and H. Vettikalladi, "Sub-THz circularly polarized horn antenna using wire electrical discharge machining for 6G wireless communications," *IEEE Access*, vol. 8, pp. 117245–117252, June 2020.
135. G. Saxena, Y.K. Awasthi, P. Jain, "High isolation and high gain super-wideband (0.33–10 THz) MIMO antenna for THz applications," *Optik-International Journal for Light and Electron Optics*, vol. 223, pp. 165335/1–10, July 2020.
136. A. Goldsmith, *Wireless Communications*. Cambridge University Press, USA, pp. 125–126, June 2012.
137. T. Harter, C. Fullner, J. N. Kemal, S. Ummethala, J. L. Steinmann, M. Brosi, J. L. Hesler, E. Brndermann, A. S. Muller, W. Freude, S. Randel, and C. Koos, "Generalized Kramers–Kronig receiver for coherent terahertz communications," *Photonic Letters*, vol. 14, pp. 601–606, Sept. 2020.
138. C. M. Cooke, K. Leong, A. Escorcia, X. B. Mei, T. W. Barton, M. A. Vega, D. L. Wu, W. R. Deal, "A 220 GHz dual channel LNA front-end for a direct detection polarimetric receiver," *Proceedings of IEEE MTT-S International Microwave Symposium (IMS)*, Boston, USA, 02–07 June 2019, pp. 508–511.
139. M. Abbasi, S. E. Gunnarsson, N. Wadefalk, R. Kozhuharov, J. Svedin, S. Cherednichenko, I. Angelov, I. Kallfass, A. Leuther, and H. Zirath, "Single-chip 220-GHz active heterodyne receiver and transmitter MMICs with on-chip integrated antenna," *IEEE Transactions on Microwave Theory and Techniques*, vol. 59, no. 2, pp. 466–478, Feb. 2011.
140. I. Dan, P. Szriftgiser, E. Peytavit, J. F. Lampin, M. Zegaoui, M. Zaknoune, G. Ducournau, and I. Kallfass, "A 300-GHz wireless link employing a photonic transmitter and an active electronic receiver with a transmission bandwidth of 54 GHz," *IEEE Transactions on Terahertz Science and Technology*, vol. 10, no. 3, pp. 271–281, May 2020.

141. C. Yi, S. H. Choi, and M. Kim, "300-GHz integrated heterodyne receiver with carrier lock-in Costas loop," *IET Electronics Letters*, vol. 56, no. 6, pp. 276–278, Mar. 2020.
142. A. Gadallah, M. H. Eissa, T. Mausolf, D. Kissinger, and A. Malignaggi, "A 300-GHz low-noise amplifier in 130-nm SiGe $SG_{13}G_3$ technology," *IEEE Microwave and Wireless Components Letters*, vol. 32, no. 4, pp. 331– 334, April 2022.
143. H. Liu, F. Meng, and H. Fu, "A d-band wideband high-gain low-noise amplifier in 55-nm CMOS," *Proceedings of 13th UK-Europe-China Workshop on Millimetre-Waves and Terahertz Technologies (UCMMT)*, Tianjin, China, 29 Aug. 2020–01 Sept. 2020, pp. 1–4.
144. J. M. P. Escudero, C. Quemada, R. Gonzalo, and I. Ederra "A millimeter-wave 4th-harmonic Schottky diode mixer with integrated local oscillator," *Applied Science*, vol. 11, no. 16, pp. 7238/1–10, Aug. 2021.
145. C. M. Grötsch, I. Dan, L. John, Sandrine Wagner, and Ingmar Kallfass, "A compact 281–319 GHz low-power downconverter MMIC for superheterodyne communication receivers," *IEEE Transactions on Terahertz Science and Technology*, vol. 11, no. 2, pp. 231–239, March 2021.
146. C. J. Lee and C. S. Park, "A d-band gain-boosted current bleeding down-conversion mixer in 65 nm CMOS for chip-to-chip communication," *IEEE Microwave and Wireless Components Letters*, vol. 26, no. 2, pp. 143–145, Feb. 2016.
147. Y. H. Xu, S. Yang, H. Li, J. M. Lv, and N. Bai, "Adaptive noise-resistant low-power ASK demodulator design in UHF RFID chips," *Electronics*, vol. 10, no. 24, pp. 3168/1–13, Dec. 2021.
148. Communications Systems Design, *White paper*, Plextek RF Integration, pp. 1–12, https://www.microwavejournal.com/ext/resources/pdf-downloads/Comms-System-Design-for-MWJ.pdf?1612492798.
149. F. Etemadi, P. Heydari, and H. Jafarkhani, "On analog QAM demodulation for millimeter-wave communications," *IEEE Transactions on Circuits and Systems II: Express Briefs*, vol. 66, no. 3, pp. 402–406, Jul. 2018.
150. J. D. Park, S. Kang, S. V. Thyagarajan, E. Alon, and A. M. Niknejad, "260 GHz fully integrated CMOS transceiver for wireless chip-to-chip communication," *Proceedings of IEEE Symposium* on VLSI Circuits (VLSIC), Honolulu, USA, 13–15 June 2012, pp. 48–49.
151. H. J. Song, J. Y. Kim, K. Ajito, N. Kukutsu, and M. Yaita, "50-Gb/s direct conversion QPSK modulator and demodulator MMICs for terahertz communications at 300 GHz," *IEEE Transactions on Microwave Theory and Techniques*, vol. 62, no. 3, pp. 600 – 609, March 2014.
152. S. P. Voinigescu, T.O. Dickson, R. Beerkens, I. Khalid, and P. Westergaard, "A comparison of Si CMOS, SiGe BiCMOS, and InP HBT technologies for high-speed and millimeter-wave ICs," *Proceedings of IEEE Topical Meeting on Silicon Monolithic Integrated Circuits in RF Systems*, 08–10 Sept. 2004, pp. 111–114.
153. R. L. Freeman, *Telecommunication System Engineering*. John Wiley & Sons, fourth-edition pp. 398–399, New Jersey, USA, 2004.
154. W. Khalil, J. Wilson, B. Dupaix, S. Balasubramanian, and G. L. Creech, "Toward millimeter-wave DACs: challenges and opportunities," *Proceedings of IEEE Compound Semiconductor Integrated Circuit Symposium (CSICS)*, La Jolla, USA, 14–17 Oct. 2012, pp. 1–4.
155. H. Al-Rubaye and G. M. Rebeiz, "W-Band direct-modulation >20-Gb/s transmit and receive building blocks in 32-nm SOI CMOS," *IEEE Journal of Solid-State Circuits*, vol. 52, no. 9, pp. 2277–2291, Sept. 2017.
156. N. Zimmermann, "Design and Implementation of a Broadband RF-DAC Transmitter for Wireless Communications," Ph.D. Electrical and Electronic Engineering Thesis, RWTH Aachen University, Germany, pp. 1–151, 2011.
157. S. M. Taleie, T. Copani, B. Bakkaloglu, and S. Kiaei, "A linear Σ–Δ digital IF to RF DAC transmitter with embedded mixer," *IEEE Transactions on Microwave Theory and Techniques*, vol. 56, no. 5, pp. 1059–1068, May 2008.

158. S. P. Voinigescu, S. Shopov, J. Bateman, H. Farooq, J. Hoffman, and K. Vasilakopoulos, "Silicon millimeter-wave, terahertz, and high-speed fiber-optic device and benchmark circuit scaling through the 2030 ITRS horizon," *Proceedings of the IEEE*, vol. 105, no. 6, pp. 1087–1104, June 2017.
159. M. Chen, L. Zhang, K. Bai, F. Li, G. Liu, H. Zhou, and Q. Chen, "100-Gbit/s DMT transmission for short-reach optical, interconnect with DAC clock leakage compensation," *IEEE Photonics Technology Letters*, vol. 31, no. 22, pp. 1791–1794, Nov. 2019.
160. L. Li, X. Zhang, D. Kong, M. Bi, S. Jia, W. Hu, and H. Hu, "Digital-analog hybrid optical access integrating 56-Gbps PAM-4 signal and 5G mm-wave signal by spectral null filling," *IEEE Journal of Lightwave Technology*, vol. 39, no. 5, pp. 1278–1288, Mar. 2021.
161. J. Kim, S. Kundu, A. Balankutty, M. Beach, B. C. Kim, S. T. Kim, Y. Liu, S. K. Murthy, P. Wali, K. Yu, H. S. Kim, C. C. Liu, D. Shin, A. Cohen, Y. Segal, Y. Fan, P. Li, and F. O'Mahony, "A 224-Gb/s DAC-based PAM-4 quarter-rate transmitter with 8-tap FFE in 10-nm finFET," *IEEE Journal of Solid-State Circuits*, vol. 57, no. 1, pp. 6–20, Jan. 2022.
162. M. Matthaiou, O. Yurduseven, H. Q. Ngo, Da. M. Jimenez, S. L. Cotton, and V. F. Fusco, "The road to 6G: ten physical layer challenges for communications engineers," *IEEE Communications Magazine*, vol. 59, no. 1, pp. 64–69, Jan. 2021.
163. S. P. Singh, *Physical Layer Design Challenges for 6G Wireless*. Y. Wu, S. Singh, T. Taleb, A. Roy, H. S. Dhillon, M. R. Kanagarathinam, and A. De, *6G Mobile Wireless Networks*. Springer, Cham Switzerland, 2021.

Chapter 3
Terahertz Communication: Standardization, Channel Modeling, and Link-Budget

3.1 Introduction

The IEEE's standardization of THz wireless communications started in early 2008 when a Terahertz Interest Group (IGthz) confirmed the following 802.15 standardization with many different parts. The critical design selections and fundamental operation prognostication by the group members are based on the cornerstone of the IEEE Task Group on "100G Wireless," i.e., TG100G in 2014 [1]. The first IEEE criterion for THz wireless communications by TG100G (IEEE Std 802.15.3d–2017) was formally certified in the Autumn of 2017 [2]. The following paragraph will describe the standardization activities and specifications of future THz wireless communication networks.

3.2 Terahertz Wireless Communication Standardizations

Diverse guiding 6G enterprises examine THz communications, containing European Commission's Horizon 2020 ICT-09 THz project assemble and Finland's 6Genesis Flagship Program (6GFP). In addition, the US Defense Advanced Research Projects Agency (DARPA) describes THz technology as specific of quadruple main research subjects that may influence society over the internet. Thus, THz-associated research has captivated severe funding, and regulation achievements have been started [3–6].

The Federal Communications Commission (FCC) has recently assigned a spectrum of 21.2 GHz within the 116–246 GHz for unauthorized service [7]. Additionally, the first IEEE 802.15.3d standardization for sub-THz communications toward 6G was approved in 2017, discussing communication point-to-point links, which can enable 100 Gb/s within a couple of centimeters (cms) up to hundreds of meters within THz channels in 252.72–321.84 GHz and BW from 2.16 GHz–69.12 GHz, including up to 69 covering channels, in 2.16 GHz multiplying groups [2]. The

U. Nissanov, G. Singh, *Antenna Technology for Terahertz Wireless Communication*,
https://doi.org/10.1007/978-3-031-35900-2_3

frequency band 275–450 GHz was approved for THz communication use for fixed service and land mobile at the World Radiocommunication Conference (WRC-2019) out of defining the necessary actions to shield passive usages, including the radio astronomy earth investigation-satellite usage [8].

The criterion contains physical (PHY) layer and medium access (MAC) designs. MAC examinations contain initial directional channel access, synchronization, and neighbor detection solutions. Examinations are in two states during the PHY layer: THz-OOK PHY and THz-SC PHY. THz-SC PHY enables high rates in backhaul and fronthaul links and data centers, where high-order modulators, demodulators, and complex signals are used, including up to 64-QAM. On the other hand, THz-OOK PHY is more adapted to short-distance (up to several meters) in intra-device and kiosk communications that can achieve tens of Gb/s. Therefore, a couple of low-rate (11/15) and high-rate (14/15) within low-density parity-check (LDPC) codes are facilitated in both states. However, within THz-OOK PHY, an additional Reed Solomon (RS) code with standard decoding excluding soft decision information is also facilitated. Both modes will have target minimum RX sensitivity signals set to $0.2 \cdot 10^{-9}$ W for BW of 2.16 GHz and 11/15 LDPC [9]. Figure 3.1 is attached, which shows the THz communication frequency plan of IEEE Std 802.15.3d–2017.

ITU-R identified operational and technical characteristics in this procedure, studied spectrum needs for THz communications, developed propagation models, introduced sharing studies within the passive services, identified applicant frequency bands, and demonstrated components for fronthaul/backhaul cellular links and wireless data centers, additionally next-generation channel coding and wireless-optical incorporation supporting data rates of Tb/s [10].

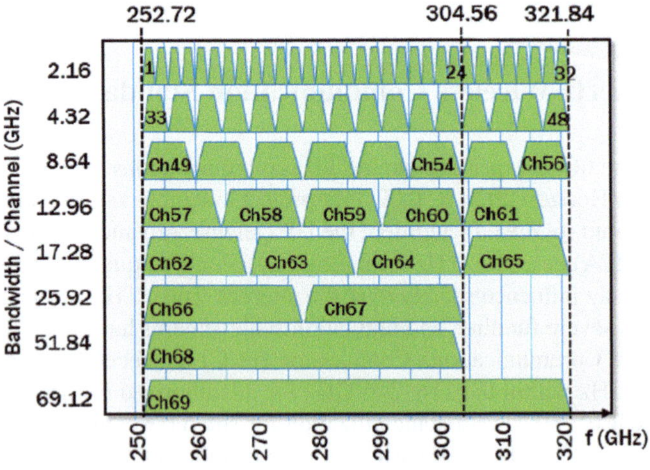

Fig. 3.1 THz communication frequency plan of IEEE Std 802.15.3d–2017 (Amendment 2) [2]

3.3 Terahertz Wireless Communication Potential Applications

3.3.1 Terahertz Chip-to-Chip Wireless Communication

Although chip-to-chip, inter-chip, and inter-board communications are currently done via wired copper connections, those links become blockage when data rates surpass 100 Gb/s. Consequently, there have been recommendations to use additional THz wireless connections or optical to change wired connections. In addition, the evolution of corresponding "nano-networks" establishes an additional potential realm for 6G. Essential specifications for these networks, besides the data rate, are energy effectiveness that needs to integrate feasible needed receiver processing, latency, and reliability. Average transmission distances are a small number of cms for LoS and NLoS conditions. So powerful multipath needs to be regarded, but the transmitter and receiver antenna need a relatively low gain because of the small range propagation. Furthermore, the completely stationary surroundings allow for a stable positioning through the modeling operation within computerized beam-steering capabilities as a selection for chronological multipoint operation beyond two chips [11].

3.3.2 Terahertz Wireless Local Area Network

The THz wireless local area network (THz-WLAN) intends to supply extremely high data rates of at least 100 Gb/s links to access points (APs). Classical operations, such as restaurants, airports, and offices, will be primarily indoors, covering 10–15 m. Even though the placement will suppose LoS links, shadowing via mobile person may source NLoS links that must be moderated using indirect transmission paths within scattering or reflections from walls through beam-steering capabilities [12].

3.3.3 Terahertz Wireless Personal Area Network

The THz wireless personal area network (THz-WPAN) intends to supply extremely high data rates of at least 100 Gb/s ad hoc links through devices at short distances below a meter, for example, via a notebook and a camera or a laptop and an external hard disk. Classical operations will be primarily indoors. Density-directivity antennas should be placed through beam-steering capabilities [13].

3.3.4 Terahertz Wireless Kiosk Downloading

The THz kiosk downloading is a particular category of THz-WPAN links, where one apparatus is connected to a fixed kiosk download terminal supplying close-proximity communications and extremely high downloads with a data rate of at least 100 Gb/s of multimedia content, for example, a 4 K hi-definition (4 K-HD) video, to a wireless mobile apparatus [14]. Average transmission distances are up to 0.5 m within certain channel features, enabling lower directive transmitters and receiver antennas without needing beam-steering capabilities. However, numerous reflections via the transmitter and receiver may further minimize the possible data rates [15].

3.3.5 Terahertz Wireless Connections Through Servers Inside a Data Center

Recently, cloud services have increased and stepped up data center construction. Usually, multiple server trays supplied within different servers, containing storage and several switches, can be set up in data centers, and it is acceptable that the threading through servers via server trays and the threading via trays will be made wireless. This wireless communication intends to supply extremely high data rates of at least 100 Gb/s. In contrast, the communication ranges begin at a few cm, considering link through servers vertically arranged in server racks, to a dozen meters, considering link through trays. It is essential to consider twain LoS and NLoS signals that assume an office model where masonries within correspondingly high reflectivity and low permeability are used [16].

3.3.6 Terahertz Backhauling and Fronthaul for Macro and Small Cell Wireless Communications

A backhaul connection is a terrestrial cellular link via the base station and an integrated network unit. On the other hand, the fronthaul connection is the terrestrial connection through the base-band unit (BBU) and a cellular base station's remote radio head (RRH). This wireless communication intends to supply extremely high data rates of at least 100 Gb/s while the communication ranges from 100 m up to about 300 m. So, it is essential to consider twain LoS and NLoS signals that assume an outdoor scenario that demands high transmitter and receiver antennas gain with electronic beam steering capabilities. Furthermore, future evolutions such as small cells' massive deployment, cloud radio access networks (C-RANs), and cooperative multipoint transmission (CoMP) can enhance the needed data rates for backhauling and fronthauling. Finally, this communication using THz wireless/cellular links may engage when fiber links are inaccessible [17].

3.3.7 Terahertz Airborne and Inter-Satellite Wireless Communications

In outer-space communication without the presence of the atmosphere, the transmission of THz is lossless and without any atmospheric attenuation. This non-terrestrial (NTN) wireless communication intends to supply extremely high data rates of at least 100 Gb/s and utilize flying systems that can fly above 99% of atmospheric water vapor to eliminate molecular absorption loss. Therefore, it is possible to reach communication ranges of a couple of km so that the THz band can be used at the inter-satellite, air-borne, and deep space wireless communication links [18, 19].

3.3.8 Terahertz Autonomous Vehicles and Unmanned Aerial Vehicles Wireless Communications

The THz may be used in the Internet of Vehicles (IoV) due to the massive increase in autonomous vehicles' data sharing, transmission, and consumption. Besides that, the THz can be used in all mobile nano cell networks, such as vehicle-to-vehicle (V2V), vehicle-to-infrastructure (V2I), shipment drones, smart railways, and unmanned aerial vehicles (UVAs). This THz wireless communication intends to supply high data rates of at least 20 Gb/s while the communication ranges up to 100 m. So, it is essential to consider the LoS with an indoor scenario for V2I, LoS, and NLoS signals that assume an outdoor scenario in this rest category, which demands high transmitter and receiver antenna gain with electronic beam steering capabilities [20].

3.3.9 Terahertz Nano-Cells Wireless Communication

The THz wireless communication can be used in medical applications for in-body via nano-machine communication through sub-mm-sized human body regions, such as in vivo applications. This THz wireless communication intends to supply exceptionally high data rates above 100 Gb/s. In contrast, the communication range is a couple of cm, meaning low-gain transmitter and receiver antennas are needed. In addition, this communication will be on-chip communication [21, 22]. Table 3.1 compares THz transmitter and receiver wireless communication antenna propagation range and gain and antenna technology with THz wireless communication applications.

Table 3.1 Comparison of needed antenna gain and propagation distance for types of THz wireless applications

THz wireless communication applications	Beam steering and MIMO capabilities	Typical propagation distance @ 100 Gb/s	Antenna gain
Chip-to-chip	No	Few cms	Low
WLAN	Yes	10–15 m	High
WPAN	No	<1 m	Moderate
Kiosk downloading	No	<0.5 m	Moderate
Server and data center	Yes	1–18 m	High
Backaul and fronthaul	Yes	100–300 m	Extremely high
Airborne and inter-satellite	Yes	Few km	Ultra extremely high
Autonomous vehicles and UVAs	Yes	50–100 m	Extremely high
Nano-cell network	Yes	Few cms	Low

3.4 Terahertz Wireless Communication Channel Properties

The foundation of radio and wireless communication applications is the transmission of electromagnetic (EM) waves. The transmitter and receiver antennas within the propagation channel form the radio channel. The designing and characterization of the radio channel are fundamental in modeling the abilities of novel wireless systems and estimating the achievement of transmitter and receiver components. The channel properties below 6 GHz have been decently designed and explored for mobile and wireless communications. The channel properties for frequencies below 100 GHz are already well researched for implementing 5G wireless, while there is insufficient knowledge because of confined measurement abilities [23]. Due to tiny wavelengths in the THz band, the propagation natures differ from the current 4G and 5G bands. Therefore, channel properties and noise should be designed first to make high-performance and trustworthy THz communications accurate [24]. So, the success in deploying future THz 6G wireless communication depends mainly on understanding how the channel properties, modeling, and noise behave at frequencies above 100 GHz.

3.4.1 The Friss Free-Space Path Loss

Free-space path loss (FSPL) or spreading loss is frequently assigned as the loss accepted by a radio frequency (RF) signal as it has been transmitted straightforwardly via a vacuum without energy reflection or absorption of proximate targets. However, FSPL considers optimal atmospheric situations: no EM energy waves are dissipated or lost, propagating far away via the radiating source, leading to lesser related power densities. FSPL describes a situated RF energy wave loss as it propagates via the air through its radiating source, where this loss is dependent on the frequency (f) of the transmitted RF signal and the distance (d) between the transmitter and receiver [25]. The FSPL can be calculated by the following equation [25]:

$$P_{RX} = P_{TX}\frac{G_{TX}\cdot G_{RX}\cdot \lambda^2}{(4\pi d)^2\cdot L}, \lambda^2 = \frac{C^2}{f^2}, FSPL = \frac{P_{TX}}{P_{RX}} = \frac{(4\pi d)^2\cdot L}{G_{TX}\cdot G_{RX}\cdot \lambda^2} \quad (3.1)$$

P_{RX} is the RX power in (W), P_{TX} is the transmitter power in (W), G_{RX} is the receiver antenna gain, G_{TX} is the transmitter antenna gain, λ is the RF signal wavelength in [m], and d is the range between transmitter and receiver in (m), L is the system or the channel losses, where $L \geq 1$, and C is the speed of the light in a vacuum, which equals to 299,792,458 m/s. Following Eq. 3.1, the received power P_{RX} reduces within the square of the distance and frequency, while the FSPL attenuation enhances proportionately with the reduction in P_{RX}.

3.4.2 *Atmospheric Gas Effect Attenuations*

(a) *Absorption of Water Vapor*

At the THz band, the primary attenuation of the transmitter signal is predictable because of meteorological attenuation, named atmospheric absorption. The spectrum under 1000 GHz is most tempting for wireless communications because, above 1000 GHz, the attenuation is stringent and significantly confines the communication propagation distance. While the wavelength of the THz wave accesses the dimensions of oxygen and water vapor molecules introduced in the atmospheric layers, so resonances are developed at precise carrier frequencies proceeding in a significant attenuation climax, which degrades the signal roughly, playing an essential part in calculations of the link budget [26]. This specific attenuation γ_0 is calculated by the ITU-R model [27]:

$$\gamma(dB/km) = \gamma_0 + \gamma_\omega = 0.1820\cdot f\cdot N''(f) \quad (3.2)$$

where γ_0 is the spectroscopic resonant behavior of water vapor and oxygen molecules γ_ω in (dB/km), and $N''(f)$ is the complex refractive index imaginary part, which can be attained from THz time-domain spectroscopy (TDS) measurements.

(b) *Dispersion*

The atmospheric conditions cause attenuation of the transmitted THz waves due to water vapor absorption. However, we also notice that the frequency-dependent absorption in the channels drastically lowers the BW of the transmitted THz impulse signal. This decrease also expands the THz data impulse signals. The THz propagating impulse signal is expanding basically because of the frequency-pendent dispersion of the refractivity of water vapor and oxygen [28, 29].

When the THz impulse signal electric field, E (0, ω), shown in the frequency domain, undergoes via the dispersive atmosphere, the electric field phase alters because of the resonance of water vapor. As a result, the output complex spectrum E (z, ω) is expressed per the multiplication of the input field within the attenuation and the phase function as [29]:

$$E(z, \omega) = E(0, \omega) \cdot e^{j\Delta k(\omega)z} \cdot e^{\alpha(\omega)z/2} \tag{3.3}$$

$$\Phi(\omega) = \Delta k(\omega)z = \beta(\omega)z \tag{3.4}$$

where β(ω) is the computed propagation vector [28], α(ω) is the attenuation factor, while z is the propagation interval. For communication channels, β(ω) is estimated by a Taylor series concerning the angular center frequency carrier ω_o of the transmitted pulse signal. Following the Taylor series, a literal expression for the estimate $\beta_A(\omega)$ is given by [29]:

$$\beta_A(\omega) = \beta(\omega_0) + \beta_1(\Delta\omega) + \frac{1}{2!}\beta_2(\Delta\omega)^2 + \frac{1}{3!}\beta_3(\Delta\omega)^3 \tag{3.5}$$

where $\Delta\omega = (\omega - \omega_0)$, and β_1, β_2, β_3 are the first, second, and third byproducts of β(ω) concerning ω, respectively, while β_2 is the group velocity dispersion (GVD). We can rewrite the approximate phase equation $\Phi_A(\omega, z) = \beta_A(\omega) \cdot z$ into [29]:

$$\Phi_A = \Phi_0 + A_1\Delta\omega + A_2\Delta\omega^2 + A_3\Delta\omega^3 \tag{3.6}$$

Therefore, we can see that the transmitted THz bit pulses via the atmosphere are broadening and transforming due to $A_2\Delta\omega^2$ and $A_3\Delta\omega^3$ phrases, while at the same time $\Phi_0 = \beta(\omega_0)z$ and $A_1\Delta\omega$ set the velocity of undeformed THz bit pulses propagation via the atmosphere.

An appropriate phase linking via the THz signal components must be restored to mitigate dispersion. For example, this can be achieved by inserting a device within the signal path within a group delay profile antipodal the atmosphere through the wanted BW. In other words, the total of the GVDs of the imported device and the atmosphere must be zero. If this is made correctly, the proportional phase shifts of the device and the atmosphere will nullify reciprocally, restoring the initial phase proportion of the transmitted THz signal and canceling the GVD atmospheric effects [30]. In contrast, Ghasempour et al. [31] proposed a technique to use a THz wireless NLoS signal using the nonlinearity of the angular dispersion in leaky antenna waveguides to achieve a higher data rate than a THz wireless LoS signal, and despite the phenomenon of dispersion.

3.4.3 Weather Effect Attenuations

(a) *Snow and Rain*

Snow and rain are other meteorological attenuations that the THz wave can encounter during propagation in an outdoor medium [32]. The ITU model supplied in [33] represents the rain attenuation, γ_R in (dB/km), experienced by the THz wave atop the frequency range up to 1000 GHz donated by the following equation [33]:

$$\gamma_R(\mathrm{dB/km}) = \mathrm{k} \cdot \mathrm{R}^{\mathrm{a}} \tag{3.7}$$

R is the rain rate in (mm/h), while (a) and (k) are frequency pendent factors determined from scattering calculations at horizontal and vertical polarizations. Since the rain drops size and shape are comparable to the wavelength order, the propagation losses can be exhibited as a function of wavelength and polarization.

(b) *Fog and Cloud*

Another meteorological attenuation that can cause fading in an outdoor medium is cloud and fog. Cloud and fog can be treated as similar atmospheric aspects, changing exclusively over the height atop the ground. The ITU model defines the couple phenomena [34], assuming that the THz wave travels entirely via a constant cloud or fog medium. The following equation can calculate the cloud or the fog attenuations γ_c [34]:

$$\gamma_c(\mathrm{f},\mathrm{T})(\mathrm{dB/km}) = \mathrm{K_l}(\mathrm{f},\mathrm{T}) \cdot \mathrm{M} \tag{3.8}$$

where $\mathrm{K_l}$ (f, T) is given by [35]:

$$\mathrm{K_l}(\mathrm{f},\mathrm{T}) = \frac{0.819 \cdot f}{\epsilon''(\eta^2 + 1)} \tag{3.9}$$

where $\mathrm{K_l}$ (f, T) is a specific cloud water attenuation factor in (dB/km)/(g/m^3) for 1–1000 GHz and a temperature range of ±40 °C, while $\eta = \frac{\epsilon'+2}{\epsilon''}$ with ϵ' and ϵ'' are set within the model given by H. Liebe et al. [36], M is the fluid water density in the cloud or fog in (g/m^3), while (f) and (T) are the THz carrier frequency and temperature, respectively. Atmospheric gas affects attenuation, and weather effect attenuation, so Fig. 3.2 is attached, which exhibits the atmosphere attenuation in (dB/km) depending on the frequency in (GHz) for a clear atmosphere at sea level (blue), rain at a rate of 50 mm/h (black), and fog with a range of vision of 50 m (red), separately [37]. For example, the peak attenuation at the frequency of 183 GHz, 325 GHz, 380 GHz, and 450 GHz is 28 dB/km, 39 dB/km, ~400 dB/km, and ~ 500 dB/km, respectively. Therefore, it is necessary to work at frequency bands below 1000 GHz with an attenuation lower than 100 dB/km. Therefore, the frequency windows are approximate: 100–117, 123–175, 190–320, 335–376, 385–443, 453–515, 627–711, and 808–902 GHz [26, 38].

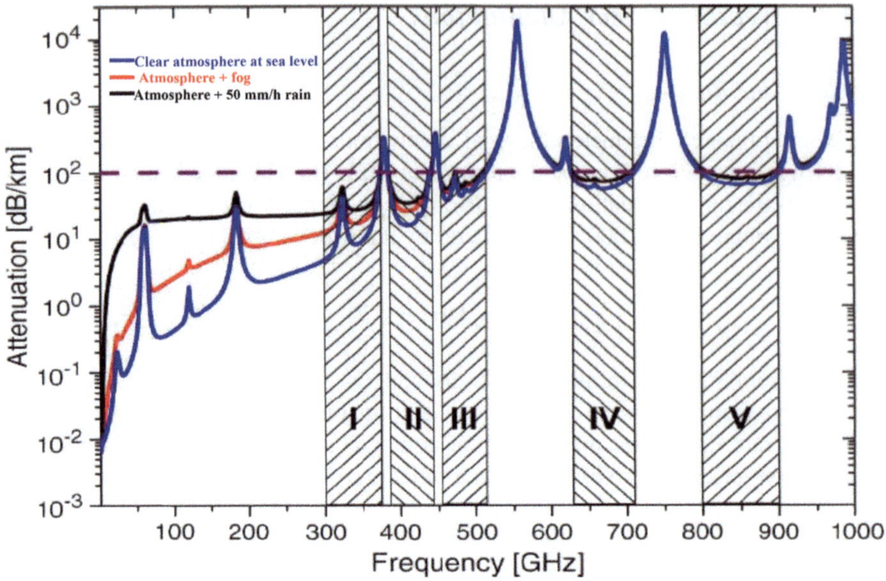

Fig. 3.2 THz band attenuation due to atmospheric gas and weather effects [15]

3.4.4 Blockage

The THz waves are significantly more affected by blockages than those at lower frequencies. These blockages are because the THz communication is based on the accessibility of line-of-sight (LoS) links because of the inferior propagation properties of the non-line-of-sight (NLoS) links [39, 40]. Therefore, humans, vehicles, buildings, and foliage can efficiently block these THz waves. An individual blockage can drive an average loss of 30 dB [40]. For instance, the reflection loss because of glass is about 11 dB at 326–350 GHz, for wood it is 18.4 dB at 350 GHz, and for building substances such as concrete, brick, and cinder block it is up to 75 dB at 326 GHz [41]. While a maximum of single tree foliage loss of about 11 dB was achieved at 142 GHz in LoS urban microcells [42].

Furthermore, the THz waves endure the individual blockage caused by human obstacles that are capable of causing attenuation of about 20–35 dB and can be affected by the THz indoor communication [43, 44]. These blockages can lower the signal power and result in electrical failure. Consequently, finding an effective solution to bypass blockages is crucial when the THz communication link gets blocked. In addition, it is necessary to consider the blockage loss in the THz communication link budget.

3.4.5 *Diffraction*

Diffraction effects are frequent in frequencies below 100 GHz, which can be depicted per the uniform theory of diffraction (UTD) or knife-edge diffraction (KED) schemes. They are significantly reduced at frequencies beyond 100 GHz because of the higher directivity antenna for LoS THz communication networks, but they may be considered multipath propagation forms, which could endow the total signal level by adding a loss between 5 and 50 dB for a frequency range of 100–2000 GHz and NLoS links [45]. These results may indicate that THz communication can be with NLoS links beside the LoS links and may cause extra interference to THz communication links and possibly inter-symbol interference (ISI).

3.4.6 *Scattering and Reflection*

Comparable to lower frequency band radiation, THz waves can penetrate via diverse nonconducting substances such as clothing, leather, paper, and plastic, and the absorption loss can be up to 2.35 dB at 326 GHz [41]. The scattering and reflection behavior of THz waves of diverse layers within diverse coarseness levels is exclusive at THz because of the tiny wavelength. Other factors influencing the magnitude of scattering include the scatterer's refractive index, shape, and size. When the layer is a coarse, diffuse scattering of the THz waves develops, resulting in an attenuation of THz waves intensity in the main lobe direction, which will originate in multipath propagation attenuation, which can be about 6–18 dB for a frequency range of 200–400 GHz, respectively [46]. So, an adequate understanding of scattering and reflection is needed to develop ultimate THz communication channel schemes. In addition, it is necessary to consider the scattering and reflection losses in the THz communication link budget.

3.4.7 *Doppler Frequency Shift*

The Doppler frequency shift effect occurs when transmitter communication transmits to a transceiver, i.e., a man with a cellular phone is located on a moving object such as a vehicle. For example, Fig. 3.3 shows a stationary cellular transmitter, which transmits to the receding/approaching vehicle transceiver with speed ($v > 0$). If the transceiver were not moving ($v = 0$), the received signal frequency by the receiver would be the same as the original transmitted frequency (f_c). However, when the transceiver moves, the frequency received by the receiver will become higher or lower than fc. That depends on the vehicle's direction of travel; for receding, the frequency will be higher than fc, while the frequency will be lower than f_c for approaching [47].

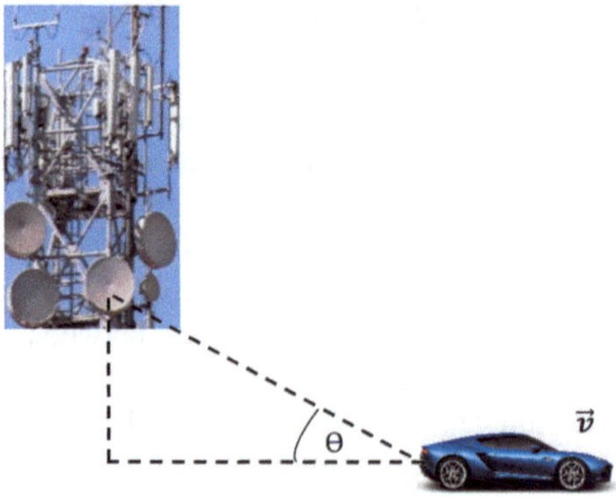

Fig. 3.3 Doppler frequency shift effect

The following equation sets the received signal f_r by the receiver [47]:

$$f_r = f_c \left(1 \pm \frac{V}{C}\right) \tag{3.10}$$

where f_c is the frequency in (Hz) of the transmitter, v is the vehicle speed in (m/s), and C is the speed of the light in a vacuum, which is 299,792,458 m/s. For Communicating with a moving object, there is a frequency error (f_d), i.e., the Doppler frequency shift effect, which is proportional to the speed v of the moving object [47]:

$$f_d = \frac{v}{C} f_c \cdot \cos \Theta \tag{3.11}$$

where f_d is the Doppler frequency shift, while Θ is the angle of the moving object from the transmitter. Furthermore, the above equation shows that the Doppler frequency shift effect is proportional to the f_c and is higher than in lower frequency bands.

As the channel time coherence is negatively related to the Doppler frequency shift effect, the channel will alter fast in time for a placed speed at THz wireless communication. Furthermore, the channel time-changing mentions that the channel information state must be refreshed frequently for the closed-loop link adjustment. Therefore, the change of channel time may enhance the overhead because of the demands for constant feedback and packed arranging of reference signals in time. Nonetheless, the overhead does not automatically enhance as the symbol duration and transmission time interval (TTI) can be lower with the enhancement of carrier frequency. So, the overhead proportion can be comparable [48]. On the other hand, the Doppler frequency shift effect may be expected to enhance as the carrier frequency enhances, causing problems that the rigor demands could dare the THz 6G wireless PHY design, particularly in moving objects in vehicular surroundings [49].

Basar et al. [50] suggested a technique to mitigate the Doppler frequency shift effect and multipath fading using reconfigurable intelligent surfaces (RISs). Without the RISs, 18 dB of the received LoS signal was lost at 250 m away, while Va et al. [49] suggested mitigating the Doppler frequency shift effect by a high directional transmitter and receiver wireless communication antennas. These techniques can be used for future THz 6G wireless communication networks.

3.4.8 Delay Spread

The delay spread (DS) calculates the expansion of propagation lagging of the communication channel multipath elements. It is generally specified in the root mean square (RMS) value, in which the channel path values compatible energy lagging of every multipath element. The significance of DS is how it influences the ISI and detects the frequency selectivity of the wireless communication networks. The channel is equivalent to ISI-free if the symbol span is sufficiently lengthy, usually a dozen times more, compared to the DS. DS characteristics are distinct in outdoor and indoor surroundings [51].

The channel's DS in (RMS) is defined by [51]:

$$\mathrm{DS} = \tau(\mathrm{RMS}) = \sqrt{\frac{\int_0^{\infty} (\tau - \overline{\tau})^2 \cdot \mathrm{A_c}(\tau)\mathrm{d}\tau}{\int_0^{\infty} \mathrm{A_c}(\tau)\mathrm{d}\tau}} \tag{3.12}$$

where $\mathrm{A_c}(\tau)$ is the power delay profile and $\overline{\tau}$ is the average channel's DS time and defined by [51]:

$$\overline{\tau} = \frac{\int_0^{\infty} \tau \cdot \mathrm{A_c}(\tau)\mathrm{d}\tau}{\int_0^{\infty} \mathrm{A_c}(\tau)\mathrm{d}\tau} \tag{3.13}$$

Wang et al. [52] proposed a technique to mitigate the DS using multiple RISs, while Va et al. [49] suggested mitigating the DS phenomena by a high directional transmitter and receiver wireless communication antennas. These techniques can be used for future THz 6G wireless communication networks. Table 3.2, which summarizes the THz channel properties with the attenuation mechanism, cause, peak attenuation, and frequency range, is attached.

Table 3.2 THz channel properties with attenuation mechanism and peak attenuation

Ref.	Attenuation mechanism	Attenuation cause	Peak attenuation	Frequency [GHz]
[37]	Weather	50 mm/h rain	~70 dB/km	100–500
[37]	Weather	Fog with 50 m visibility	~30 dB/km	100–500
[41]	Blockage	Glass, wood	11–18.4 dB	326–350
[41]	Blockage	Concrete, brick, and cinder block	75 dB	326
[42]	Blockage	Tree foliage	11 dB	140
[42]	Blockage	Human obstacles	25–30 dB	>100
[43, 44]	Diffraction	Any obstacles	5–50 dB	100–2000
[41]	Penetration	Nonconducting substances such as clothing, leather, paper, and plastic	2.35 dB	326
[46]	Scattering and reflection	Rough surfaces	6–18 dB	200–400
[50]	Doppler frequency shift effect	Moving object	25 dB variation of the received signal	>100

3.5 Terahertz Wireless Communication Channel Modeling

3.5.1 Deterministic Model

Several significant schemes exist for THz channel propagation studies for time, frequency, and spatial domains in a place surrounding. First is the deterministic method, based on the ray-tracing (RT) model. The RT scheme can be applied for multipath channel scheme models, using the assumptions of geometric optics to track the propagation of line-of-sight (LoS), non-LoS (NLoS), diffracted, scattered, and reflected EM waves. This approach in the THz communication signals, which have a very short wavelength, permits precise modeling. While RT is selected to design the channel, the EM waves' propagation medium is entirely assigned by objects' shapes, sizes, and materials. It is a fact that this scheme interferes with getting each outcome if any modification in the scheme is introduced. The complexity increases exponentially when the environment's size becomes more significant. Still, it gives precise data about the channel beneath any conditions [53]. In spatial and temporal channel characteristics, the RT scheme is used in THz indoor communication and at the THz data center [54, 55]. In the THz kiosks application, the frequency, time, and spatial domain channel parameters are designed for any RT type to analyze the THz channel feathers [56].

(a) *General Ray-Tracing Model*

General ray-tracing (GRT) can forecast the EM radiation field power and DS for several indoor and outdoor construction structures and antenna settings. In this model, the construction database, such as location, height, dielectric values, and transmitter and receiver positions, should be defined precisely [57, 58]. The GRT scheme utilizes geometrical optics to track the LoS propagation of the reflected signal components and signals via construction-expanded scattering and diffraction. The LoS signal and reflected paths typically supply the most influential elements of the received signal because scattering and diffraction losses are incredibly high. Still, these other multipath components can predominate in regions near diffracting or scattering layers, typically blocked from the LoS and reflecting waves [59].

(b) *Joined Path Loss and Shadowing Fading Model*

Most mobile communication systems operate in complex propagation environments that FSPL cannot accurately model. Over the years, several path loss models have been developed to predict path loss in typical wireless environments such as large urban macro-cells, urban microcells, and, more recently, inside buildings. These models are mainly based on empirical measurements over a given distance, frequency range, and geographical area or building [60]. Apart from path loss, there is also channel shadowing fading, which shows the wireless signal's attenuation due to many variables. For example, fading may occur from atmospheric attenuation, obstructions, time-domain variations (e.g., due to receiver motion), multipath interference effects, and more [61]. Accordingly, the following combined model for path loss and shadowing fading as a function of distance is generally used for system design and is based on an experimental model or RT model in a low-frequency band, and the ratio of received to the transmitted power in (dB) is given by [59]:

$$\frac{P_{RX}}{P_{TX}}(\mathrm{dB}) = 10\log K - 10\cdot\gamma\log\frac{d}{d_0} + \psi(\mathrm{dB}) \tag{3.14}$$

K, d, and d_0 are reference constants associated with antenna characteristics, the average channel attenuation, the transmit distance, and the reference distance for the antenna farfield. γ is the path-loss exponent depending on the environment, which can be between 2 and 4 for a simple propagation environment. While the value of γ for other complicated environments can be attained through a minimum mean square error (MMSE) fit experimental measurements. ψ(dB) is a Gauss-distributed arbitrary variable within an average zero and variance of $\sigma^2_{\psi(\mathrm{dB})}$. This model is only valid at transmission distance $d > d_0$, where d_0 is 1–10 m indoors and beyond 10 m outdoors. The value of $K < 1$ is sometimes set to the FSPL at a distance d_0 and equal to [59]:

$$K(\mathrm{dB}) = -20\log\frac{4\pi d_0}{\lambda} \tag{3.15}$$

For ideal LoS channel $\gamma = 2$, $\psi = 0$ dB corresponds to zero shadowing effect and average space path loss. NLoS fading requires even more complications due to the Rician or Rayleigh statistical variations of the received signal [62].

3.5.2 Stochastic Model

Stochastic channel models have defined radio channel outcomes via different components, including distance-pendent path loss, fast fading, and large-scale shadowing. THz stochastic channel models demand multiple datasets to remove probability distributions, so stochastic schemes acquire the mean of the environmental outcomes in place of a concrete environment scheme model, which is contradictory to the deterministic scheme. However, only several upper mmWave and THz radio channel models exist, and those that do, are not noticeably common [63]. Some stochastic scheme models are suggested in [64, 65], where the path loss and noise variance is quick for THz in vivo communication applications due to the human tissues containing substances within variant molecular characteristics. Path loss relies on distance, frequency, and tissue dielectric loss in cases like these. Furthermore, the spreads like cells and particles originate the loss in signal energy [64, 65]. Finally, studies [66, 67] have centered on nanodevice THz communications in the human corpus.

(a) *Stochastic Channel Design Form Plan*

The essential characteristics that a THz stochastic channel pattern design shall involve are path loss (PL), 3D angular spreads (AS) with elevation and azimuth realm, DS, power angle profile (PAP), power delay profile (PDP), several multipath components (MPCs), K-factor (KF), auto-correlation function (ACF), channel impulse response (CIR), time of arrival (ToA), spatial cross-polarization function (CCF), and angle of arrival (AoA). Pulling the critical channel constants and evaluating their statistical elements are crucial to stochastic channel patterns [68, 69].

The evaluations of the channel specifications firmly rely on the precision of the extracted rays and clusters. Nevertheless, it is regularly demanding to recognize rays and clusters from the experimental measurements. Standard procedures are conceived across preprocessing channel measurement outcomes, following identification algorithms and vision checkups. Nevertheless, there are unpredicted user-specified specifications in the analyzing algorithms to be manually readjustment, which makes the extracted outcome of the standard procedures unpredictable, and lacks stringent cause explanation [70–72].

3.5.3 Terahertz Channel Simulators

In a practical wireless communication channel, the LoS propagation signal, NLoS propagation signals, multipath, shadowing, and fading; therefore, the mathematical complexity is high in analyzing the THz wireless communication channel and finding a model close to a practical model. In addition, THz wireless communication will include many communication environments, such as indoor channels, outdoor channels, moving object channels, beamforming channels, and MIMO channels. Furthermore, the characteristics of the material firmly influence THz channel feathers in the density and the surroundings of these materials. So, setting up a standard THz wireless communication channel model is complex [24].

Consequently, unique simulators/solvers are used to find the models suitable for THz wireless communication in each environment. For example, using Remcom Wireless InSite for analyzing indoor, urban outdoor, moving vehicles or aircraft, and MIMO beamforming channels using RT models [73, 74]. Another example is TeraMIMO in matrix laboratory (MATLAB) simulator for 3D LoS and NLoS channels propagation for indoor and outdoor surroundings, beamforming MIMO channel. In addition, TeraMIMO includes channel statistics depending on the Doppler shift effect, water vapor absorption, DS in (RMS), and coherence BW [75].

3.6 Terahertz Link-Budget

The link budget of 6G backhaul and fronthaul THz wireless communication for 100–500 GHz will be described in this paragraph. Shannon's capacity equation is defined as the maximal transmitted error-free data rate (C), for a point-to-point scheme, for a band-limited Gaussian noisy channel [76, 77]:

$$C = B \cdot \log_2(1 + SNR) \tag{3.16}$$

$$SNR_{min}(dB) = P_{RX} - (N_0 + 10 \log B + N_F + S_M) \tag{3.17}$$

The SNR_{min} is the minimum signal-to-noise ratio of the receiver, P_{RX} is the received power at the receiver in [dB], N_0 is the noise density in (dBm/Hz), B is the BW in (Hz) of the receiver, and the transmitter, N_F is the total noise figure of the receiver in (dB), and S_M is the system margin in (dB).

The link budget of the 6G channel is defined as [77, 78]:

$$P_{RX} = P_{TX} + G_{TX} + G_{RX} + 20 \cdot \log\left(\frac{\lambda_0}{4 \cdot \pi \cdot d}\right) - L_{Atm+Rain}(f_0) \cdot d - \alpha \cdot L_r - L_{ex} \tag{3.18}$$

$$P\,(dBm) = 10 \cdot \log\left(\frac{P}{1 \cdot 10^{-3}}\right) \tag{3.19}$$

The P_{RX} is the received power at the receiver in (dBm), and P_{TX} is the transmitter's output power in (dBm). G_{TX} and G_{RX} are antenna gains of the transmitter and receiver antennas in (dB), appropriately. d and λ_0 are the propagation distance and free-space wavelength in (m), appropriately. $L_{Atm\,+\,Rain}(f_0)$ is atmospheric absorption loss due to the water vapor and with rain at rates of 50 mm/h at the frequency f_0. L_{ex} is the excess loss that is not included in the FSPL. L_r is the reflection loss; when $\alpha = 0$, the rays propagate in LoS, and $\alpha = 1$, the rays propagate in NLoS.

Table 3.3 6G channel parameters

Parameter	Value
Transmitter power (P_{TX})	10 dBm (=10 mW)
Propagation distance (d)	100–300 m
Frequency (f_O)	100–500 GHz
Wavelength (λ_O)	0.6–3 mm
Atmospheric absorption loss with 50 mm/h rain ($L_{Atm\ +\ Rain}(f_O)$)	70 dB/km
The bandwidth of the receiver (B=BW)	6.48 GHz and 25.92 GHz
Total spectral efficiency	4 bit/s/Hz
Data rate (C)	~26 Gb/s and ~ 104 Gb/s
The total noise figure of the receiver (N_F)	10 dB
Noise density (N_O)	−174 dBm/Hz
SNR_{min} of the receiver	10 dB
System margin (S_M)	10 dB

There is a need to use the Eqs. 3.16, 3.17, 3.18, and 3.19 to find the minimum needed transmitter and receiver antennas gain at 100–500 GHz. While the propagation distance of the transmitted and received signal is d = 100–300 m for 6G backhaul and fronthaul wireless communication [17], and the impedance BW of the transmitter and the receiver is 6.48 GHz (3 × 2.16 GHz) and 25.92 GHz (12 × 2.16 GHz) [2] with total spectral efficiency of at least four bit/s/Hz [79], so the data rate will be about 26 Gb/s and 104 Gb/s where the minimum data rate needed for 6G wireless communication is 20–100 Gb/s [80, 81]. Furthermore, we will assume that the receiver parameters are equal to the following: $N_0 = -174$dBm/Hz [77], $N_F = 10$ dB, $S_M = 10$ dB, $SNR_{min} = 10$ dB [38], while the transmitter power is at least $P_{TX} = 10$ dBm [1, 82, 83], $G_{TX} = G_{RX} = G$ = antennas gain, when the atmospheric absorption loss due to the water vapor with the effect of rain at rates of 50 mm/h and the channel parameters are: $L_{Atm\ +\ Rain}(f_O) = 70$dB/km [15], $L_{ex} = 0$ dB, $\alpha = 0$ for LoS propagation. Table 3.3 summarizes the parameters of 6G channel propagation.

Figure 3.4 shows the obtained results on suitable graphs of the needed transmitter and the receiver antenna's gain in (dB) vs. the frequency in (GHz) for propagation distances of 100–300 m, and BW of 6.48–25.92 GHz corresponds to data rates of 26–104 Gb/s, respectively. From Fig. 3.4, it may have been shown that the transmitter and the receiver antenna's gain should be between 31.8 and 38.8 dB for the frequency range of 100–500 GHz, respectively, for a propagation distance of 100 m and BW of 6.48 GHz corresponds to a data rate of 26 Gb/s. While from Fig. 3.4 it may have been shown that the transmitter and the receiver antenna's gain should be between 38.3 and 45.3 dB for the frequency range of 100–500 GHz, respectively, for a propagation distance of 200 m and BW of 6.48 GHz corresponds to a data rate of 26 Gb/s. Moreover, it may have been shown that the transmitter and the receiver antennas gain should be between 43.6 and 50.6 dB for the frequency range of 100–500 GHz, respectively, for a propagation distance of 300 m and BW of

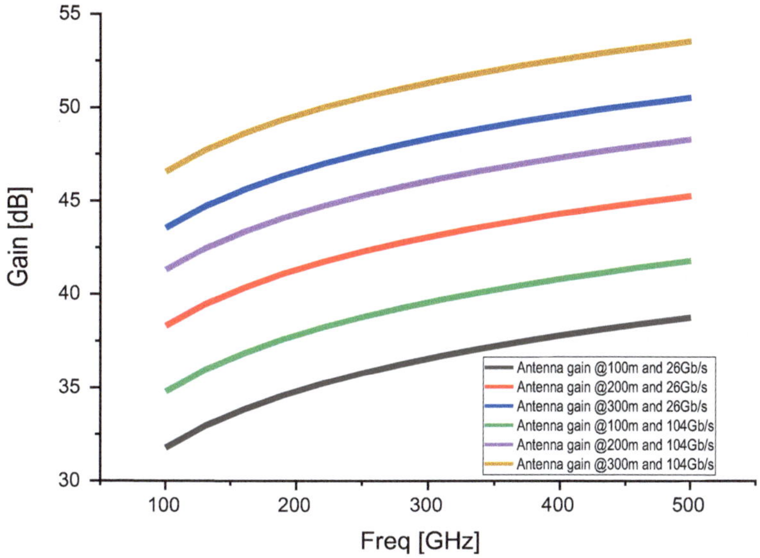

Fig. 3.4 The needed transmitter and the receiver antennas gain vs. the frequency for propagation distances of 100–300 m and data rates of 26–104 Gb/s

6.48 GHz corresponds to a data rate of 26 Gb/s. Furthermore, from Fig. 3.4, it may have been shown that the transmitter and the receiver antenna gain should be 34.8–41.8 dB for the frequency range of 100–500 GHz, respectively, for a propagation distance of 100 m and BW of 25.92 GHz corresponds to a data rate of 104 Gb/s. While from Fig. 3.4, it may have been shown that the transmitter and the receiver antenna's gain should be 41.3–48.3 dB for the frequency range of 100–500 GHz, respectively, for a propagation distance of 200 m and BW of 25.92 GHz corresponds to a data rate of 104 Gb/s. Furthermore, it may have been shown that the transmitter and the receiver antenna's gain should be between 46.6 and 53.6 dB for the frequency range of 100–500 GHz, respectively, for a propagation distance of 300 m and BW of 25.92 GHz corresponds to a data rate of 104 Gb/s. To conclude this section, the transmitter and the receiver antenna gains should be between 31.8 and 53.6 dB for the frequency range of 100–500 GHz, for a propagation distance of 100–300 m, and BW of 6.48–25.92 GHz corresponds to data rates of 26–104 Gb/s, namely, these needed transmitter and the receiver antennas gains corresponds to reported [84, 85]. In addition, with each increase in the THz transmitted signal propagation distance by 100 m at a given data rate and a given frequency, the gain for the transmitter and the receiver antennas should be increased by about 5.2–6.6 dB.

3.7 Open Research Challenges

Based on this chapter, there are several open research challenges in THz channel properties and modeling: (1) Flexible and accurate THz channel models. The THz wireless communications models used based on deterministic channels based on RT schemes and stochastic models are insufficient. So, we need to integrate statistical and deterministic modeling schemes to achieve high accuracy and efficiency under severe multipath fading. Furthermore, we need to consider THz UM-MIMO channel modeling and THz beam steering channel modeling with high power efficiency and spectrum for future 6G THz wireless communications. Finally, further investigation of outdoor and non-stationary channel scenarios remains an open issue. (2) Effective THz channel measurement systems. The unique properties of THz waves drove flexible and advanced systems to compete with significant path loss, ultra-large BW within the high frequency, and short-time coherence. Based on such properties, a considerable number of THz channel measurements should be managed for diverse supported by THz wireless communications scenarios, such as UVAs, nano-communication, backhaul communication, fronthaul communication, airborne and satellite communication, indoor communication, outdoor communication, virtual reality (VR), augmented reality (AR), mixed reality (MR), and extended reality (XR). (3) Material characterization. Setting material parameters such as permittivity and conductivity for frequency beyond 100 GHz can assist in characterizing the THz propagation channel, which depends on multipath scattering behavior and absorption losses. (4) AI-enabled channel modeling. THz wireless communication with AI and ML can potentially address complicated channel scenario problems. However, AI-enabled channel modeling demands considerable data input to the software algorithms. This big data is used to prepare these software algorithms and make effective forecasts. Thus, the computational complication can be significantly reduced because the predictions are precise, and the software algorithms are powerful for modeling THz diverse scenarios. (5) Improved channel sounding. Improved channel sounding at THz via a 3D milieu permits a temporal and spatial synthesis to predict and characterize THz wave propagation behavior, such as the ToA, Doppler frequency shifts, DS, and AoA. Knowing these parameters implements the forecast of a THz wireless communications system's execution and remarkably affects the future THz wireless communications standardization processes.

3.8 Summary

This chapter discusses the THz wireless communication standardization, potential applications, channel properties, modeling, and link budget. The THz potential communication applications can be chip-to-chip THz wireless communication, THz-WLAN, THz-WPAN, THz-kiosk downloading, THz wireless connections via servers inside a data center, THz backhauling and fronthaul for macro and small cells wireless communications, THz airborne, and inter-satellite wireless communications, THz autonomous vehicles, UAV's wireless communications, and, finally, THz nano-cells wireless communication. The wireless propagation distance can vary between a few centimeters to several kilometers in indoor and outdoor communication environments. At the same time, the THz have some channel properties such as the FSPL, e.g., spreading loss, atmospheric absorption of the water vapor loss, and communication proposes we need to operate at frequency bands below 1000 GHz that atmosphere absorption loss needs to be below 100 dB/km. Therefore, the frequency windows are approximately 100–117, 123–175, 190–320, 335–376, 385–443, 453–515, 627–711, and 808–902 GHz.

Furthermore, the THz channel properties are snow and rain, fog and cloud, blockage, diffraction, scattering, reflection, Doppler frequency shift effect, and DS, where all of these properties attenuate the THz wave signals. So, to make a reliable THz wireless backhauling and fronthaul communication, and based on the THz link budget and THz channel properties, there is a need to develop THz high-gain antennas with gains of at least 31.8–53.6 dB to facilitate reliable 6G LoS wireless communication with a data rate of 20–100 Gb/s and propagation distance of 100–300 m for 100–500 GHz. The THz wireless communication channel modeling can be a deterministic or stochastic model based on the RT scheme. Researchers are using THz channel simulators because setting up a standard THz wireless communication channel model is complex for the reasons described in this chapter and above.

References

1. V. Petrov, T. Kürner, and I. Hosako, "IEEE 802.15.3d: first standardization efforts for sub-terahertz band communications toward 6G," *IEEE-Communications Magazine*, vol. 58, no. 11, pp. 28–33, Nov. 2020.
2. IEEE Std 802.15.3d–2017, "IEEE Standard for High Data Rate Wireless Multimedia Networks – Amendment 2: 100 Gb/s Wireless Switched Point-to-Point Physical Layer," pp. 1–55, Oct. 2017, https://standards.ieee.org/ieee/802.15.3d/6648/
3. C. Paoloni, A. Alexiou, O. Bouchet, A. Davy, V. Ermolov, T. Kürner, B. Napier, and O. Sahin, "ICT beyond 5G cluster: seven H2020 for future 5G," TU Braunschweig University, pp. 1–2, 2018, https://doi.org/10.24355/dbbs.084-201908211514-0
4. http://www.h2020-dream.eu/ict-09-2017-cluster/
5. M. Katz, M. M. Blue, and M. L. Aho, "6Genesis Flagship Program: building the bridges towards 6G-enabled wireless smart society and ecosystem," *Proceedings of IEEE 10th Latin-American Conference on Communications (LATINCOM)*, Guadalajara, Mexico, 14–16 Nov. 2018, pp. 1–6.
6. H. Elayan, O. Amin, B. Shihada, R. M. Shubair, and M. Alouini, "Terahertz band: The last piece of RF spectrum puzzle for communication systems," *IEEE Open Journal of the Communications Society*, vol. 1, pp. 1–32, Nov. 2019.
7. Federal Communications Commission, Federal Register Rules and Regulation, vol. 84, no. 107, pp. 25685–25692, June 2019, https://www.govinfo.gov/content/pkg/FR-2019-06-04/pdf/2019-10925.pdf
8. T. Kürner and A. Hirata, "On the impact of the results of WRC 2019 on THz communications," *Proceedings of IEEE in 3rd International Workshop on Mobile Terahertz Systems (IWMTS)*, Essen, Germany, 01–02 July 2020, pp. 1–3.

9. M. Shehata, K. D. Wang, J. Webber, M. Fujita, T. Nagatsuma, and W. Withayachumnankul, "IEEE 802.15.3d-compliant waveforms for terahertz wireless communications," *IEEE Journal of Lightwave Technology*, vol. 39, no. 24, pp. 7748–7760, Dec. 2021.
10. Deliverable D7.3 Standardisation Activities – Final Release, TERRANOVA Project, pp. 1–43, July 2017, https://ict-terranova.eu/wp-content/uploads/2020/06/D7.3.pdf.
11. T. Kürner and S. Priebe, "Towards THz communications - status in research, standardization, and regulation," *Journal of Infrared, Millimeter, and Terahertz Waves*, vol. 35, pp. 53–62, Aug. 2014.
12. V. Petrov, T. Kürner, and I. Hosako, "IEEE 802.15.3d: first standardization efforts for sub-terahertz band communications towards 6G," [Online] arXiv, pp. 1–6, Dec. 2020, https://arxiv.org/pdf/2011.01683
13. C. Yi, D. Kim, S. Solanki, J. H. Kwon, M. Kim, S. Jeon, Y. C. Ko, and I. Lee, "Design and performance analysis of THz wireless communication systems for chip-to-chip and personal area networks applications," *IEEE Journal on Selected Areas in Communications*, vol. 39, no. 6, pp. 1785 – 1796, June 2021.
14. H. J. Song; H.i Hamada, and M. Yaita, "Prototype of KIOSK data downloading system at 300 GHz: Design, technical feasibility, and results," *IEEE Communications Magazine*, vol. 56, no. 6, pp. 130 – 136, June 2018.
15. M. A. Salhi, T. K. Ostmann, and T. Schrader, "Propagation channel measurements in the mm and sub-mm wave range for different indoor communication scenarios," *Journal of Infrared, Millimeter, and Terahertz Waves*, vol 42, pp. 357–370, Mar. 2021.
16. Report ITU-R SM.2352-0, Technology Trends of Active Services in the Frequency Range 275–3000 GHz, pp. 1–25, June 2015, https://www.itu.int/dms_pub/itu-r/opb/rep/R-REP-SM.2352-2015-PDF-E.pdf
17. F. Liu, Y. Cui, C. Masouros, J. Xu, T. X. Han, Y. C. Eldar, and S. Buzzi, "Integrated sensing and communications: toward dual-functional wireless networks for 6G and beyond," *IEEE Journal on Selected Areas in Communications*, vol. 40, no. 6, pp. 1728–1767, June 2022.
18. F. A. Dicandia, N. J. G. Fonseca, M. Bacco, S. Mugnaini, and S. Genovesi, "Space-air-ground integrated 6G wireless communication networks: A review of antenna technologies and application scenarios," *Sensor*, vol. 22, no. 9, pp. 3136/1–34, Apr. 2022.
19. M. Civas and O. B. Akan, "Terahertz wireless communications in space," *ITU Journal on Future and Evolving Technologies*, vol. 2, no. 7, pp. 1–8, Oct. 2021.
20. C. Zhang, K. Ota, J. Jia, and M. Dong, "Breaking the blockage for big data transmission: Gigabit road communication in autonomous vehicles," *IEEE Communications Magazine*, vol. 56, no. 6, pp. 152–157, Jun. 2018.
21. K. Yang, A. Pellegrini, M. Munoz, A. Brizzi, A. Alomainy, and Y. Hao, "Numerical analysis and characterization of THz propagation channel for body-centric nano-communications," *IEEE Transactions on Terahertz Science and Technology*, vol. 5, no. 3, pp. 419–426, 2015.
22. F. Lemic, S. Abadal, W. Tavernier, P. Stroobant, D. Colle, E. Alarcón, J. M. Barja, and J. Famaey, "Survey on terahertz nano communication and networking: A top-down perspective," *IEEE Journal on Selected Areas in Communications*, vol. 39, no. 6, pp. 1506–1643, April 2021.
23. RF Enabling 6G – Opportunities and Challenges from Technology to Spectrum, *White paper 6G Research Vision*, no. 13, pp. 1–68, University of Oulu, Apr. 2021.
24. K. Tekbiyik, A. R. Ekti, G. K. Kurt, and A. Gorcin, "Terahertz band communication systems: challenges, novelties, and standardization efforts," *Physical Communication*, vol. 35, pp. 100700/1–28, Apr. 2019.
25. O. D. Oyeleke, S. Thomas, O. I. Bismark, P. Nzerem, and I. Muhammad, "Absorption, diffraction, and free space path losses modeling for the terahertz band," *International Journal of Engineering and Manufacturing*, vol. 1, pp. 54–65, Feb. 2020.
26. Demos Serghiou, M. Khalily, Tim W C Brown, and R. Tafazolli, "Terahertz channel propagation phenomena, measurement techniques, and modeling for 6G wireless communication applications: a survey, open challenges, and future research directions," *IEEE Communication Surveys & Tutorials*, vol. 24, no. 4, pp. 1957–1996, Fourth Quarter 2022.

27. Attenuation by Atmospheric Gases, International Telecommunication Union ITU-R Recommendation P.676-10, pp. 1–22, Sep. 2013, [online] available at: https://www.itu.int/dms_pubrec/itu-r/rec/p/R-REC-P.676-10-201309-S!!PDF-E.pdf
28. M. Mandehgar, Y. Yang, and D Grischkowsky, "Experimental confirmation and physical understanding of ultra-high bit rate impulse radio in the THz digital communication channels of the atmosphere," *Journal of Optics*, vol. 16, pp. 094004/1–17, July 2014.
29. M. Mandehgar and D. R. Grischkowsky, "Understanding dispersion compensation of the THz communication channels in the atmosphere," *IEEE Photonics Technology Letters*, vol. 27, no. 22, pp. 2387–2390, Nov. 2015.
30. K. Strecker, "Group Velocity Dispersion Management in Terahertz Wireless Communication Channels," M.Sc. Electrical and Electronic Engineering Dissertation, Oklahoma State University, USA, pp. 1–52, May 2020.
31. Y. Ghasempour, Y. Amarasinghe, C. Y. Yeh, E. Knightly, and D. M. Mittleman, "Line-of-sight and non-line-of-sight links for dispersive terahertz wireless networks," *APL Photonics*, vol. 6, pp. 041304/1–8, March 2021.
32. J. Ma, J. Adelberg, R. Shrestha, L. Moeller, and D. M. Mittleman, "The effect of snow on a terahertz wireless data link," *Journal of Infrared Millimeter Terahertz Waves*, vol. 39, pp. 505–508, March 2018.
33. Specific attenuation model for rain prediction methods, International Telecommunication Union ITU-R Recommendation P.838-3, pp. 1–8, Mar. 2005, [online] available at: https://www.itu.int/dms_pubrec/itu-r/rec/p/R-REC-P.838-3-200503-I!!PDF-E.pdf.
34. Specific attenuation model for clouds and fog prediction methods, International Telecommunication Union ITU-R Recommendation P.840-3, pp. 1–7, 1997, https://www.itu.int/dms_pubrec/itu-r/rec/p/R-REC-P.840-3-199910-S!!PDF-E.pdf.
35. G. A. Siles, J. M. Riera, and P. García-del-Pino, "Atmospheric attenuation in wireless communication systems at millimeter and THz frequencies," *IEEE Antennas and Propagation Magazine*, vol. 57, no. 1, pp. 48 – 61, Feb. 2015.
36. H. Liebe, G. Hufford, and M. Cotton, "Propagation modeling of moist air and suspended water/ice particles at frequencies below 1000 GHz," *Proceedings of AGARD, 52nd Specialists Meeting of the Electromagnetic Wave Propagation Panel*, May 1993, pp. 3-1–3-11.
37. T. Schneider, A. Wiatrek, S. Preubler, M. Grigat, and R. P. Braun, "Link budget analysis for terahertz fixed wireless links," *IEEE Transactions on Terahertz Science and Technology*, vol. 2, no. 2, pp. 250–256, March 2012.
38. H. Tataria, M. Shafi, A. F. Molisch, M. Dohler, H. Sjöland, and F. Tufvesson, "6G wireless systems: vision, requirements, challenges, insights, and opportunities," *Proceeding of the IEEE*, vol. 109, no. 7, pp. 1166–1199, July 2021.
39. F. Boccardi, R. W. Heath, A. Lozano, T. L. Marzetta, and P. Popovski, "Five disruptive technology directions for 5G," *IEEE Communications Magazine*, vol. 52, no. 2, pp. 74–80, Feb. 2014.
40. S. Tripathi, N. V. Sabu, A.K. Gupta, and H. S. Dhillon, *Millimeter-Wave and Terahertz Spectrum for 6G Wireless*. Ed by Y. Wu et al. *6G Mobile Wireless Networks: Computer Communications and Networks*. Springer, Cham, Feb. 2021.
41. Tersence, "Attenuation of terahertz radiation in various materials," [Online], Feb. 2018, http://terasense.com/news/attenuation-terahertz/
42. Y. Xing and T. S. Rappaport, "Propagation measurements and path loss models for sub-THz in urban microcells," *Proc. of the IEEE International Conference on Communications* (ICC 2021), Montreal, QC, Canada, 14–23 June 2021, pp. 1–6.
43. I. A. Hemadeh, K. Satyanarayana, M. El-Hajjar, and L. Hanzo, "Millimeter-wave communications: physical channel models, design considerations, antenna constructions, and link-budget," *IEEE Communications Surveys and Tutorials*, vol. 20, no. 2, pp. 870–913, Dec. 2017.
44. M. A. Salhi, T. K. Ostmann, and T. Schrader, "Propagation channel measurements in the mm and sub-mm wave range for different indoor communication scenarios," *Journal of Infrared, Millimeter, and Terahertz Waves*, vol. 42, pp. 357–370, March 2021.

45. J. Kokkoniemi, P. Rintanen, J. Lehtomaki, and M. Juntti, "Diffraction effects in terahertz band - measurements and analysis," *Proceedings of IEEE Global Communications Conference* (GLOBECOM), Washington DC, USA, 04–08 December 2016, pp. 1–6.
46. J. Ma, R. Shrestha, W. Zhang, L. Moeller, and D. M. Mittleman, "Terahertz wireless links using diffuse scattering from rough surfaces," *IEEE Transactions on Terahertz Science and Technology*, vol. 9, no. 5 pp. 463–470, Sept. 2019.
47. S. Farahani, *ZigBee Wireless Networks and Transceivers. Chapter 5- RF Propagation, Antennas, and Regulatory Requirements. Elsevier, Amsterdam*, pp. 171–206, Sep. 2008.
48. J. Jeon, K. Muhammad, J. Cho, G. Xu, I. Na, and J. Zhang, "Design considerations for terahertz wireless communication systems," *Proceedings of IEEE Wireless Communications and Networking Conference Workshops* (WCNCW), Seoul, S. Korea, 06–09 April 2020, pp. 1–5.
49. V. Va, T. Shimizu, G. Bansal, and R. W. Heath Jr., "Millimeter wave vehicular communications: A survey," Foundations and Trends in Networking, vol. 10, no. 1, pp. 1–113, 2015.
50. E. Basar, "Reconfigurable intelligent surfaces for Doppler effect and multipath fading mitigation," *Front. Comms. Net.,* vol. 2, pp. 672857/1–12, May 2021.
51. S. Priebe, M. Jacob, and T. Kürner, "Angular and RMS delay spread modeling in view of THz indoor communication systems," *Radio Science*, vol. 49, pp. 242–251, Feb. 2014.
52. K. Wang, C. T. Lam, and B. K. Ng, "Positioning information based high-speed communications with multiple RISs: Doppler mitigation and hardware impairments," *Applied Sciences*, vol. 12, pp. 7076/1–25, July 2022.
53. K. Guan, H. Yi, D. He, B. Ai, and Z. Zhong, "Towards 6G: paradigm of realistic terahertz channel modeling," *IEEE China Communications*, vol. 18, no. 5, pp. 1–18, May 2021.
54. B. Peng, S. Rey, T. Kurner, "Channel characteristics study for future indoor millimeter and submillimeter wireless communications," *Proceedings of IEEE 10th European Conference on Antennas and Propagation (EuCAP)*, Davos, Switzerland, 10–15 April 2016, pp. 1–5.
55. S. Nie and I. F. Akyildiz, "Three-dimensional dynamic channel modeling and tracking for terahertz band indoor communications," *Proceedings of IEEE 28th Annual International Symposium on Personal, Indoor, and Mobile Radio Communications (PIMRC)*, Montreal, Canada 08–13 October 2017, pp. 1–5.
56. D. He, K. Guan, A. Fricke, B. Ai, R. He, Z. Zhong, A. Kasamatsu, I. Hosako, and T. Kurner, "Stochastic channel modeling for kiosk applications in the terahertz band," *IEEE Transactions on Terahertz Science and Technology*, vol. 7, no. 5, pp. 502–513, July 2017.
57. R. Piesiewicz, T. K. Ostmann, N. Krumbholz, D. Mittleman, M. Koch; J. Schoebel, and T. Kurner, "Short-range ultra-broadband terahertz communications: concepts and perspectives," *IEEE Antennas and Propagation Magazine*, vol. 49, no. 6, pp. 24–39, Dec. 2007.
58. F. Sheikh, B. Salah, D. Lessy, and T. Kaiser, "Unexplored aspects in THz ray-tracing," *Proceedings of 4th International Workshop on Mobile Terahertz Systems (IWMTS)*, Essen, Germany, 05–06 July 2021, pp. 1–7.
59. A. Goldsmith, *Wireless Communications*. Cambridge University Press, USA, pp. 35–79, 2005.
60. J. F. O'Hara, S. Ekin, W. Choi, and I. Song, "A perspective on terahertz next-generation wireless communications," *Technologies*, vol. 7, pp. 43/1–18, June 2019.
61. E. N. Papasotiriou, A. A. A. Boulogeorgos, K. Haneda, M. F. de-Guzman, and A. Alexiou, "An experimentally validated fading model for THz wireless systems," *Scientific Reports*, vol. 11, no. 18717, pp. 1–14, Sep. 2021.
62. S. Li and L. Yang, "Performance analysis of dual-hop THz transmission systems over α-μ fading channels with pointing errors," *IEEE Internet of Things Journal*, vol. 9, no. 14, pp. 11772–11783, July 2022.
63. C. Han and Y. Chen, "Propagation modeling for wireless communications in the terahertz band," *IEEE Communications Magazine*, vol. 56, no. 6, pp. 96–101, June 2018.
64. J. Bian, C.X. Wang, X. Gao, X. You, and M. Zhang, "A general 3D non-stationary wireless channel model for 5G and beyond," *IEEE Transactions on Wireless Communications*, vol. 20, no. 5, pp. 3211–3224, May 2021.

65. Z. D. Taylor, R. S. Singh, D. B. Bennett, P. Tewari, C. P. Kealey, N. Bajwa, M. O. Culjat, A. Stojadinovic, H. Lee, J. P. Hubschman, E. R. Brown, and W. S. Grundfest, "THz medical imaging: in-vivo hydration sensing," *IEEE Transactions on Terahertz Science and Technology*, vol. 1, no. 1, pp. 201–219, Sept. 2011.
66. K. Yang, A. Pellegrini, M. O. Munoz, A. Brizzi, A. Alomainy, and Y. Hao, "Numerical analysis and characterization of THz propagation channel for body-centric nano-communications," *IEEE Transactions on Terahertz Science and Technology*, vol. 5, no. 3, pp. 419–426, May. 2015.
67. H. Elayan, R. M. Shubair, J. M. Jornet, and P. Johari, "Terahertz channel model and link budget analysis for intrabody nanoscale communication," *IEEE Transactions Nanobioscience*, vol. 16, no. 6, pp. 491–503, Sep. 2017.
68. J. Wang, C. X. Wang, J. Huang, and H. Wang, "A novel 3D space-time-frequency non-stationary channel model for 6G THz indoor communication systems," *Proceedings of IEEE Wireless Communications and Networking Conference (WCNC)*, Seoul, S. Korea, 25–28 May 2020, pp. 1–7.
69. T. Kürner, D. M. Mittleman, and Tadao Nagatsuma, *THz Communications: Paving the Way Towards Wireless Tbps. Springer, Cham*, Switzerland, 2022.
70. D. He, K. Guan, A. Fricke, B. Ai, R. He, Z. Zhong, A. Kasamatsu, I. Hosako, and T. Kurner, "Stochastic channel modeling for kiosk applications in the terahertz band," *IEEE Transactions on Terahertz Science and Technology*, vol. 7, no. 5, pp. 502–513, July 2017.
71. J. Kokkoniemi, A. A. A. Boulogeorgos, M. U. Aminu, J. Lehtomaki, A. Alexiou, and M. Juntti, "Stochastic analysis of indoor THz uplink with co-channel interference and phase noise," *Proceedings of IEEE International Conference on Communications Workshops (ICC Workshops)*, Dublin, Ireland, 07–11 June 2020, pp. 1–6.
72. Z. Hossain, C. Mollica, and J. M. Jornet, "Stochastic multipath channel modeling and power delay profile analysis for terahertz-band communication," *Proceedings of the 4th ACM International Conference on Nanoscale Computing and Communication*, Washington DC, USA, 27–29 Sept. 2017, pp. 1–7.
73. https://www.remcom.com/wireless-insite-em-propagation-software/33
74. F. Sheikh, N. Zarifeh, and T. Kaiser, "Terahertz band: Channel modeling for short-range wireless communications in the spectral windows," *IET Microwaves, Antennas, and Propagation*, vol. 10, no. 13, pp. 1435 – 1444, July 2016.
75. S. Tarboush, H. Sarieddeen, H. Chen, M. H. Loukil, H. Jemaa, M. S. Alouini, and T. Y. Al-Naffouri, "TeraMIMO: A channel simulator for wideband ultra-massive MIMO terahertz communications," *IEEE Transactions on Vehicular Technology*, vol. 70, no. 12, pp. 12325 – 12341, Dec. 2021.
76. T. T. Ha, *Theory and Design of Digital Communication Systems*. Cambridge University Press, New York, USA, pp. 127–146, 2011.
77. H. J. Song and T. Nagatsuma, *Handbook of Terahertz Technologies: Devices and Applications*. Taylor & Francis Group, UK, 2015.
78. Z. Xu, "Antenna Designs and Channel Modeling for Terahertz Wireless Communications," Ph. D. Electric and Electronic Engineering Thesis, University of Victoria, Canada, 2016.
79. M. F. Hermelo, P. T. B. Shih, M. Steeg, A. Ngoma, and A. Stöhr, "Spectral efficient 64-QAM-OFDM terahertz communication link," *Optics Express*, vol. 25, no. 16, pp. 19360–19370, Mar. 2017.
80. W. Tong and P. Zhu, *6G The Next Horizon: From Connected People and Things to Connected Intelligence*. Cambridge University Press, Apr. 2021.
81. 6G: The next hyper-connected experience for all, *Samsung 6G Vision White Paper*, pp. 1–42, Jul. 2020. https://cdn.codeground.org/nsr/downloads/researchareas/6G%20Vision.pdf.
82. J. Nishizawa, P. Płotka, T. Kurabayashi, and H. Makabe, "706-GHz GaAs CW fundamental-mode TUNNETT diodes fabricated with molecular layer epitaxy," *Current Topics in Solid-State Physics,* vol. 5, no. 9, pp. 2802–2804, July 2008.

83. S. Chakraborty, A. Acharyya, A. Biswas, and A. K. Kundu, "Multi-stage-multi-iterative optimization algorithm for design optimization of multi-quantum well Terahertz IMPATT sources," *Proceeding of IEEE VLSI Device Circuit and System (VLSI DCS)*, Kolkata, India, 18–19 July 2020, pp. 129–133.
84. A. Shafie, N. Yang, C. Han, J. M. Jornet, M. Juntti, and T. Kürner, "Terahertz communications for 6G and beyond wireless networks: Challenges, key advancements, and opportunities," *IEEE Networks*, pp. 1–8, Sept. 2022.
85. I. F. Akyildiz, C. Han, Z. Hu, S. Nie, and J. M. Jornet, "Terahertz band communication: An old problem revisited and research directions for the next decade," *IEEE Transactions on Communications*, vol. 70, no. 6, pp. 4250–4285, June 2022.

Chapter 4
Terahertz Antenna Technologies for 6G Communication Systems

4.1 Introduction

For reliable THz 6G wireless communication, we need to design and fabricate a high-gain terahertz antenna for transmitters and receivers due to the lack of high-power THz sources [1, 2] and to compensate for the high losses caused by FSPL, atmospheric absorption caused by water vapor molecular, and other loss mechanisms such as rain, fog, blockage, and diffraction [3–9] to allow the signal propagation at a distance of least 100 m for THz 6G fronthaul and backhaul wireless communication networks [10]. At the same time, the second important parameter is the BW of the antenna to allow a data rate of 20–100 Gb/s for THz 6G wireless communication networks [11, 12]. However, these two parameters are always a trade-off when increasing the antenna's gain, always the BW of the antenna is reduced, and vice versa. Therefore, it is necessary to make compromises between the minimum needed transmitter and receiver antenna's gain and the minimum needed transmitter and receiver antenna's BW according to the demand.

4.2 Terahertz Antenna Technologies

The THz antennas are made in various configurations and sizes, containing angle arrays reflectors, pyramidal cavities within dipoles, dielectric lens planar antennas, microstrip planar antennas, bowtie dipoles, photoconductive antennas, THz horn antennas, and graphene-based antennas [13, 14]. In addition, THz antennas are classified into three groups pending their content processing: metallic antennas, such as horn antennas; dielectric antennas, such as planar microstrip antennas based on dielectric laminates and dielectric lens antennas; and state-of-the-art material antennas based on graphene or carbon nanotube (CNT) [15].

U. Nissanov, G. Singh, *Antenna Technology for Terahertz Wireless Communication*,
https://doi.org/10.1007/978-3-031-35900-2_4

4.2.1 Terahertz Horn Antenna

Horn antennas are a type of metallic antennae, which is designed to operate in the THz.

The horn antenna is extensively used as a feed component for satellite tracking, communication dishes, and radio astronomy. Besides, it is used as a feed for lens and reflector antennas and serves as a worldwide standard for calibration and gains measurements of different high-gain antennas. The horn antenna is one of the most used antennas at the THz band because of its simplicity in fabrication, significant gain, versatility, easiness of excitation, and superior comprehensive rendition. Some types of horn antennas include E-plane horn antennas, H-plane horn antennas, circular/conical horn antennas, and corrugated and pyramidal horn antennas [16, 17]. Figure 4.1a, b is attached and shows the E-plane horn antenna, whereas Fig. 4.1a shows the front view of the E-plane horn antenna, and Fig. 4.1b shows the E-plane view of the E-plane horn antenna [17].

The horn antenna can be examined as an aperture antenna. The fields at the horn's aperture can be found by examining the horn as a radial waveguide. It can be presented that if the fields of the feed waveguide are those of its most important TE_{10} mode and the horn longitude are sizeable compared to the aperture dimensions, the lowest order mode fields at the aperture of the E-plane horn antenna are set by [17]:

$$E'_x = H'_y = E'_z = 0 \tag{4.1}$$

$$E'_y(x', y') \approx E_1 \cos\left(\frac{\pi}{a} x'\right) e^{-j\left[k(y')^2/(2r_1)\right]} \tag{4.2}$$

$$H'_z(x', y') \approx j \cdot E_1 \left(\frac{\pi}{k \cdot a \cdot \eta}\right) \sin\left(\frac{\pi}{a} x'\right) e^{-j\left[k(y')^2/(2r_1)\right]} \tag{4.3}$$

$$H'_x(x', y') \approx \frac{E_1}{\eta} \cos\left(\frac{\pi}{a} x'\right) e^{-j\left[k(y')^2/(2r_1)\right]} \tag{4.4}$$

$$r_1 = r_e \cos \psi_e \tag{4.5}$$

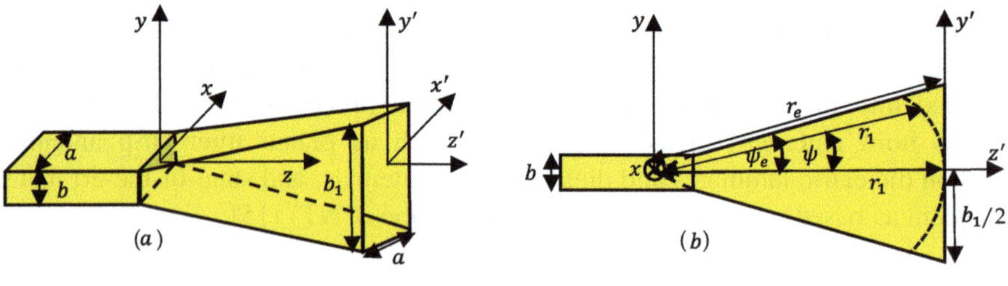

Fig. 4.1 E-plane horn antenna, (**a**) Front view, (**b**) E-plane view. (Inspired by [17]

F_1 is a constant, k is the phase constant, a and b_1 ($b_1 > a$) are aperture dimensions, and η is the wave impedance. The directivity of this antenna $D_{E\text{-plane}}$ is set by [17]:

$$D_{E\text{-plane}} = \frac{64 \cdot a \cdot r_1}{\pi \cdot \lambda \cdot b_1} \left[C^2 \left(\frac{b_1}{\sqrt{2\lambda \cdot r_1}} \right) + S^2 \left(\frac{b_1}{\sqrt{2\lambda \cdot r_1}} \right) \right] \quad (4.6)$$

where C(X) and S(X) are the Fresnel integrals, which are equal to [17]:

$$C(x) = C\left(\frac{b_1}{\sqrt{2 \cdot \lambda \cdot r_1}} \right) = \int_0^{\frac{b_1}{\sqrt{2 \cdot \lambda \cdot r_1}}} \cos\left(\frac{\pi}{2} t^2 \right) dt \quad (4.7)$$

$$S(x) = S\left(\frac{b_1}{\sqrt{2 \cdot \lambda \cdot r_1}} \right) = \int_0^{\frac{b_1}{\sqrt{2 \cdot \lambda \cdot r_1}}} \sin\left(\frac{\pi}{2} t^2 \right) dt \quad (4.8)$$

Because the THz wave's frequency is so large, the horn antenna's dimensions are so defined that preparing the horn's tip end is demanding, mainly when designing antenna arrays, and the preparation technique's difficulty donates to high costs and minimal output. Furthermore, since the bottom of a complicated horn pattern is difficult to fabricate, a primary horn antenna in a tapered/conical shape is mainly used, lowering the price and intricacy whereas keeping the antenna's efficiency [16, 18].

Aqlan et al. [19] designed a tapered circularly polarized THz horn antenna at 300 GHz. This antenna contains a tapered horn antenna with a circular polarizer disk and waveguide feed. This antenna was validated with a fabricated prototype THz antenna. The measured impedance BW for $S_{11} < -15$ dB and the peak gain were 60 GHz (270–330 GHz) and 18.4 dB, respectively. In comparison, Lee et al. [20] designed three types of THz horn antennas at 240 GHz. The first antenna was a simple horn antenna; the second was a horn antenna with a waveguide; the third was a horn antenna with a waveguide and a dielectric lens to enhance the gain of the simple horn antenna. These antennas were validated with fabricated prototype THz antennas. The measured impedance BW for $S_{11} < -13$ dB and the peak gain for the first, second, and third antennas were 40 GHz (220–260 GHz), 26 dB, 40 GHz (220–260 GHz), 32.5 dB, and 40 GHz (220–260 GHz), 39.5 dB, respectively.

4.2.2 *Terahertz Dielectric Lens Antenna*

Dielectric lens antennas are treated as quasi-optical systems, changing the waveguide horn for a planar printed antenna on a broad dielectric and a silicon lens where the lens matter delays the traveling wave and generate a plane wavefront from spherical. Compared to waveguide horn antennas, the advantages of these hybrid antennas are easy integration, low SLLs, low loss, and low fabrication cost by the micro-machined technique. In most designs, lens dimensions are significant in wavelength, enabling quasi-optical design techniques. Large lenses enable large BW, limited only by the feed BW. Dielectric lens antennas are a reliable alternative to reflector antennas because of their supporting configuration and back-feed. They do not obstruct the antenna cavity at the THz band to supply a directive radiation beam pattern, also multiple-beam or single-beam [18, 21, 22].

The lens classification is founded on triple unequal physical features: (a) The position of the feed is proportionate to the lens frame (at direct connection or far from the lens), (b) Refractive index contour (nonuniform or stepped), and (c) Number of refraction layers [23]. Thus, a dielectric lens is invaluable for attaining high directivity and efficiency than a planar microstrip antenna at the THz band. However, when planar microstrip antennas are printed over chunky dielectric laminates, they may endure power loss because of unwanted surface modes generated inside the laminate. Because of the lens geometry, these unwanted surface modes are not developed using a dielectric lens as a laminate. Instead, if the planar microstrip antenna is printed on the backside of a high dielectric lens, the radiation will abundantly go to the highest dielectric material, and with a lens contour, the radiation will be transmitted to the free space [24, 25]. Figure 4.2a shows the shaped integrated dielectric lens antenna, and Fig. 4.2b presents the incidence wave from the lower index matter, while Fig. 4.2c shows the incidence wave from higher index matter.

Using the geometrical optics (GO) principle and considering that the interface within a couple of dielectric media can be examined locally plane if the reflection of an incident plane wave develops in the identical matter within a similar incident and reflects angles [26]. The following Snell's law controls the refraction [27]:

$$\mathrm{n}_1 \sin(\theta_{\mathrm{inc}}) = \mathrm{n}_2 \sin(\theta_{\mathrm{trans}}) \tag{4.9}$$

where n_1 and n_2 are the refraction indexes of any matter and θ_{inc} and θ_{trans} are the incidence and transmission angles specified regarding the interface normal (Fig. 4.2b, c). If a couple of matters are available within the identical magnetic permeability of air, then $n_1 = \sqrt{\varepsilon_{r1}}$ and $n_2 = \sqrt{\varepsilon_{r2}}$ where ε_{r1} and ε_{r2} are the proportional electric permittivity of each matter. Generally, the lens layer has a haphazard elliptical pattern, although within an extensive curvature radius equalized to the wavelength at each point. So, it is acceptable to express at spherical coordinates centered at the feed center phase for an axially symmetric lens, taking the pattern [23]:

$$\frac{\mathrm{dr}_1}{\mathrm{d}\eta} = \frac{\mathrm{r}_1(\eta)\sin(\theta - \eta)}{\mathrm{n}_1/\mathrm{n}_2 - \cos(\theta - \eta)} \tag{4.10}$$

The $r_1(\eta)$ describes the unspecified lens contour during the output angle function, $\theta(\eta)$ is the other unspecified by several design conditions involving phase, polarization, or amplitude. Due to the lenses' large size proportional to wavelength, dissipation losses are further critical. The material permittivity is a complex-valued function of the form $\varepsilon_r(1 - j\tan\delta)$. A spherical wave propagating within the lens matter has the pattern [23]:

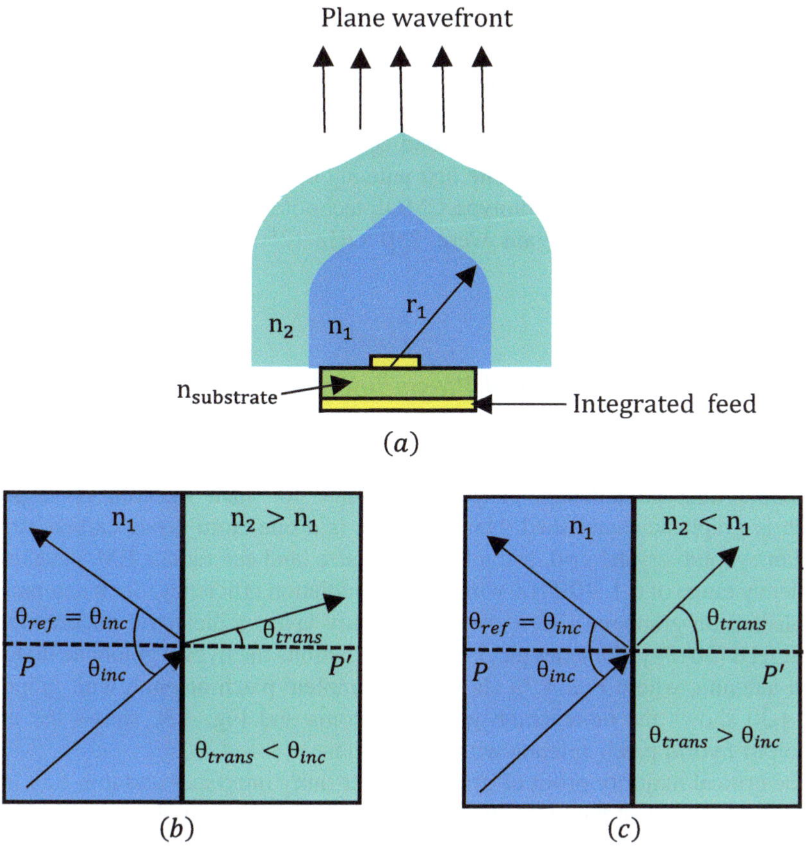

Fig. 4.2 Dielectric lens antenna and plane wave incident at an interface between two dielectrics, (**a**) Shaped integrated dielectric lens antenna, (**b**) Incidence wave from the lower index matter, (**c**) Incidence wave from the higher index matter [23]

$$E_0(\theta, \varphi)\frac{e^{-j2\pi\sqrt{\varepsilon_r(1-j\tan\delta)}\frac{r}{\lambda}}}{r} \tag{4.11}$$

So the combined dissipation loss depends on lens size, pattern, and feed illumination function. For low-loss matters, the dissipation loss (L) in a homogeneous lens can be calculated for the maximum and minimum radius (r) of the lens using [23]:

$$L(dB) = 27.3\sqrt{\varepsilon_r}\frac{r}{\lambda}\tan\delta \tag{4.12}$$

Campo et al. [28] designed a low dielectric elliptical lens antenna at 180 GHz. This antenna contains a leaky-wave feed with a dielectric grating wedge (GW) and plastic elliptical lenses. This antenna was experimentally validated within a fabricated prototype THz antenna. The measured impedance BW, maximum steering angles, and peak gain were 80 GHz (140–220 GHz), ±25°, and 34.5 dB, respectively. In comparison, Berkel et al. [29] designed two types of dielectric lens antennas at 375 GHz. The first antenna was fed with a co-planar waveguide (CPW) and dielectric silicon lens, while the second antenna was with a double leaky slot center-fed silicon lens antenna. The first antenna with a CPW-fed lens antenna was validated within fabricated prototype CMOS technology. The measured impedance BW and maximum peak gain were 250 GHz (250–500 GHz) and 26.5 dB, respectively.

4.2.3 Terahertz Nano and Plasmonic Antennas

The THz wireless communications at the nanoscale can be achieved using a plasmonic material antenna. A good candidate for such material is graphene, which is graphene-established. Nano-antenna is a one-atom-dense carbon stratum in a honeycomb crystal grid, just a few μm in size, and can radiate EM waves at the frequency range of 0.1–10 THz, with a higher radiation efficiency. Such antennas are established on graphene sheets over a flat tinny layer, a dielectric material, and a signal feed [30–34]. For example, Fig. 4.3a–c exhibits the hybrid graphene dielectric patch antenna, where Fig. 4.3a shows the complete patch antenna with graphene, Fig. 4.3b shows the honeycomb graphene atoms and Fig. 4.3c shows the cross-section of hybrid patch antenna with the graphene.

The critical majority profit of graphene over tinny interfaces and thin-tiny layers is the ability to adjust the graphene conductivity using a gate voltage firmly E_{bais} or chemical doping instantaneous, non-homogeneously, and locally. By using diverse magnitudes of E_{bais} at diverse points over the single graphene sheet, certainly wanted conductivity formats can be created. Graphene's conductivity directly depends on its unique band configuration and parameters such as chemical doping, Fermi energy (chemical potential), relaxation time, electron mobility, environmental temperature, operating frequency, and the lack of magnetic field bias. The graphene can enable surfaces' Plasmon polaritons (SPPs) modes at THz frequencies, where SPPs modes subsist on the connection of a positive ε_r laminate (such as Si, SiO2) and a negative

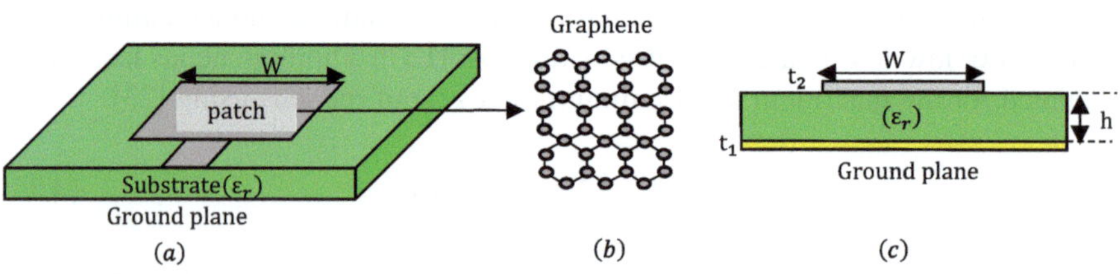

Fig. 4.3 (**a**) The complete hybrid patch antenna with the graphene, (**b**) The honeycomb graphene atoms, and (**c**) The cross-section of the hybrid patch antenna with graphene

ε_r laminate (such as graphene) and scale down different THz wireless communication networks. The design of antennas by utilizing graphene supplies proper matching, high miniaturization, decent total efficiency, and reconfiguration abilities [34, 35].

The conductivity of graphene can be modeled at THz frequencies only by exploiting the intra-band concept of the Kubo equation [36]:

$$\sigma = -j\frac{e^2 \cdot k_B \cdot T}{\pi h^2(\omega - j\tau^{-1})} \ln\left\{2\left[1 + \cosh\left(\frac{\mu_c}{k_B \cdot T}\right)\right]\right\} \tag{4.13}$$

where τ is the graphene electron's relaxation time, and T is the temperature. While μ_c is the chemical potential of graphene with a unit of electron-volt (eV), e is the charge of an electron, h is the reduced constant of Planck's, k_B is the constant of Boltzmann's, and ω is the angular frequency. This equation is applicable if $h\omega < 2|\mu_c|$, valid at the frequency of 0.1–10 THz for medium μ_c. The graphene sustains the spread of SPPs at those frequencies, precisely, at a free-space standing thin graphene layer of a σ conductivity. The characteristic impedance Z_C and dispersion relation of the maintained TM plasmons may be calculated by [37, 38]:

$$k_{SPP} = k_0\sqrt{1 - \left(\frac{2}{\eta \cdot \sigma}\right)^2} \tag{4.14}$$

$$Z_C = \frac{k_{SPP}}{\omega \cdot \varepsilon_0 \cdot \varepsilon_{eff}} \tag{4.15}$$

where k_{SPP} is the wavenumber of SPPs, while k_0 and η are the impedance and free-space wavenumber, ε_o is the vacuum permittivity, and ε_{eff} is the effective permittivity of the surrounding matter/substrate.

Graphene can trick out its composite conductivity, which can quickly be achieved using graphene's field reaction by directly executing an exterior electrostatic bias field orthogonal to the specimen. The bias inserts holes or electrons on the thin layer, therefore adjusting the chemical potential of graphene. This adjustment can be high-speed and signed up hundreds of GHz, enabling reconfiguration capabilities in all graphene-established components and antennas. So the graphene-established THz array antenna will achieve high gain and good radiation characteristics, and the radiation shape of an antenna array can be changed by dominating the gate potentials on any antenna. So it removes the necessity for phase shifters or switches for beam-steering [39–42].

Metasurfaces established on brassy nanoparticles can inspect the polarization, phase, amplitude, and circular-angular sweep of EM waves to trick out the propagation of EM waves. Still, the common brassy meta-surfaces cannot be applicable to be well applicant at Sub-mm wavelengths. Therefore, graphene is the best suitable for SPPs modes at THz frequency by tuning the graphene antennas μ_c. The capability to control the magnitudes and phases of reflection coefficients mostly fetches paramount flexibility in beam-steering exploiting the metasurfaces [43].

Dash et al. [44] designed a THz ultra-wideband (UWB) bowtie graphene plasmonic antenna for 1350 GHz. The achieved simulated BW and the peak directivity was 4600 GHz (400–5000 GHz) and 4.25 dBi. Tripathi et al. [45] designed a graphene plasmonic patch antenna for 300 GHz based on RT Duroid microstrip laminate. The obtained simulated peak directivity and BW was 7.47 dBi and about 40 GHz.

Following [30–45], further research is needed for graphene plasmonic technology to mature into THz applications.

4.2.4 Terahertz Microstrip Planar Antenna

Microstrip antenna has the advantages of conforming to planar and non-planar surfaces and low fabrication cost when using printed-circuit-board (PCB) technology. Moreover, they are very versatile in impedance, resonant frequency, pattern, impedance, and polarization; it also suppresses spurious radiation and parasitic substrate waves with the right design. However, at the same time, the disadvantages are their low efficiency, low power-handling ability, poor scan performance, insufficient polarization purity, low gain, and highly narrow frequency BW that is usually 2–5% [46, 47].

The microstrip antennas include a thin copper strip/radiating patch located above the dielectric substrate of height (h) and metallic ground plane of a thickness (t), as shown in Fig. 4.4a, b.

Figure 4.4a illustrates the basic configuration of the rectangular microstrip antenna, while Fig. 4.4b illustrated the side view of the microstrip antenna and its fringing effects. The sizes of the radiator are restricted over the width and length. So, the EM fields at the boundaries of the radiator cross fringing. The total fringing effects are proportional to the radiating patch's size and the substrate's height. The fringing effects influence derives from electric field lines that ample the antenna dimensions after stimulation. The leading factor of the fringing is the feed's position and the antenna's width. Fringing fields support radiation and give an additional directivity. Still, the resonant frequency is changed over the wanted frequency [48], while a dielectric substrate with permittivity ε_r separates the ground plane and the patch.

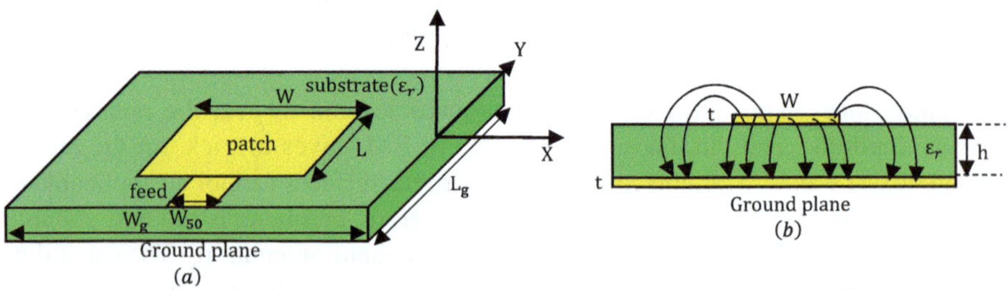

Fig. 4.4 (**a**) Basic configuration rectangular microstrip antenna, and (**b**) Side view. (Inspired by [46])

(a) *Terahertz Microstrip Rectangular Planar Antenna Design Equations*

For most implementations, the microstrip substrate is electrically thin ($h << \lambda_o$), where λ_o is the free-space wavelength, and the fields inside the microstrip antenna are quasi-transverse electromagnetic (TEM). The width W_x of the THz rectangular planar microstrip antenna is equal to [49]:

$$W_x = \frac{c}{2 \cdot f_r} \cdot \sqrt{\frac{2}{(\varepsilon_r + 1)}} \tag{4.16}$$

where c is the speed of the light in a vacuum, while f_r is the frequency of the resonance of the THz rectangular planar microstrip antenna. The effective dielectric constant ε_{reff} is equal to [49]:

$$\varepsilon_{r_{eff}} = \frac{\varepsilon_r + 1}{2} + \frac{\varepsilon_r - 1}{2} \cdot \left(1 + 12\frac{h}{W_x}\right)^{-1/2}, \ \frac{W_x}{h} > 1 \tag{4.17}$$

The effective length L_{eff} of the THz rectangular planar microstrip antenna is equal to [49]:

$$L_{eff} = \frac{c}{2 \cdot f_r \sqrt{\varepsilon_{r_{eff}}}} \tag{4.18}$$

Due to the fringing effects, the THz rectangular planar microstrip antenna exhibits more than its actual size [48]. Like its length, the THz rectangular planar microstrip antenna sizes have been expanded on any end by a distance ΔL. The extension of the length is equal to [49]:

$$\Delta L = 0.412 \cdot h \cdot \frac{\varepsilon_{r_{eff}} + 0.3}{\varepsilon_{r_{eff}} - 0.258} \cdot \frac{(W_x/h) + 0.264}{(W_x/h) + 0.8} \tag{4.19}$$

The length of the patch L_x of the THz rectangular planar microstrip antenna after consideration of the fringing effects is equal to [49]:

$$L_x = L_{eff} - 2 \cdot \Delta L \tag{4.20}$$

The dimensions of the ground plane (W_g, L_g) of the THz rectangular planar microstrip antenna are equal to at least [49]:

$$L_g = 6 \cdot h + L_x \tag{4.21}$$

$$W_g = 6 \cdot h + W_x \tag{4.22}$$

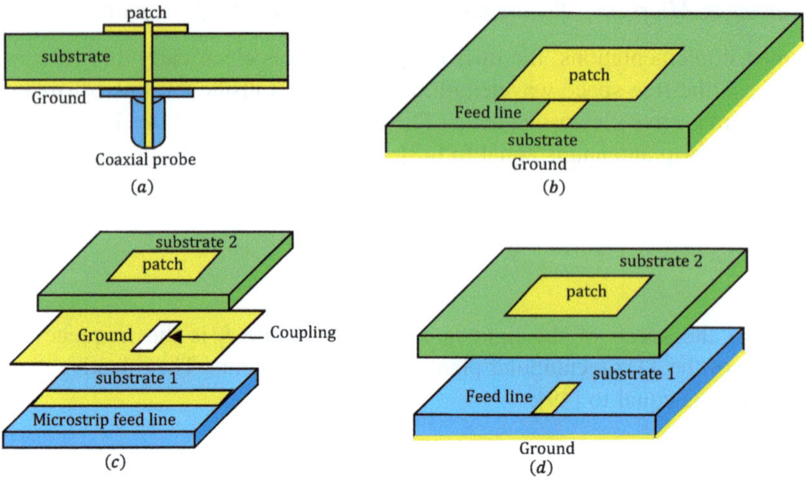

Fig. 4.5 Microstrip antenna feeding techniques, (**a**) Coaxial probe feed, (**b**) Microstrip transmission line feed, (**c**) Aperture coupling feed, and (**d**) Proximity coupling feed

The width of the THz rectangular planar microstrip antenna feed line W_{50} to supply a characteristic impedance of 50 Ω is equal to [49]:

$$W_{50} = h\left(\frac{377}{50\sqrt{\varepsilon_r}} - 2\right) \tag{4.23}$$

Similarly, the feed line length L_f of the THz rectangular planar microstrip antenna is equal to [49]:

$$L_f = (2 \cdot M + 1)\lambda, M = 0, \pm 1, \pm 2, \ldots \pm N \tag{4.24}$$

where M is an integer and λ is the wavelength.

(b) *Terahertz Microstrip Planar Antenna Feeding Techniques*

The potential techniques used to feed the microstrip antenna, as shown in Fig. 4.5, are classified as (a) Coaxial probe, (b) Microstrip transmission line, (c) Aperture coupling, and (d) Proximity coupling [50–52].

The coaxial probe feed is shown in Fig. 4.5a, and the microstrip transmission line is shown in Fig. 4.5b, while the aperture and proximity coupling are shown in Fig. 4.5c, b, appropriately. The classification of feeding techniques for the planar microstrip antenna with their performance is presented in Table 4.1. From this table, it is well shown that the proximity-coupled feeding technique allows high BW.

Table 4.1 Classification of feeding types for THz planar microstrip antennas

Characteristics	Coaxial feed (Not been used at THz)	Microstrip line feed	Aperture coupled	Proximity coupled
Fabrication	Solder and drilling needed	Easy	Alignments required	Alignments required
Configuration	Non-planar	Co-planar	Planar	Planar
Spurious feed radiation	Excess	Excess	Excess	Excess
Reliability	Poor because of soldering	Better	Good	Good
Polarization cleanness	Inferior	Inferior	Inferior	Inferior
Impedance matching	Simple	Simple	Simple	Simple
Impedance BW	2–5%	2–5%	2–5%	13%

Ma et al. [53] demonstrated a couple of THz antennas fed by a substrate-integrated waveguide (SIW). The first was a horn antenna, while the second was a rectangular planar patch antenna. Both of the antennas were experimentally verified. The peak measured gain and impedance BW for the first and second antennas were 15.46 dB, 27 GHz (129.5–156.5 GHz), 7.26 dB, and 29.1 GHz (125.3–154.4 GHz), respectively. Nissanov et al. [54] demonstrated a couple of THz antennas to increase the gain. The first was a planar microstrip array antenna, while the second was a planar array microstrip antenna with frequency selective surfaces (FSSs) for more gain enhancement. The simulated peak gain and impedance BW for the first and second antennas were 19.6 dB, 22 GHz (114.8–136.8 GHz), 21.55 dB, and 21.9 GHz (114.9–136.8 GHz), respectively. In addition, the first and the second antenna simulation results were compared with simulation with other design antenna solvers, and a good agreement was achieved. The performance summary of these antennas discussed above is presented in Table 4.2 according to the discussed literature.

4.3 Terahertz Microstrip Antenna Design Considerations

Due to the conductor and substrate losses, the potential demanding selection of the suitable microstrip laminate is gravely noticeable at mmWave and THz wave frequencies. All the PCBs have twain dielectric losses and conduction losses; however, the former losses are correlated with the laminate, and later losses are resistive and generated by the conductive materials applied in the PCBs. In a conductor, there are several mechanisms of energy dissipation. At high frequencies, in the Joule effect, in which the electrical resistance of the conductor induces ohmic

Table 4.2 THz antenna classification and performances

References	Antenna technology	Freq. range (GHz)	Measured BW (GHz)	Measured max. gain (dB)
[19]	THz tapered horn	270–330	60	18.4
[20]	THz horn	220–260	40	26
	THz horn with waveguide			32.5
	THz horn with waveguide and dielectric lens			39.5
[28]	THz elliptical lens	140–220	80	34.5
[29]	THz CPW with silicon lens	250–500	250	29.5[a] (26.5)
	THz double leaky slot with silicon lens			26.5[a]
[44]	THz UWB graphene bowtie	400–5000	4600[a]	4.25[a]
[45]	THz graphene planar patch	~480–500 and ~ 560–580	40[a]	7.47[a]
[53]	THz horn with SIW	129.5–156.5	27	15.46
	THz planar patch with SIW	125.3–154.4	29.1	7.26
[54]	THz planar microstrip array	114.8–136.8	22[b]	19.6[b]
	THz planar microstrip array with FSSs	114.9–136.8	21.9[b]	21.55[b]

Notes:
[a]Simulated
[b]Simulated and compared to a different type of antenna solver/simulator

heating on the PCB, there is a change in the current distribution along the conductor that involves the skin effect [55, 56]. The skin effect increases the effective resistance relative to the square root of frequency; this effect is typically considered for frequencies above 2 GHz [57].

The substrate loss is dependent on the dielectric polarization: when the signal is applied, the dipole moment formed in the dielectric starts to oscillate at the operating frequency; this energy dissipation is an additional source of signal attenuation at high frequencies. A complex permittivity was defined to include this effect. Materials can be classified following their complex-valued permittivity $\varepsilon(\omega) = \varepsilon' - j\varepsilon''$, upon the comparison of their real permittivity ε' and imaginary permittivity ε'' components. The dielectric constant D_k or relative permittivity $\varepsilon_r(=D_k)$ is $\varepsilon_r = \varepsilon'/\varepsilon_0$, where the real part of the permittivity ε' defines the material's capability to bond the electric energy or allow the electric field. Where ε_0 is the vacuum permittivity. The dissipation factor $D_f(=\tan\delta)$ represents the loss of the material, and it defines as $\tan\delta = \frac{\varepsilon''}{\varepsilon'} = \frac{\sigma}{\omega\varepsilon'}$, when σ is the material conductivity, and ω is the angular frequency of the wave at (rad/s) [56].

The attenuation constant α_D in (dB/m),due to dielectric loss is expressed as [55, 56]:

$$\alpha_D = 27.3 \frac{\varepsilon_r}{\sqrt{\varepsilon_{r_{eff}}}} \cdot \frac{(\varepsilon_{r_{eff}} - 1)}{(\varepsilon_r - 1)} \cdot \frac{\tan\delta}{\lambda_0} \tag{4.25}$$

The Eq. (4.25) exhibits the dielectric loss, which is directly proportional to the signal's frequency, the ε_r, and tanδ of the substrate. At higher frequencies (lower wavelength), the attenuation is higher, while low dielectric substrate or lower tanδ substrate ensures low attenuation. The properties of a dielectric substrate are sensitive to changes in temperature, and the dielectric losses increase with an increase in temperature. The conduction loss α_C at (dB/m) equals [55, 56]:

$$\alpha_C = 6.1 \cdot 10^{-5} \cdot \frac{\left(R_s' \cdot Z_0 \cdot \varepsilon_{r_{eff}}\right)}{h} \left[\frac{W'}{h} + \frac{0.667 \frac{W'}{h}}{\frac{W'}{h} + 1.444}\right] \cdot \Lambda, \quad \text{for } \frac{W}{h} \geq 1 \tag{4.26}$$

Z_0 is the characteristic impedance in [Ω] of the microstrip line and equals to [55, 56]:

$$Z_0 = \frac{120\pi}{\sqrt{\varepsilon_{r_{eff}}}\left\{\frac{W}{h} + 1.393 + 0.667 \ln\left(\frac{W}{h} + 1.444\right)\right\}}, \quad \text{for } \frac{W}{h} \geq 1 \tag{4.27}$$

Λ is a constant dependent on the microstrip thickness (h) and equals to [55, 56]:

$$\Lambda = 1 + \frac{h}{W'}\left[1 - \frac{1.25t}{\pi W} + \frac{1.25}{\pi} \ln\left(\frac{2h}{t}\right)\right], \quad \text{for} \frac{W}{h} \geq \frac{1}{2\pi} \tag{4.28}$$

W' is the effective microstrip thickness (h) and expressed as [55, 56]:

$$\frac{W'}{h} = \frac{W}{h} + \frac{1.25t}{\pi h}\left[1 + \ln\left(\frac{2h}{t}\right)\right], \quad \text{for } \frac{W'}{h} \geq \frac{1}{2\pi} \tag{4.29}$$

R_s' is the effective surface resistance in (Ω) of the conductor and equals to [55, 56]:

$$R_s' = \frac{1}{\delta}\left\{1 + \frac{2}{\pi} \tan^{-1}\left[1.4\left(\frac{R_{RMS}}{\delta}\right)^2\right]\right\} \tag{4.30}$$

where δ is the RF current skin-depth effect, and R_{RMS} is the conductor roughness's root-mean-square (RMS) value. Finally, the total loss α_T is the sum of the conduction loss α_C, dielectric loss α_D, radiation loss α_R, and leakage loss α_L [55, 56]:

$$\alpha_T = \alpha_C + \alpha_D + \alpha_R + \alpha_L \tag{4.31}$$

The radiation loss α_R is due to many issues related to energy radiating off the substrate. The radiation loss is (1) frequency-dependent- if the working frequency of the substrate increases, the radiation loss increases, (2) substrate thickness-dependent- if the thickness increases, the radiation loss increases, and (3) dielectric constant-dependent if the ε_r increase the radiation loss decreases. The leakage loss α_L is primarily due to electrical leakage through the dielectric betwixt conductor layers.

The microstrip arrays that work at the THz frequencies needing transmission lines in the feed network have a longitude of about 50–400μm, so high precision with tight tolerances associated with fabrication and high-cost fabrication technology is needed. Following PCB technology, the minimal widths/gaps of the line that can be etched at the microstrip substrate are 0.05 mm = 50μm, while the declared etching accuracy is about ±10μm [58]. For example, if we need that to etch a rectangular radiator patch on the microstrip substrate with a width of 50μm, the fabricated patch can be 50 ± 10μm, meaning that the minimum etched patch can be 40μm (−10μm possible etching machine accuracy). On the other hand, the maximum etched patch can be 60μm (+10μm possible etching machine accuracy). These deviations can alter the designed THz microstrip antenna parameters after fabrication. The fabrication complications that are regularly linked with printed antennas at the THz frequencies is to incorporate these antennas with silicon and gallium arsenide (GaAs) for the active electronics substrates, which typically have very high ε_r (e.g., $\varepsilon_r = 12.4$ for GaAs); as an outcome, surface waves are facilely excited in the laminate [59, 60].

Numerous radiators are frequently stationed together in the microstrip array antenna to attain an excessively directive pattern, and if we use large corporate/hybrid fed networks will increase the losses. An extensive array containing large antenna feed lines at corporate/hybrid fed at power dividers have discontinuities that introduce undesired radiation/spurious emission and interfere with its regular operation; these discontinuities are formed in the transition betwixt antenna and the transmission line, also at mmWave and THz waves there are a significant feed line losses because of large inductance ($\approx\omega L$), therefore a careful selection of the feed line must be made to prevent it. The requirement to reduce this problem is to use a low-loss feed mesh, which is simple to manufacture and diminishes radiation losses in the feed line. Large arrays are needed to attain excessively directive radiation patterns and cause a more significant mutual coupling, increasing the surface waves in the substrate's losses [61].

The BW restriction can be solved by applying a thick substrate. As the frequency magnifies, the thickness of the laminate in wavelengths magnifies, and it can likely have a laminate that is less or equal to $\lambda_0/4$ thick. Thicker laminates are either used to moderate the fabrication endurances but will cause surfaces waves losses. A thicker laminate leans to capture surface waves that do not increase the wanted radiation pattern. Instead, their radiated power is weighted an antenna loss due to surface wave losses [62, 63].

4.3.1 Rule of Thumbs for Choosing Terahertz Microstrip Antenna Substrates

Numerous guidelines help to reduce surface wave propagation problems, substrate and conductor losses, and radiation losses, such as:

1. For removing intolerable resonances through diverse circuit planes in PCBs, a general rule by applying a laminate height/thickness (h) that is slimmer than the one-eighth wavelength at the maximal working frequency ($f_{max} = c/\lambda_{max}$) and λ_{max} is the wavelength of the maximum working frequency. At the same time, c is the speed of light in the vacuum and the speed of EM waves [64, 65]:

$$h \leq \lambda_{max}/8 \tag{4.32}$$

As we can see, when the frequency increases, the wavelength decreases. These resonances arise betwixt a laminate's metal layers and can disrupt the desired signal spread of the quasi-TEM waves via the planar microstrip transmission lines. These resonances can disturb the wanted wave spread for PCBs and create surface waves. In contrast, the surface waves reduce antenna efficiency, the antenna gain, and the transmission signal's propagation distance, which is the primary concern at THz antennas; they distort radiation patterns and increase the mutual coupling in an antenna array configuration. In addition, this parasitic radiation reduces the antenna efficiency

2. A typical way is to lower the substrate height/thickness (h) to about one-tenth of the dielectric wavelength (ε_r) at the maximal working frequency (f_{max}) to lower the power radiated from surface waves [64, 65]:

$$h \leq \lambda_{max}/10\sqrt{\varepsilon_r} \tag{4.33}$$

A proper offset is achieved with [64, 65]:

$$\varepsilon_r \leq 4, \text{and } \tan\delta < 0.01 \tag{4.34}$$

3. The width of the electrical conductor (t) is also connected to the thickness of a PCB substrate, but a slimmer substrate will use a limited conductor width. Excessive conductor width (t) can cause spurious-mode resonances, interfering with the desired signal propagation through the microstrip conductors. To mitigate surface waves problems, the conductor width (t) should be one-eighth wavelength or lesser at the maximal working frequency (f_{max}) [64, 65]:

$$t \leq \lambda_{max}/8 \tag{4.35}$$

4. Copper with surface roughness is frequently applied for bettered joining to the dielectric laminate in PCBs. Nevertheless, this roughness can make higher losses. Therefore, as a guideline, when the skin-depth effect (δ) for a wanted frequency is parred or lesser than the copper surface roughness, the surface waves of the conductor will be meaningful [66]:

$$(\delta \leq \text{Surface roughness}) \tag{4.36}$$

For example, for the frequency of 110 GHz, the skin depth is about 0.2μm RMS so the copper surface roughness at the microstrip laminate must be lower than 0.2μm RMS so that the conductor losses will decrease.

4.3.2 *Example for Terahertz Recommended Microstrip Substrates*

The THz recommended microstrip substrates are used experimentally up to the frequency of 100 GHz [67–77], at least, which have $\varepsilon_r \leq 4$, and $\tan\delta < 0.01$. Table 4.3 summarizes the THz recommended microstrip substrates and their feathers.

Table 4.3 THz recommended microstrip substrates and their feathers

Substrate type	Substrate height/ thickness (h) (μm)	Copper/ gold thickness (t) (μm)	ε_r measured @ (GHz)	tanδ measured @ (GHz)	Working up to (GHz)
Standard electrodeposited (ED) Rogers RO3003 [67, 68]	127	Copper (8.75/17.5)	3 ± 0.04 @ 10	0.001 @ 10	100
			3.16 @ 100	0.0019 @ 100	
Rolled copper Rogers RO3003 [69]	127	Copper (8.75/17.5)	3 ± 0.04 @ 10	0.001 @ 10	100
			3.02 @ 100	0.0019 @ 100	
Taconic TLY-5 [70, 71]	127	Copper (17.5)	2.2 ± 0.02 @ 10	0.0009 @ 10	500
			2.252–2.254 @ (325–500)	0.00943–0.0111 @ (325–500)	
Isola Astra MT77 [72, 73]	127	Copper (17.5)	3 @ 10–20	0.0017 @ 10–20	170
			2.82 ± 0.05 @125	0.0025 ± 0.0005 @125	
BCB CYCLOTENE 3000 Series Polymer [74–77]	30	Gold (2)	2.65 @ 10	0.002 @ 10	312
			2.6 @ 65	0.009 @ 65	
			2.54 @ 1500	0.01 @ 1500	

4.4 Third Dimensions Commercial Terahertz Antenna Design Simulators

Some antenna design simulators are full-wave electromagnetic (EM) commercial simulators used to compute the EM performance of a third-dimensions (3D) structure used to design analysis and optimize THz antenna modeling. Examples of such simulators are Computer Simulation Technology Microwave Studio (CST MWS), Ansys High-Frequency Structure Simulator (HFSS), and multiphysics COMSOL [78–80].

4.4.1 CST MWS

The full-wave 3D EM commercial CST MWS simulator contains different solvers that analyze the 3D structures, which use different mathematical methods such as the time domain solver, frequency domain solver, and integral equation solver. The time-domain solvers such as the finite integration technique (FIT) solver is based on the finite-difference-time-domain (FDTD) technique and the transmission-line-matrix (TLM) technique [78]. The time domain solvers can compute broadband simulations in a single run by describing Maxwell's equations on time-grid space [81]. In contrast, the frequency domain solver is based on the finite element method (FEM). The FEM solver is based on a mathematical algorithm for finding approximate solutions to integral and partial differential equations by dividing the computational domain into singular small patches and finding local solutions that satisfy the differential equation inside the boundary of this patch so that the frequency domain solver can compute all ports simultaneously [82]. In comparison, the integral equation solver analyzes electrically large structure models. The asymptotic of the calculation region is lower than the object's boundaries, leading to a linear equation system within fewer unidentified volume solvers [83].

4.4.2 Ansys HFSS

Ansys HFSS is a full-wave 3D EM commercial simulator for designing and simulating high-frequency electronic devices such as filters, RF or microwave components, antennas, PCBs, and ICs. In addition, this simulator analyzes EM compatibility (EMC), EM interference (EMI), and radio frequency interference (RFI) within composite surroundings. The Ansys HFSS contains different solvers, such as memory matrix solvers, time and frequency domain solvers based on FEM solver, and integral equation solvers based on the method of moments (MoM) [79].

The MoM is a full-wave accurate numerical method for solving open-edge EM issues. This solver allows us to analyze EM wave propagation, scattering, and radiation issues within comparatively short computation and moderate computing systems. Furthermore, it solves the integral model of Maxwell's equations in contrast to differential solvers such as FEM or FDTD techniques [84, 85].

4.5 Terahertz Antenna Design Validation Techniques

The results attained within the predictions within full-wave 3D EM antenna design solvers must be verified to trust the simulation results and use the antenna in natural environments. There are a couple of significant, accepted ways of validating the wanted mmWave or THz band antenna structure's design and simulation results:

1. Software comparison/validation by applying two different 3D full-wave EM commercial design solver works within varying simulation methods. For example, we can design and simulate the wanted antenna within the CST MWS simulator [78] that operates within a FIT solver and uses the FDTD method. We can use the Ansys HFSS simulator [79] to verify the wanted antenna's simulation results within the FEM approach. For the simulation results to coincide, leastwise one of the simulations should be done with an adaptive mesh algorithm or equivalent algorithm.

In addition to simulator comparisons, it is necessary to carry out additional simulations due to the deviation of the production machines, which is $\pm 10 \mu m$ [58]. The antenna dimensions are small for mmWave and THz frequencies due to the small wavelength of the antenna size. So the accuracy of these machines can alter the dimensions of the fabricated antennas. The comparison of the simulators must be made with the stated and measured values of ε_r and $\tan\delta$ because these values affect the simulation results and can cause a deviation between the results and the measurement results of the antenna at mmWave and THz frequencies, as done in [86, 87]. If these values are unknown, some deviation in these parameters should be considered in the simulators; for example, a deviation of $\pm 5\%$ of ε_r and increasing the $\tan\,\delta$ the simulations should be performed after considering this deviation within the simulation solver. These parameters are frequency depended and can alter the measured results of the mmWave and THz antennas, as done in [88]. So the simulation results will be accurate and close to products that should be attained at experimental validation with a mmWave and THz prototype antenna.

2. Another way is by experimental validation by manufacturing the original mmWave or THz band prototype antenna and measuring the results attained within a THz vector network analyzer (VNA) installed in an anechoic chamber. To compare the simulation results via the measurement results to get an acceptable agreement between them as done in reference [89, 90].

In academia, respecting a mmWave or THz band, it is additionally authorized to lower the high payment of fabrication and evaluation of the mmWave or THz archetype antenna, reaching about a dozen thousand US dollars (USD) [91, 92], and downscale it to a low-frequency unit-cell antenna. Initially, a 3D full-wave EM antenna simulator should simulate the low-frequency antenna design. Consequently, it should be fabricated as a downscale low-frequency unit-cell archetype antenna and measured to verify its simulation results within the measurements as long as the ultimate verification of the mmWave or THz band antenna simulation results has been done in [93].

4.5.1 Fabrication of Terahertz Antenna

As described, at THz frequencies, the dimensions of the antenna are small because of the short wavelength; therefore, manufacturing machines with as high etching accuracy as possible is needed. Furthermore, in addition, the etching accuracy deviation of the manufacturing machines is about ±10μm [58], which affects the measurement results of the electrical parameters of the THz antenna due to different fabricated antenna sizes vs. designed THz antenna. Therefore, improving THz antennas is connected to the growth of manufacturing technology. The designed THz antennas can be fabricated, depending on the operating frequency, by commercially available computer numerical control (CNC) machining, electrical discharge machining (EDM), printed-circuit-board (PCB), focused ion beam (FIB), photolithography, nano-imprint lithography, and electron beam lithography. Additionally, deep reactive ion etching (DRIE), micromachining, laser milling, and 3D printing technologies [94–96]. However, due to the fabrication accuracy tolerance, the PCB and 3D printing processes can only handle THz antennas up to 220–325 GHz. Therefore, micromachining is used beyond this band, while high-end CNC milling machines can have an acceptable tolerance accuracy of ±1μm [97].

The lithography technology using the FIB technique can be used for the evolution of THz spiral antennas and 3D printing, which is frequently used for printing THz lenses and horn antennas because of the superiority of rapid prototyping at a low expense, miniaturization, and high accuracy [98, 99]. In addition, one of the advantages of silicon DRIE microfabrication is that it is a clustering process that facilitates us to micromachine up to thousands of THz components of one wafer, contradictory to the traditional serial milling machining technique [100].

4.5.2 *Measurement of Terahertz Antenna*

(a) *Vector Network Analyzer*

Vector network synthesis is a confirmed method for deciding the radio frequency (RF) reflection and transmission features of apparatus, circuits, or linear time-invariable networks. First, a VNA sets the frequency-pendent and complex-valued scattering parameters assigned for an n-port network considering measurands. Then, these scattering parameters are used to obtain component features such as mixer attenuation filter transmission, antenna features such as antenna factor, gain, or transmission channel features such as channel impulse reaction after a conversation of measurement results within the time domain [101, 102]. The major components of a VNA are the generator within the source switch that supplies the excitation signal. This test set divides backward and forward traveling EM waves, and the measurement and reference channel receivers are essential to accurately measure the scattering parameters as corresponding quantities. In addition, the coaxial connector typically determines the upper-frequency limit of a VNA system used, for example, 2.92 mm (40 GHz), 2.4 mm (50 GHz), or 1.85 mm (67 GHz) [103]. Finally, a set of frequency converters, for example, VNA extender (VNAX) specific to the respective band, will be used to extend the available frequency range to the waveguide bands to the THz frequencies, which can cover the frequency range of 50–1100 GHz, while the possible VNAX configuration can be full two-port, through-reflect, through and one port which is illustrated in Fig. 4.6 [104].

(b) *Anechoic Chamber*

Compact antenna test ranges (CATRs), VNA, AUT holders, 3D position controllers within robotic arms, and VNAX are situated in anechoic chambers. These chambers are impermeable regions, shielded and coated by EM absorbing matter within a

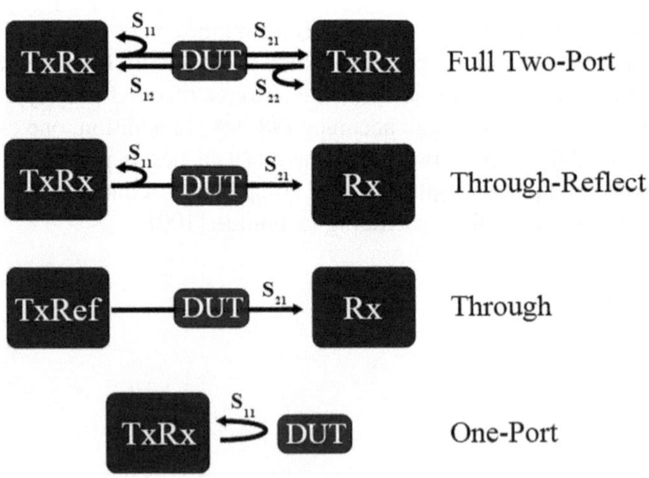

Fig. 4.6 The VNAX possible configurations [104]

dielectric constant near one to refrain reflections that mimic free space propagation because of the absorption of the radiation absorbing material (RAM) within about – 50 dB reflectivity. These anechoic chambers are used for antenna under test (AUT), radar cross-section (RCS), radiated emissions, and EMC measurements. The significant superiorities of applying anechoic chambers in antenna measures are all weather conditions, the controlled surroundings, and the nonappearance of RFIs. On the other hand, the significant deficiency is the sizeable maximum dimension causing an ample, expensive unique space [105].

(c) *Measurements Methods*

The VNA impedance calibration is the first process before testing the AUT by standard port calibration of the short-open-load (SOL) adapters. After this calibration is finished, the calibrated probe can be connected to the AUT via the AUT holder for return loss (S_{11}) measurements, which can be recorded within the VNA. The 3D radiation patterns at farfield positioning can be measured via the VNA by changing the robotic arms, which hold a standard RX horn antenna, through 3D position controllers and a personal computer (PC) for comparison. A gain comparison with THz standard horn antenna within known gain is being used to reference reliable 3D radiation pattern measurements and AUT gain measurements [106, 107].

(d) *Antenna Field Regions*

Space bounded THz antenna is generally divided into the following sections: (a) Reactive near-field, (b) Radiating near-field ("Fresnel"), and (c) Farfield ("Fraunhofer") regions as shown in Fig. 4.7 [108].

The radiation-reactive near-field region is the area of the sources and the adjacent environment, where there are reactive fields, which the energy includes but does not spread in free space. For the majority of antennas, the external border of this region is

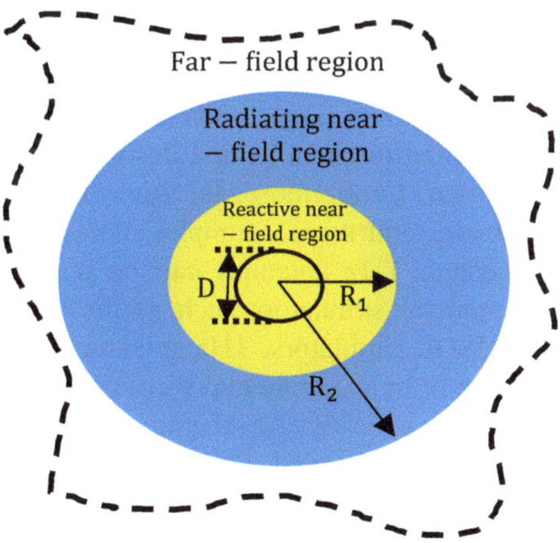

Fig. 4.7 Field regions of the THz antenna

valid at a distance of $r < R_1 < 0.62\sqrt{(D^3/\lambda)}$. D is the largest dimension from the antenna plane, and λ is the antenna's wavelength. The radiating near-field ("Fresnel") region is specified, where the developed radiation fields are with real energy. Their shape varies depending on the distance from the antenna, and their wavefront is not planar. This region is valid at a distance of $R_1 < r < R_2 = (2D^2)/\lambda$. Farfield ("Fraunhofer") region is specified, where there is a clean planar wave, which includes an E-field and an H-field that are orthogonal to each other and perpendicular to the direction of THz wave propagation. This region exists at the distance of $r > R_2 > (2D^2)/\lambda$.

4.6 Open Research Challenges

Based on this chapter, THz antenna technologies have several open challenges: (1) More research and developments on THz graphene antenna technologies. (2) Additional analysis and evolutions on THz planar microstrip antenna, including planar horn antenna and dielectric lens planar antenna. (3) Further investigation and improvements on coaxial feed beyond 110 GHz.

4.7 Summary

This chapter discusses THz antenna technologies for 6G communication system networks and validation techniques. The THz antenna technologies are classified into three groups depending on their content processing: metallic antennas, dielectric antennas, and state-of-the-art materials antennas. We can use the THz horn antenna, THz dielectric lens antennas, THz nano, and Plasmonic antennas based on graphene, and THz planar microstrip for the THz wireless communication antennas. In addition, the THz rectangular microstrip antenna design equations, feeding techniques of the microstrip antenna, and design considerations of THz microstrip antennas are presented. Furthermore, the rule of thumb for choosing THz microstrip antenna laminates and some examples of recommended microstrip antenna laminates are presented. THz antennas can be designed and simulated within full-wave EM commercial simulators such as the CST MWS, Ansys HFSS, and multiphysics COMSOL simulators. THz antenna design validation can be within a solver comparison or by experimental verification, which is the best way of validation, but this technique is costly. At the end of this chapter, some proper fabrication of the THz antennas is presented. Furthermore, the VNA and anechoic chamber are presented, which are needed for the experimental verification of THz antennas.

References

1. D. Pavlidis, *Fundamentals of Terahertz Devices and Applications. John Wiley & Sons*, Chapter 11, pp. 447–477, July 2021.
2. M. J. W. Rodwell, "Beyond 5G: 100–340GHz transistor, IC, and system design," *Short Course: Device Research Conference*, University of Michigan, USA, 23 June 2019, pp. 1–68.
3. O. D. Oyeleke, S. Thomas, O. I. Bismark, P. Nzerem, and I. Muhammad, "Absorption, diffraction, and free space path losses modeling for the terahertz band," *International Journal of Engineering and Manufacturing*, vol. 1, pp. 54–65, Feb. 2020.
4. D. Serghiou, M. Khalily, T. W. C. Brown and R. Tafazolli, "Terahertz channel propagation phenomena, measurement techniques and modeling for 6G wireless communication applications: A survey, open challenges, and future research directions," *IEEE Communications Surveys & Tutorials*, vol. 24, no. 4, pp. 1957–1996, Fourth quarter 2022.
5. Attenuation by Atmospheric Gases, International Telecommunication Union ITU-R Recommendation P.676-10, pp. 1–22, Sep. 2013, [online] available at: https://www.itu.int/dms_pubrec/itu-r/rec/p/R-REC-P.676-10-201309-S!!PDF-E.pdf
6. M. Mandehgar and D. R. Grischkowsky, "Understanding dispersion compensation of the THz communication channels in the atmosphere," *IEEE Photonics Technology Letters*, vol. 27, no. 22, pp. 2387–2390, Nov. 2015.
7. F. Boccardi, R. W. Heath, A. Lozano, T. L. Marzetta, and P. Popovski, "Five disruptive technology directions for 5G," *IEEE Communications Magazine*, vol. 52, no. 2, pp. 74–80, Feb. 2014.
8. Tersence, "Attenuation of terahertz radiation in various materials," [Online], Feb. 2018, http://terasense.com/news/attenuation-terahertz/
9. J. Kokkoniemi, P. Rintanen, J. Lehtomaki, and M. Juntti, "Diffraction effects in terahertz band - measurements and analysis," *Proceedings of IEEE Global Communications Conference (GLOBECOM)*, Washington DC, USA, 04–08 December 2016, pp. 1–6.
10. F. Liu, Y. Cui, C. Masouros, J. Xu, T. X. Han, Y. C. Eldar, and S. Buzzi, "Integrated sensing and communications: toward dual-functional wireless networks for 6G and beyond," *IEEE Journal on Selected Areas in Communications*, vol. 40, no. 6, pp. 1728–1767, June 2022.
11. W. Tong and P. Zhu, "6G The Next Horizon: From Connected People and Things to Connected Intelligence," Cambridge University Press, pp. 1–462, Apr. 2021.
12. 6G: The next hyper-connected experience for all, *Samsung 6G Vision White Paper*, pp. 1–42, Jul. 2020. https://cdn.codeground.org/nsr/downloads/researchareas/6G%20Vision.pdf.
13. T. K. Ostmann and T. Nagatsuma, "A review on terahertz communications research," *Journal of Infrared, Millimeter, and Terahertz Waves*, vol. 32, no. 2, pp. 143–171, Jan. 2011.
14. S. K. Tripathi and A. Kumar, "Performance Comparison of Graphene Terahertz Antenna and Copper Terahertz Antenna," In book: Advances in Computational Intelligence and Communication Technology: Proceedings of Springer CICT, Singapore, pp. 489–497, 2019.
15. Y. He, Y. Chen, L. Zhang, S. W. Wong, and Z. N. Chen, "An overview of terahertz antennas," *Proceeding of IEEE China Communications*, vol. 17, no. 7, pp. 124–165, July 2020.
16. X. Chen and X. Liu, "Terahertz Antennas and Measurement," In book: Handbook of Antenna Technologies, Springer Nature, Singapore, pp. 620–659, 2016.
17. C. A. Balanis, "Horn Antennas," In book: Antenna Theory Analysis and Design, 4th Edition, NJ: John Wiley & Sons, pp. 719–782, 2016.
18. G. Chattopadhyay, M. Alonso-delPino, N. Chahat, D. G. Ovejero, C. Lee, and T. Reck, "Terahertz Antennas and Feeds," In book: Aperture Antennas for Millimeter and Sub-Millimeter Wave Applications, Springer International Publishing, Switzerland, pp. 335–406, 2018.
19. B. Aqlan, M. Himdi, L. L. Coq, and H. Vettikalladi, "Sub-THz circularly polarized horn antenna using wire electrical discharge machining for 6G wireless communications," *IEEE Access*, vol. 8, pp. 117245– 117252, June. 2020.

20. J. N. Lee, Y. K. Cho, J. H. Jung, and S.B. Hyun, "High-gain sub-terahertz lens horn antenna with a metal guide," *IET Electronics Letters*, vol. 54, no. 14, pp. 689–691, July 2020.
21. D. F. Filipovic, S. S. Gearhart, and G. M. Rebeiz, "Double-slot antennas on extended hemispherical and elliptical silicon dielectric lenses," *IEEE Transactions on Microwave Theory and Techniques*, vol. 41, no. 10, pp. 1738–1749, Oct. 1993.
22. L. E. G. Munoz, S. Preu, S. Malzer, G. H. Dohler, J. Montero-de-Paz, R. Gonzalo, David G. Ovejero, D. Segovia-Vargas, D. Lioubtchenko, and A. V. Raisanen, *Principles of Emission of THz Waves.* In book: *Semiconductor Terahertz Technology Devices and Systems at Room Temperature Operation*, John Wiley & Sons, pp. 90–93, 2015.
23. C. A. Fernandes, E. B. Lima, and J. R. Costa, "Dielectric Lens Antennas," In book: Handbook of Antenna Technologies, Springer, Singapore, pp. 1001–1064, 2016.
24. U. Nissanov, G. Singh, and A. Akinola, "Low sidelobe levels (SLLs) microstrip antennas for THz bio-sensors and THz communications," *Sensors International*, vol. 2, pp. 100097/1–10, April 2021.
25. G. M. Rebeiz, "Millimeter-wave and terahertz integrated circuit antennas," *Proceedings of the IEEE*, vol. 80, no. 11, pp. 1748–1770, Nov. 1992.
26. G. Molesini, "Geometrical Optics," In book: Encyclopedia of Condensed Matter Physics, pp. 257–267, 2005.
27. M. Born and E. Wolf, "Principles of Optics," Cambridge University Press, 7th edition, UK, p. 36–38, 1999.
28. M. A. Campo, G. Carluccio, S. Bruni, and N. Llombart, "Dielectric gratings enhancing the field of view in low dielectric permittivity elliptical lenses," *IEEE Transactions on Antennas and Propagation,* vol. 69, no. 11, pp. 7308–7322, Nov. 2021.
29. S. V. Berkel, E. S. Malotaux, C. D. Martino, M. Spirito, D. Cavallo, A. Neto, and N. Llombart, "Wideband double leaky slot lens antennas in CMOS technology at submillimeter wavelengths," *IEEE Transactions on Terahertz Science and Technology*, vol. 10, no. 5, pp. 540–553, Sep. 2020.
30. S. Kavitha, K. V. S. S. S. Sairam, and A. Singh, "Graphene plasmonic nano-antenna for terahertz communication," *S N Applied Science*, vol. 4, pp. 114/1–11, Mar. 2022.
31. I. F. Akyildiz, J. M. Jornet, and C. Han, " TeraNets: Ultra-broadband communication networks in the terahertz band," *IEEE Wireless Communications*, vol. 21, no. 4, pp. 130–135, Aug. 2014.
32. E. Lallas, "Key roles of plasmonics in wireless THz nano communications-a survey," *MDPI*, vol. 9, no. 5488, pp. 1–34, Dec. 2019.
33. I. Laser, C. Kremers, D. N. Chigrin, J. M. Jornet, M. C. Lemme, A. Cabellos-Aparicio, and E. Alarcon, "Characterization of graphene-established nano-antennas in the terahertz band," *Proceedings of IEEE 6th European Conference on Antennas and Propagation (EUCAP)*, Prague, Czech Republic, 26–30 March 2012, pp. 1–5.
34. Y. Luo, Q. Zeng, X. Yan, T. Jiang, R. Yang, J. Wang, Y. Wu, Q. Lu, and X. Zhang, "A graphene-established tunable negative refractive index metamaterial and its application in dynamic beam-tilting terahertz antenna," *Microwave and Optical Technology Letters*, vol. 61, no. 12, pp. 2266–2672, Dec. 2019.
35. G. Bansal, A. Marwaha, and A. Singh, "A graphene-established multiband antipodal Vivaldi nano-antenna for UWB applications," *Journal of Computational Electronics*, vol. 19, pp. 709–718, Feb. 2020.
36. G. W. Hanson, "Dyadic Green's functions and guided surface waves for a surface conductivity model of graphene," *Journal of Applied Physics*, vol. 103, no. 06430, pp. 0643021–0643028, Jan. 2008.
37. C. L. Wang, Y. Q. Wang, H. Hu, D. J. Liu, D. L. Gao, and L. Gao, "Reconfigurable sensor and nano-antenna by graphene-tuned fano resonance," *Optics Express*, vol. 27, no. 24/25, pp. 35925–35935, Nov. 2019.
38. L. Cui, J. Wang, and M. Sun, "Graphene plasmon for optoelectronics," *Reviews in Physics*, vol. 6, pp. 100054/1–20, June 2021.

39. M. H. Mubarak, S. Hara, I. Watanabe, and A. Kasamatsu, "Beam steering in graphene-based sub-THz dipole phased array antenna," *Proceedings of IEEE International Symposium on Radio-Frequency Integration Technology (RFIT)*, Hiroshima, Japan, 02–04 Sep. 2020, pp. 154–156.
40. D. Yang, "Analysis on the conductivity of graphene-based composite material," *IOP Publishing Journal of Physics: Conference Series*, vol. 2011, no. 012002, pp. 1–8, July 2021.
41. X. Fu, F. Yang, C. Liu, X. Wu, and Tie Jun Cui, "Terahertz beam steering technologies: from phased arrays to field-programmable metasurfaces," *Advanced Optical Material*, vol. 8, no. 3, pp. 1900628/1–22, 2020.
42. A. Singh, M. Andrello, N. Thawdar, and J. M. Jornet, "Design and operation of a graphene-based Plasmonic nano-antenna array for communication in the terahertz band," *IEEE Journal on Selected Areas in Communications*, vol. 38, no. 9, pp. 2104–2117, Sep. 2020.
43. I. Soltani, T. Soltani, and T. Aguili, "Electromagnetic modeling of a reconfigurable graphene-established metasurface using MoM-GEC method for 3D digital holography," *Journal of Electromagnetic Waves and Applications*, vol. 33, no. 13, pp. 1661–1680, June 2019.
44. S. Dash and A. Patnaik, "Graphene Plasmonic bowtie antenna for UWB THz application," *Proceedings of IEEE 24th National Conference on Communications (NCC)*, Hyderabad, India, 25–28 Feb. 2018, pp. 1–4.
45. S. K. Tripathi and A. Kumar, "Performance Comparison of Graphene Terahertz Antenna and Copper Terahertz Antenna," in X. Z. Gao, S. Tiwari, M. Trivedi, and K. Mishra, (eds) "Advances in Computational Intelligence and Communication Technology," Advances in Intelligent Systems and Computing, vol 1086. Springer, Singapore, pp 489–497, 2021.
46. T. A. Milligan, *Microstrip Antennas*. In book: *Modern Antenna Design*. IEEE Press and John Wiley & Sons Inc., Second edition, NJ, pp. 285–335, 2005.
47. A. Kishk, *Advancement in Microstrip Antennas with Recent Applications*. InTech, Croatia, pp. 1–383, 2013.
48. A. Elrashidi, K. Elleithy, and H. Bajwa, "Performance Analysis of a microstrip printed antenna conformed on cylindrical body at resonance frequency 4.6 GHz for TM_{01} mode," *Procedia Computer Science*, vol. 10, pp. 775–784, 2012.
49. C. A. Balanis, *Microstrip, and Mobile Communications Antennas*. In book: *Antenna Theory Analysis and Design*. John Wiley & Sons 4th Edition, NJ, pp. 783–874, 2015.
50. Y. Hang and K. Boyle, *Antennas from Theory to Practice*. Wiley Telecom: UK, 2008.
51. D. M. Pozar, "A review of aperture coupled microstrip antennas: history, operation, development, and applications," The University of Massachusetts, Amherst, USA, May 1996, https://www.ecs.umass.edu/ece/pozar/aperture.pdf
52. J. R. James and P. S. Hall, *Handbook of Microstrip Antennas*. IET Electromagnetics Waves Series 28, UK, pp. 1–1312, 1989.
53. C. Ma, S. Ma, L. Dai, Q. Zhang, H. Wang, and H. Yu, "Wideband and high-gain D-band antennas for next-generation short-distance wireless communication chips," *IEEE Transactions on Antennas and Propagation*, vol. 69, no. 7, pp. 3700–3708, July 2021.
54. U. Nissanov, G. Singh, E. Gelbart, and N. Kumar, "Highly directive microstrip array antenna with FSS for future generation cellular communication at THz band," *Wireless Personal Communications*, vol. 118, no. 1, pp. 599–617, Jan. 2021.
55. G. Kouemou, "Radar Technology," InTechOpen, Croatia, pp. 297–317, 2010.
56. D. Liu, B. Gaucher, U. Pfeiffer, and J. Grzyb, *Advanced Millimeter-Wave Technologies Antennas, Packaging, and Circuits*. John Wiley and Sons, UK, pp. 49–80, 2009.
57. S. Nagai, T. Hayashi, and A. Sanada, "Measurements of Anomalous skin effect in 1 THz band," *Proceedings of IEEE MTT-S International Microwave Symposium Digest (MTT)*, Seattle, USA, 02–07 June 2013, pp. 1–3.
58. J. Francey and T. Bateman, "PCB technology requirements for millimeter-wave interconnect and antenna," *The PCB Magazine*, pp. 36–42, Apr. 2017.
59. C. Chen, K. Kaj, X.g Zhao, Y. Huang, R. D. Averitt, and X. Zhang, "On-demand terahertz surface wave generation with microelectromechanical-system-based metasurface," *Optica*, vol. 9, no. 1, pp. 17–25, Jan. 2022.

60. P. Fernández, A. D. Corso, and A. Baldereschi, "Polarized Wannier functions: Ab-initio study of the dielectric properties of silicon and gallium arsenide," *Physical Review B*, vol. 58, no. 12, pp. R7481–R7483, Sep. 1998.
61. J. D. Preez and S. Sinha, *Millimetre-Wave Antennas Configurations and Application*. Springer Signals and Communication Technology, Switzerland, pp. 45–46, 2016.
62. D. M. Pozar, "Considerations for millimeter wave printed antennas," *IEEE Transactions on Antennas and Propagation*, vol. 31, no. 5, pp. 740–747, Sep. 1983.
63. F.K. Schwering, "Millimeter wave antennas," *Proceedings of IEEE*, vol. 80, no. 1, pp. 92–102, Jan. 1992.
64. L. K. S. Granados, "Antennas for Millimeter-Wave Applications," Ph.D. Thesis, Electrical and Electronic Engineering, Santa María Tonantzintla University, Puebla, Mexico, pp. 1–89, 2016.
65. U. Nissanov, G. Singh, and A. Akinola, "Sixth-generation (6G) microstrip antenna with high-gain," *International Journal on Communications Antenna and Propagation*, vol. 11, no. 4, pp. 279–287, Aug. 2021.
66. S. Hinaga, M. Y. Koledintseva, P. K. R. Anmula, and James L. Drewniak "Effect of conductor surface roughness upon measured loss and extracted values of PCB laminate dissipation factor," *IPC APEX EXPO Conference*, Las Vegas USA, 31 March – 2 April 2009, pp. 1–14, http://web.mst.edu/~marinak/files/My_publications/Papers/S_Hinaga-IPC2009_Formatted.pdf
67. G. F. Hamberger, U. Siart, and T. F. Eibert, "A broadband low-loss WR10 waveguide to microstrip line transition with T-shaped probe," *Progress in Electromagnetics Research Letters (PIER-L)*, vol. 73, pp. 17–22, Jan. 2018.
68. O. Khan, J. Pontes, X. Li., and C. Waldschmidt, "A wideband variable-width microstrip grid array Antenna," *Proceedings of IEEE 44th European Microwave Conference*, Rome, Italy, 6–9 Oct. 2014, pp. 1–5.
69. M. S. Esfahlan and I. Tekin, "77 GHz PCB patch antenna," *URSI-TURKİYE*, Ankara, Turkey, 1–3 Sep. 2016, pp. 1–3.
70. Z. W. Miao, Z. C. Hao, Y. Wang, B. B. Jin, J. B. Wu, and W. Hong, "A 400-GHz wideband high-gain quartz-established single-layered folded reflectarray antenna for terahertz," *IEEE Transactions on Terahertz Science and Technology*, vol. 9, no. 1, pp. 78–88, Nov. 2018.
71. Y. W. Wu, Z. C. Hao, M. C. Tao, X. Wang, and J. S. Hong, "A simple and accurate method for extracting super wideband electrical properties of the printed circuit board," *IEEE Access*, vol. 7, pp. 57321–57331, April 2019.
72. A. Lamminen, M. Lahti, D. del-Rio, J. Säily, J. F. Sevillano, and V. Ermolov, "Characterization of interconnects on multilayer high-frequency PCB for D-band," *Proceedings of IEEE 2nd 6G Wireless Summit (6G SUMMIT)*, 17–20 March 2020, pp. 1–6.
73. A. Lamminen, P. Roux, J. Ala-Laurinaho, J. Säily, M. Moretto, M. G. L. Frecassetti, and V. Ermolov, "Technologies for D-band links with beam steering functionality," *Proceedings of IEEE Asia-Pacific Microwave Conference (APMC)*, 10–13 Dec. 2019, pp. 1–4.
74. W. Choe and J. Jeong, "A broadband THz on-chip transition using a dipole antenna with an integrated balun," *Electronics*, vol. 7, no. 10, pp. 236/1–12, Oct. 2018.
75. S. Costanzo, A. Borgia, I. Venneri, and G. Di-Massa, "Millimeter-waves configurations on benzocyclobutene dielectric substrate," Radio Engineering, vol. 20, no. 4, pp. 785–789, Dec. 2011.
76. S. Costanzo, I. Venneri, G. Di Massa, and A. Borgia, "Benzocyclobutene as substrate material for planar millimeter-wave structures: Dielectric characterization and application," *Journal of Infrared, Millimeter, and Terahertz Waves*, vol. 31, no. 66, pp. 66–77, July 2009.
77. A. Borgia, "New Materials and Technologies for Compact Antennas and Circuits at Millimeter Frequency," Ph.D. Thesis, Electrical and Electronic Engineering, University Della Calabria, Italy, pp. 1–81, 2012.
78. https://www.3ds.com/products-services/simulia/products/cst-studio-suite/
79. https://www.ansys.com/products/electronics/ansys-hfss
80. https://www.comsol.com/

81. R. Marklin, *The Finite Integration Techniques as a General Tool to Compute Acoustic, Electromagnetic, Elastodynamic, and Coupled Wave Fields*. IEEE Press, New York, USA, pp. 201–244, 2002.
82. https://www.comsol.com/multiphysics/finite-element-method
83. https://space.mit.edu/RADIO/CST_online/mergedProjects/3D/special_overview/frequency_domain_solver_overview_-_meshtype__surface_mesh.htm
84. Y. Zhang, T. K. Sarkar, M. Taylor, and H. Moon, "Solving MoM problems with million-level unknowns using a parallel out-of-core solver on a high-performance cluster," *Proceedings of IEEE Antennas and Propagation Society International Symposium*, North Charleston, USA, 01–05 June 2009, pp. 1–4.
85. Y. Zhang, J. Porter, M. Taylor, and T. K. Sarkar, "Solving challenging electromagnetic problems using MoM and a parallel out-of-core solver on high-performance clusters," *Proceedings of IEEE AP-S & USNC/URSI Symposium*, San Diego, USA, July 2008, pp. 1–4.
86. U. Nissanov, "6G Rotman lens D-band beam-steering microstrip antenna," *Nature Journal of Computational Electronics*, pp. 1–14, Jan. 2022.
87. U. Nissanov and G. Singh, "mmWave/THz reconfigurable ultra-wideband (UWB) microstrip antenna," *Progress in Electromagnetic Research C*, vol. 111, pp. 207–224, April 2021.
88. U. Nissanov and G. Singh, "Beamforming D-band phased array microstrip antennas," *Elsevier Sensors International*, vol. 3, pp. 100196/1–14, July 2022.
89. B. Aqlan, M. Himdi, H. Vettikalladi, and L. Coq, "A 300 GHz low-cost high-gain fully metallic Fabry-Perot cavity antenna for 6G terahertz wireless communications," *Scientific Reports*, vol. 11, no. 7703, pp. 1–10, April 2021.
90. H. Kim, W. Choe, and J. Jeong, "A Terahertz CMOS V-shaped patch antenna with defected ground structure," *Sensors*, vol. 18, no. 8, pp. 2432/1–13, July 2018.
91. M. A. Jamshed, A. Nauman, M. A. B. Abbasi, and S. W. Kim, "Antenna selection and designing for THz applications: Suitability and performance evaluation: A Survey," *IEEE Access*, vol. 8, pp. 113246–113261, June 2020.
92. U. Nissanov, G. Singh, and A. Akinola, "Low sidelobe levels (SLLs) microstrip antennas for THz bio-sensors and THz communications," *Elsevier Sensors International*, vol. 2, no. 100097, pp. 1–10, April 2021.
93. R. Xu, S. Gao, B. S. Izquierdo, C. Gu, P. Reynaert, A. Standaert, G. J. Gibbons, W. Bosch, M. E. Gardiner, and D. Li, "A review of broadband low-cost and high-gain low-terahertz antennas for wireless communications applications," *IEEE Access*, vol. 8, pp. 57615–57629, Mar. 2020.
94. A. Abohmra, "Terahertz Antenna Design for Future Wireless Communication," Ph.D. Thesis, Electrical and Electronic Engineering, University of Glasgow, Scotland, pp. 1–137, 2021.
95. F. Laermer, S. Franssila, L. Sainiemi, and Kai Kolari, *Deep Reactive Ion Etching*. M. Illi, M. Paulasto-Krockel, M. Petzold, H. Theuss, T. Motooka, and V. Lindroos, (Eds.), *Handbook of Silicon Based MEMS Materials and Technologies*. Elsevier, pp. 349–361, 2010.
96. S. Lavadiya and V. Sorathiya, *Terahertz Antenna: Fundamentals, Types, Fabrication, and Future Scope*. S. Das, N. Anveshkumar, J. Dutta, and A. Biswas, (Eds.), *Advances in Terahertz Technology and Its Applications*. Springer Nature, Singapore, pp. 128–130, 2021.
97. C. H. Chan, "THz antennas – Design, fabrication, and testing," *Proceeding of IEEE 13th European Conference on Antennas and Propagation (EuCAP)*, Krakow, Poland, 31 March 2019–05 April 2019, pp. 1–4.
98. B. Zhang, Y. X. Guo, H. Zirath, and Y.P. Zhang, "Investigation on 3-D-printing technologies for millimeter-wave and terahertz applications," *Proceedings of IEEE*, vol. 105, no. 4, pp. 723–736, Apr. 2017.
99. L. Guo, H. Meng, L. Zhang, and J. Ge, "Design of MEMS on-chip helical antenna for THz application," *Proceedings of IEEE MTT-S International Microwave Workshop Series on Advanced Materials and Processes for RF and THz Applications (IMWS-AMP)*, Chengdu, 20–22 July 2016, pp. 1–4.

100. C. Lee, G. Chattopadhyay, E. Decrossas, A. Peralta, I. Mehdi, C. A. L. Sevillano, M. A. Del Pino, and N. Llombart, "Terahertz antenna arrays with silicon micromachined-based microlens antenna and corrugated horns," *Proceedings of IEEE International Workshop on Antenna Technology (iWAT)*, Seoul, S. Korea, 04–06 March 2015, pp. 70–73.
101. Understanding the Fundamental Principles of Vector Network Analysis, Keysight application notes, pp. 1–17, https://www.keysight.com/us/en/assets/7018-06841/application-notes/5965-7707.pdf
102. A. V. Räisänen, J. A. Laurinaho, T. Crowe, S. Pivnenko, M. Sierra Castañer, and V. Viikari, *Antenna Measurements at Millimeter and Submillimeter Wavelengths.* A. Boriskin and R. Sauleau, *Aperture Antennas for Millimeter and Sub-Millimeter Wave Applications.* Springer International Publishing, Switzerland, pp. 412–430, 2018.
103. T. Kleine-Ostmann, *Vector Network Analyzer (VNA).* T. Kürner, D. M. Mittleman, T. Nagatsuma, eds., *THz Communications.* Springer Series in Optical Sciences, vol 234. Springer, Cham. pp. 29–34, 2022.
104. https://www.vadiodes.com/en/products/vector-network-analyzer-extension-modules.
105. L. H. Hemming, "Electromagnetic Anechoic Chambers: A Fundamental Design and Specification Guide," Wiley-IEEE Press, 2002.
106. L. L. Coq, N. Mézières, P. Leroy, and B. Fuchs, "Some contributions for antenna 3D farfield characterization at terahertz," *Sensors*, vol. 21, no. 4, pp. 1438/1–11, Feb. 2021.
107. Z. Zheng, Y. Zhang, and J. Mao, "Testing method of a millimeter-wave/THz farfield antenna measurement setup," *Proceedings of IEEE 9th Asia-Pacific Conference on Antennas and Propagation (APCAP)*, Xiamen, China, 04–07 Aug. 2020, pp. 1–2.
108. http://elearningatria.files.wordpress.com/2013/10/ece-vi-antennas-and-propagation-10ec64-notes.pdf.

Chapter 5
Terahertz Passive Components for 6G Communication

5.1 Introduction

Three-dimensional (3D) printed mmWave and THz components employed for future-generation communication (6G) systems are classified into active and passive devices used to examine the performances of the communication networks. Due to the geometric tolerance of current 3D printing technologies, defined within the needed gate length of mmWave transistors, it is exceptionally demanding to print a transistor of f_{max} and f_T decreasing in the mmWave and THz wave regimes of the spectrum. The majority of 3D printed mmWave and THz wave passive components are perceived as frequency selective surfaces (FSSs), electromagnetic bandgap (EBG) structures, Wilkinson power dividers, T-junction power dividers, substrate integrated waveguide (SIW), planar radiator antennas, lenses antennas, and horn antennas to name a few [1]. However, this chapter discusses the THz power dividers and the THz silicon integrated waveguide (SIW).

5.2 Terahertz Power Dividers

The antenna with an incredible gain is an essential component for 6G wireless communication networks to mitigate atmospheric path loss due to resonance with water vapor and oxygen molecules, which is one of the potential concerns in the THz regime of the spectrum [2–5]. Thus, an array of antennas has been exploited to enhance the microstrip antenna gain and directivity compared to the unit-cell radiator. Therefore, developing a feeding network with minimum loss power dividers, such as the T-junction power splitter and Wilkinson power splitter [6, 7], is essential.

U. Nissanov, G. Singh, *Antenna Technology for Terahertz Wireless Communication*,
https://doi.org/10.1007/978-3-031-35900-2_5

The power dividers are passive components used for power combining or power splitting. It generally provides an equal power division ratio (0.5) with the same phase output signals. However, an unequal power division can also be performed. When equal power dividers are used in the corporate fed or hybrid fed array antennas, the sidelobe levels (SLLs) can be up to −13.5 dB, which is occasionally insufficient, and significantly higher SLLs can reduce the THz transmitter's efficiency. Reducing the THz transmitter's efficiency limits its power and reduces the THz transmit signal's propagation distance, which is the main challenging problem of THz wireless communication. Further, unequal power dividers are generally used to lower the SLLs. Subsequently, an amplitude weighing/tapering method is required to reduce the SLLs, such as Binomial amplitude, Dolph-Tschebyscheff, and Taylor n-bar amplitude weighing distribution techniques [8].

5.2.1 Wilkinson Power Divider

The Wilkinson power divider is an electrical circuitry that can attain isolation within the output ports while maintaining matched output ports. Thus, apart from output ports, the power reflected is dissipated [6, 7]. The basic Wilkinson 1 to 2 symmetrical power dividers have three ports generally used for the microstrip array antennas to split the input signal equally or comb the power. The outputs and input of the Wilkinson power divider have the impedance of $Z_0 = 50\ \Omega$, while L_q is the quarter-wave transformer ($\lambda/4$) that changes the impedance of the load to another value. Thus, impedance matching is possible and has the impedance of $\sqrt{2}Z_0 = 70.71\ \Omega$. Every power-divider is included a 100-Ω resistor, which sets up the matching and isolation of two output ports. Besides that, the Wilkinson power splitter/combiner should consider the phases' performances and power/magnitude variance in the output ports. Figure 5.1 exhibits the basic configuration of the Wilkinson 1 to 2 symmetrical power divider.

To compute the widths of the microstrip lines W_{50}, $W_{70.71}$ and the length of the the microstrip L_q that equivalent to 50 Ω, 70.71 Ω, which is a quarter-wave transformer ($\lambda/4$),appropriately. We can use the CST MWS Macros functions to calculate analytic line impedance as presented in Fig. 5.2a–c.

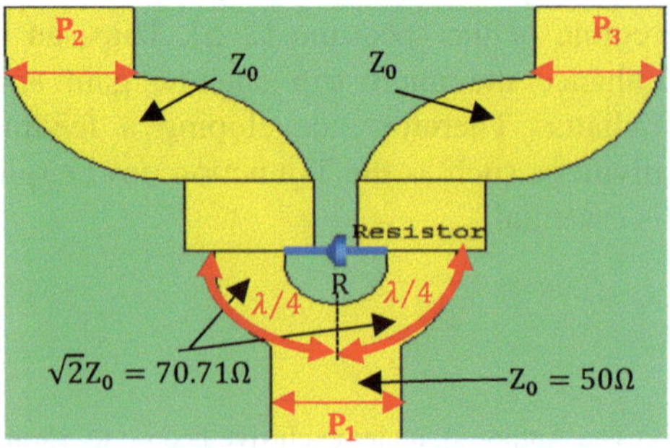

Fig. 5.1 Wilkinson 1 to 2 symmetrical power divider

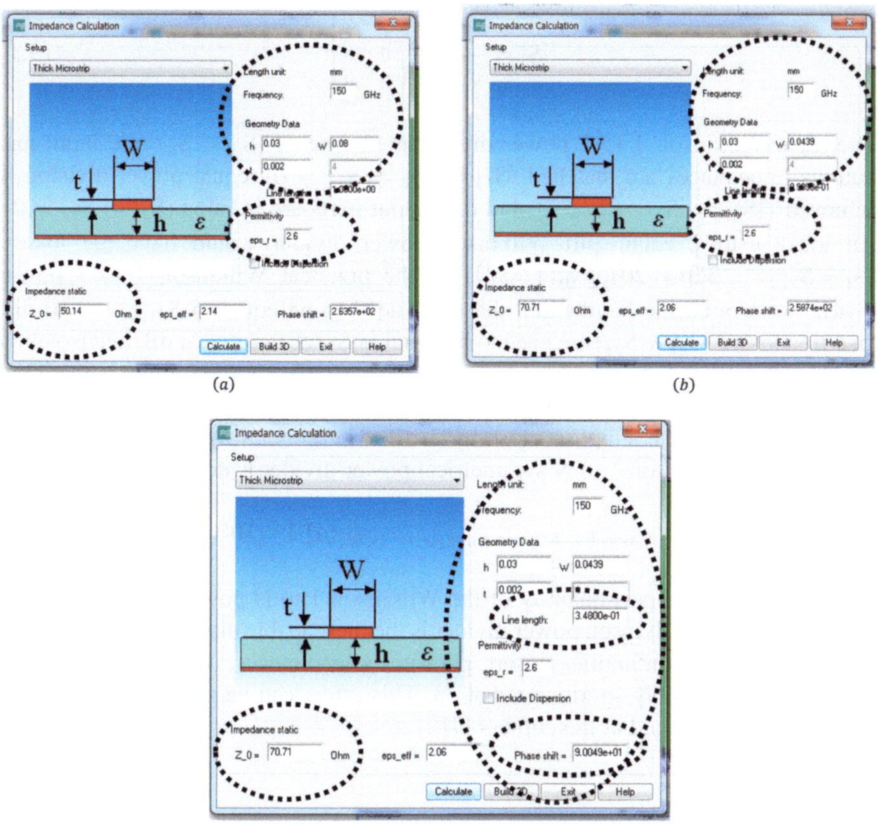

(a) (b) (c)

Fig. 5.2 Computation using CST MWS macros, (**a**) The width of W_{50}, (**b**) Width of $W_{70.71}$, and (**c**) The length of λ/4

Figure 5.2a exhibits the computed result of the width of W_{50} using CST MSW macros. It has been shown that the widths of $W_{50} = 0.08$ mm for microstrip substrates with the structure parameters ($h = 0.03$ mm, $t = 0.002$ mm, $\varepsilon_r = 2.6$, $f = 150$ GHz). Further, Fig. 5.2b exhibits the calculated result of the width of $W_{70.71}$. It has been shown that for microstrip substrates with the parameters of ($h = 0.03$ mm, $t = 0.002$ mm, $\varepsilon_r = 2.6$, $f = 150$ GHz) the width of $W_{70.71} = 0.0439$ mm. Fig. 3.2c exhibits the calculated result of the length of λ/4 (L_q). It has been illustrated that for the microstrip substrates with the parameters of ($h = 0.03$ mm, $t = 0.002$ mm, $\varepsilon_r = 2.6$, $f = 150$ GHz) the width is λ/4(L_q = Line length = 0.348 mm) for phase-shifting of 90° (Phase shift = 90.05°), which is a potential requirement for impedance matching. The scattering parameters for the Wilkinson 1 to 2 symmetrical power divider at the desired frequencies are placed by [6, 7]:

$$[S] = \frac{-j}{\sqrt{2}} \begin{bmatrix} 0 & 1 & 1 \\ 1 & 0 & 0 \\ 1 & 0 & 0 \end{bmatrix} \tag{5.1}$$

The ideal Wilkinson 1 to 2 power divider is reciprocal ($S_{ij} = S_{ji}$), the input and outputs impedances are matched ($S_{11} = S_{22} = S_{33} = 0$), equal power division is achieved $\left(|S_{21}| = |S_{31}| = 1/\sqrt{2}\right)$, and the output ports are isolated ($S_{23} = S_{32} = 0$). An ideal 1 to 2 equal-split Wilkinson power divider would have the losses: $S_{21} = S_{31} = -3\text{dB} = 20\log_{10}\left(1/\sqrt{2}\right)$. At the practical Wilkinson 1 to 2 power divider, at around the bandwidth, the (S_{11}) need to be equal to $S_{11} \leq -10$ dB, insertion losses (S_{21} & S_{31}) are need to be equal to:S_{21} & $S_{31} \leq -4$ dB, and isolation S_{32} & S_{23} are need to be equal to: S_{32} & $S_{23} \leq -10$ dB. By serial linking, the input power can be sorted to each N-number of outputs while multiplying the output ports number enhances the supposed loss by 3 dB. The general formula of theoretical losses of the Wilkinson 1 to N symmetrical power divider is defined by [6]:

$$\text{Loss}_{\text{Wilkinson 1 to N symmetrical power divider}}(\text{dB}) = 10\ \log_{10}\text{N} \tag{5.2}$$

N is the sum output port quantity of the Wilkinson 1 to N power divider. Another problem with the Wilkinson power divider is the inescapable discontinuities at bends because such discontinuities insert parasitic capacitances, which can cause a mismatch and possibly spurious radiation. This effect can be mitigated by making a smooth "swept" bend at the corners [7].

5.2.2 *T-Junction Equal and Unequal Power Divider*

The T-junction power divider is a three-port network that can be used to combine or split power. At practical T-junction 1 to 2 equal power divider, at around the BW, the (S_{11}) need to be equal to $S_{11} \leq -10$ dB, insertion losses (S_{21} & S_{31}) need to be equal to about S_{21} & $S_{31} \leq -4$ dB. As shown in Fig. 5.3, the input power to the matched T-junction equal power divider for the intersection voltage of U is [9]:

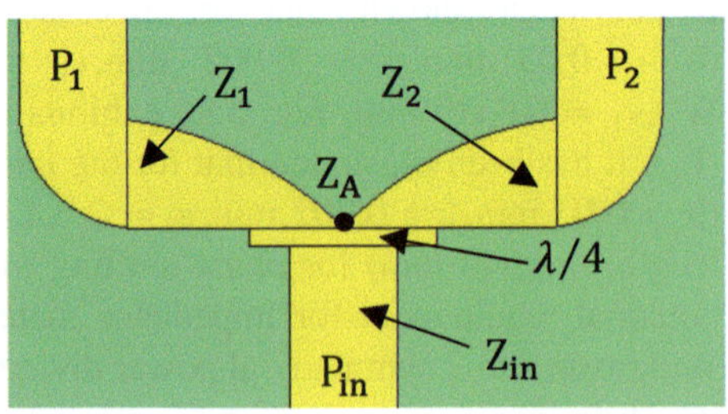

Fig. 5.3 The T-junction equal power divider

$$P_{in} = 0.5\frac{U^2}{Z_0} \tag{5.3}$$

The input and output impedance of the T-junction power divider is equal to $Z_0 = Z_{in} = Z_1 = Z_2 = 50\ \Omega$. The point A impedance is $Z_A = Z_1 \parallel Z_2 = 25\ \Omega$, so the quarter-wave transformer (λ/4) need to be within an impedance of $Z_{\lambda/4} = \sqrt{Z_A \cdot Z_0} = \sqrt{25 \cdot 50} = 35.35\ \Omega$ for impedance matching. The output powers at either lateral intersections are [9]:

$$P_1 = 0.5\frac{U^2}{Z_0}, P_2 = 0.5\frac{U^2}{Z_0} \tag{5.4}$$

Due to the lossless model being used, the total output power is [9]:

$$P_{out} = P_1 + P_2 = P_{in} \tag{5.5}$$

to achieve the weighing power distribution of the corporate or hybrid fed microstrip antenna. In addition, to diminish the microstrip array antenna's SLLs, the microstrip T-junction power divider is used using the lossless model [10–12], as shown below. For an unequal power splitting, as seen in Fig. 5.4, Eqs. (5.3), (5.4), and (5.5) can be used:Thus, for an unequal power splitting, the outputs power is [9]:

$$P_1 = K \cdot P_{in} \tag{5.6}$$

$$P_2 = (1 - K) \cdot P_{in}; 0 < K < 1 \tag{5.7}$$

The impedances at any side of the matched intersection can be attained by combining Eqs. (5.3), (5.4), (5.5), (5.6), and (5.7) by [9]:

$$Z_1 = \frac{Z_0}{K} \tag{5.8}$$

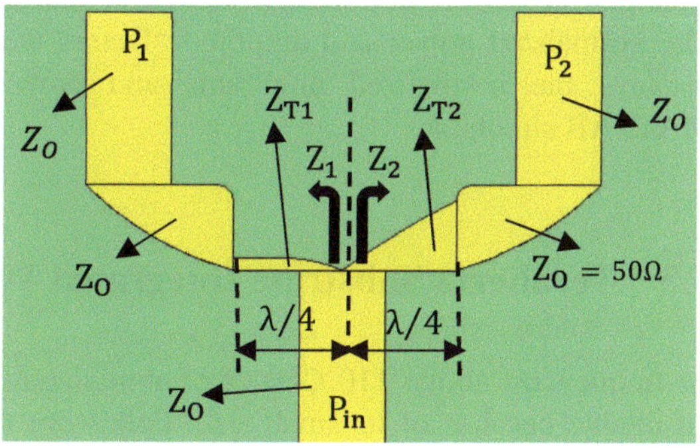

Fig. 5.4 The form of the unequal 1 to 2 T-junction power divider

Table 5.1 Normalized power and amplitude factors

Port number	Amplitude factors	Normalized amplitude factors	Normalized power factors
2, 9	0.4205	$V_2 = V_9 = 0.29116$	$P_2 = P_9 = 0.08477$
3, 8	0.8209	$V_3 = V_8 = 0.56845$	$P_3 = P_8 = 0.32313$
4, 7	1.2056	$V_4 = V_7 = 0.83485$	$P_4 = P_7 = 0.69697$
5, 6	1.4441	$V_5 = V_6 = 1$	$P_5 = P_6 = 1$

$$Z_2 = \frac{Z_0}{1 - K} \tag{5.9}$$

$$\frac{P_1}{P_2} = \frac{Z_2}{Z_1} = \frac{K}{K - 1} \tag{5.10}$$

The quarter-wave transformers Z_{T1} and Z_{T2} are computed to match the power divider's outputs within an input impedance of Z_0 to the wanted impedance at the intersection as [9]:

$$Z_{T2} = \sqrt{Z_0 \cdot Z_2} \tag{5.11}$$

$$Z_{T1} = \sqrt{Z_0 \cdot Z_1} \tag{5.12}$$

In the real world, the T-junction power divider is not lossless, and the output power is: $P_{out} = P_{in} - P_{loss}$ that equals the total of the radiation, ohmic, and dielectric losses. The power ratio is: P_1/P_2, like the lossless mode, as displayed in Fig. 5.4, the microstrip structure is practically symmetrical within the couple sides of the intersection, and losses mode may be used.

The Taylor n-bar amplitude weighting distribution [8] within the desired SLL of −30 dB was chosen for the 1 to 8 T-junction unequal power splitter outputs. The W = taylorwin (8, 50, −30) function within the matrix laboratory (MATLAB) simulator has been applied to find the normalized amplitude factors. The following equation has been applied to find the normalized power factors [9]:

$$P = V^2 \tag{5.13}$$

The normalized power and amplitude factors were uploaded in Table 5.1, which displays the normalized amplitude and power factors computed within the MATLAB simulator.

5.3 Terahertz Substrate Integrated Waveguide

At frequencies above 110 GHz, a "Subminiature version A" (SMA) connector is unsuitable because of its small size. Following the "IEEE Standard for Precision Coaxial Connectors (DC to 110 GHz)," a 1.0 mm connector can be applied with

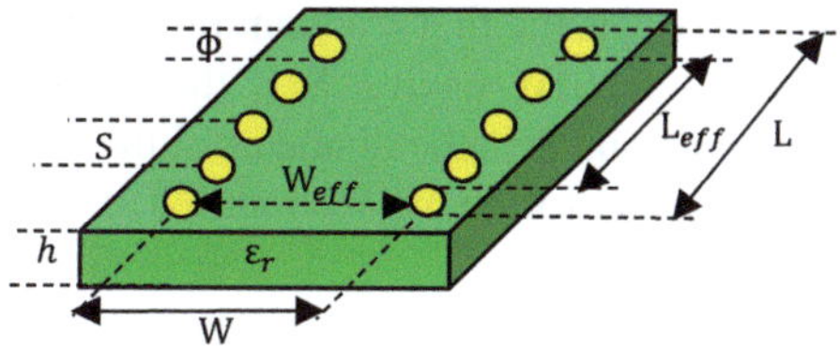

Fig. 5.5 The fundamental topology of the SIW

several connector types: "The laboratory precision connector (LPC) and the general precision connector (GPC)" [13].

So, microstrip devices are inefficient for tremendous frequency because the wavelength is petite, and microstrip apparatus manufacture demands tight tolerances. Waveguide devices are preferred at high frequencies, but they suffer from the disadvantage of higher weight and volume and high fabrication cost. SIW technique has been established to design mmWave, microwave, and THz elements. SIW is the integrated oblong shape waveguide built-in regular PCBs laminate that places into a couple of isolated brassy via-hole layouts and works under the same principles as standard oblong air-cavity waveguides. SIW has the same improvements as conventional brassy air-cavity waveguides, like superior maximum frequency, Q-factor, and selectivity [14].

Additionally, it has weightlessness, low profile, conformability to an arched or planar layer, and easy assembly with SoC designs. However, SIW has some disadvantages because of its higher frequencies. In addition, SIW is further sensitive to substrate losses [15].

Furthermore, compared to traditional 3D rectangular waveguides, slimmer dielectrics substrate inhibits TM modes from resonating. Therefore, only TE mode can efficiently propagate through SIW cavities. Figure 5.5 exhibits the configuration of the fundamental topology of the SIW waveguide established on a dielectric substrate with its physical magnitudes [16, 17].

The design rules the rectangular SIW established upon TE_{10}. The subsequent mathematical formulations provide the first resonant frequency mode for the SIW cavity [18, 19]:

$$f_{TE_{10}} = \frac{c}{2\pi\sqrt{\mu_r \varepsilon_r}} \sqrt{\left(\frac{\pi}{W_{eff}}\right)^2} \tag{5.14}$$

$$\phi \leq \frac{\lambda_g}{5} \tag{5.15}$$

$$S \leq 2\phi \tag{5.16}$$

$$L_{eff} = L - \frac{\phi^2}{0.95 \cdot S} \tag{5.17}$$

$$W_{eff} = W - \frac{\phi^2}{0.95 \cdot S} \tag{5.18}$$

where $f_{TE_{10}}$ is the first resonant mode of the cavity, W and L are the width and length of a single SIW cavity, S is the via spacing, ε_r is the dielectric constant of the substrate, and ϕ is the diameter of the via's. The guided wavelength λ_g in the SIW is given by the following formula [20]:

$$\lambda_g = \frac{c}{\sqrt{\varepsilon_r\left(f^2 - f_{TE_{10}}^2\right)}} \tag{5.19}$$

where c is the velocity of EM waves in free space, and f is the resonant frequency of the SIW. These equations give a good approximation of the TE field propagation in the SIW. In addition, a waveguide to microstrip transition needs to be used to connect the radiation patch element to the SIW. They should have minimum voltage standing wave ratio (VSWR) properties, low loss, and broadband [19].

5.4 Problem Formulation

In most microwave and mmWave systems, splitting and combining RF signals is vital. Notably, in high-power applications like radars and transmitters, an individual PA generally cannot achieve the needed RF power level, notably at the THz frequencies. Combining the output power beyond one amplifier is needed. On another side, a high gain antenna array is needed to mitigate the lack of a THz high power transmitter that needs 6G wireless communication networks, which also demand a power splitter/divider [21–25].

In the microwave and mmWave spectrum, power splitter/dividers have been designed within various configurations such as Wilkinson divider, T-junction divider, and Y-junction divider. These power dividers were embedded using microstrip laminates [26, 27], waveguides [28, 29], SIW [30, 31], and coplanar waveguides (CPW) [32, 33].

However, conventional, unalloyed laminate-based splitter/dividers usually introduce significant insertion loss at mmWave frequencies [26, 27]. In contrast, waveguide power divider structures have a low insertion loss. In addition, waveguide-based dividers usually endure unmatched ports and insufficiency of isolation. This problem, consecutively, shows that the output ports are concerned with everyone, and either reflection originating at a particular output port will be coupled into another output port [28, 29].

So at THz frequencies, these problems are more significant and need to be mitigated by proper techniques, namely, at THz frequencies, low loss, broadband THz power divider, and broadband, matched, and highly isolated THz SIW needs to be designed and fabricated as done in [34, 35].

5.5 Related Work

Barrera et al. [36] demonstrated a 1 to 4 Wilkinson divider in 28 nm CMOS technology for 140 GHz. This divider was fabricated, and isolation, insertion loss, and impedance BW measurement results were achieved at −20 dB, 2.5 dB, and 40 GHz (105–145 GHz), respectively. Treizebre et al. [37] demonstrated a 1 to 3 Wilkinson power divider based on planar Goubau lines (PGLs) for 240 GHz. This divider was fabricated, and isolation, insertion loss, and impedance BW measurement results were achieved: −15– (−30) dB, 15–23 dB, and 130 GHz (140–270 GHz), respectively. While Reichel et al. [38] demonstrated a broadband 1 to 2 T-junction variable power divider based on a parallel plate waveguide (PPWG) for 240 GHz. This divider was fabricated, and transmission and impedance BW measurements were achieved: 100% and 150 GHz (150–300 GHz), respectively. In contrast, Kumar et al. [39] demonstrated a 1 to 2 Y-junction divider based on SIW technology for 60 GHz. The simulated insertion loss results and impedance BW were achieved: 3.4 dB and 11 GHz (57–68 GHz), respectively.

Aloui et al. [40] demonstrated a SIW annular ring slot multi-band antenna based on graphene for THz within three separate bands. This design was within Ansys HFSS (FEM solver) and verified with a comparison within CST MWS (FIT solver). The impedance BW and gain of the three bands simulated results were 540 GHz, 131 GHz, 100 GHz, 7 dB, 6 dB, and 10 dB, respectively. At the same time, the resonance was within 8–16 THz. Dhwaj et al. [41] demonstrated a SIW antenna on-chip (AoC) for 550 GHz. This antenna was fabricated within 65 nm CMOS technology, and the measured results of the gain and impedance BW were 2.8 dB and 20 GHz (540–560 GHz), respectively. In contrast, Althuwayb et al. [42] demonstrated an AoC based on SIW with metasurface for 486 GHz. The simulated gain and impedance BW results were 7.4 dB and 50 GHz (450–500 GHz), respectively. The performance summary of these THz passive components discussed above is presented in Table 5.2 according to the related work.

5.6 Simulation Results and Discussion

5.6.1 THz Wilkinson Power Divider

(a) *THz Wilkinson 1 to 2 Equal Power Divider*

The designed THz Wilkinson 1 to 2 equal power divider is based on rolled copper Rogers RO3003 laminate, which has the following parameters of a copper thickness (t), substrate height (h), loss tangent (tan δ), and dielectric constant ε_r of 8.75 μm, 127 μm, (0.0019, 3.02) @ 100 GHz, respectively, for the resonance frequency of 123.8 GHz. The simulation results would be closest to the 90–138 GHz frequency range simulation, so ε_r and tanδ were taken; the manufacturer did it at the highest frequency (100 GHz) upon the microstrip substrate's measurements. This substrate

Table 5.2 THz passive components' classification and performances according to related work in this chapter

References	THz passive technology	Freq. nom. (GHz)	Measured BW (GHz)	Measured isolation (dB)	Measured insertion loss (dB)
[36]	1 to 4 Wilkinson divider	140	40 (105–145)	−20	2.5
[37]	1 to 3 Wilkinson divider based on PGLs	240	130 (140–270)	−15–(−30)	15–23
[38]	1 to 2 T-junction variable divider-based PPWG	240	150 (150–300)	n/a	n/a
[39]	1 to 2 T-junction divider-based SIW	60[a]	11[a] (57–68)[a]	n/a	3.4[a]

References	THz passive technology	Freq. nom. (GHz)	Measured BW (GHz)	Measured max. gain (dB)
[40]	Annular ring slot antenna-based SIW	(8000–16,000)[a]	(500, 131, 100)[a]	(7, 6, 10)[a]
[41]	AoC-based SIW	550	20 (540–560)	2.8
[42]	AoC-based SIW	486[a]	50[a] (450–500)[a]	7.4[a]

Notes
[a]Simulated
[b]Simulated and compared to a different type of antenna solver/simulator

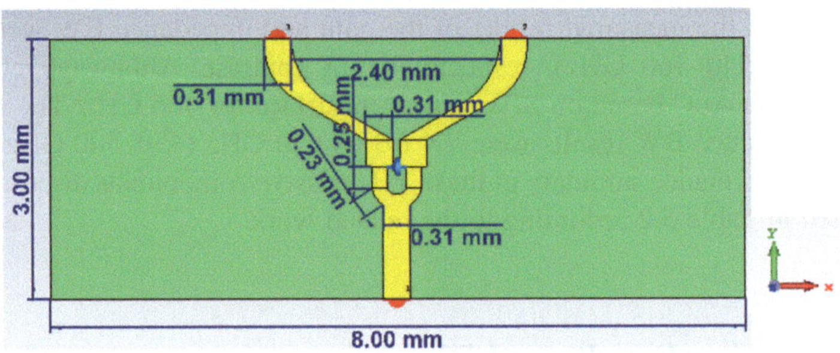

Fig. 5.6 The designed THz Wilkinson 1 to 2 equal power divider

is electrically ultra-thin ($h = 127\ \mu m \le \lambda_{max}/8 = 272\mu m$ and $t = 8.75\ \mu m \le \lambda_{max}/8 = 272\mu m$) for the sake of suppressing the surface wave in the THz band, when the Lq and W_{50}, and $W_{70.71}$ were set to 480 μm, 308 μm, and 164 μm, respectively. While the isolation resistor was within 100 ohms. The CST MWS was the designed simulator. Figure 5.6 shows the designed THz Wilkinson 1 to 2 equal power divider.

Figure 5.7 shows the simulated S-parameters of the designed THz Wilkinson 1 to 2 equal power divider. From Fig. 5.7, it has been shown that $BW(S_{11} \le -10dB)$ of this Wilkinson power divider is >48.24 GHz. Furthermore, it has been shown that

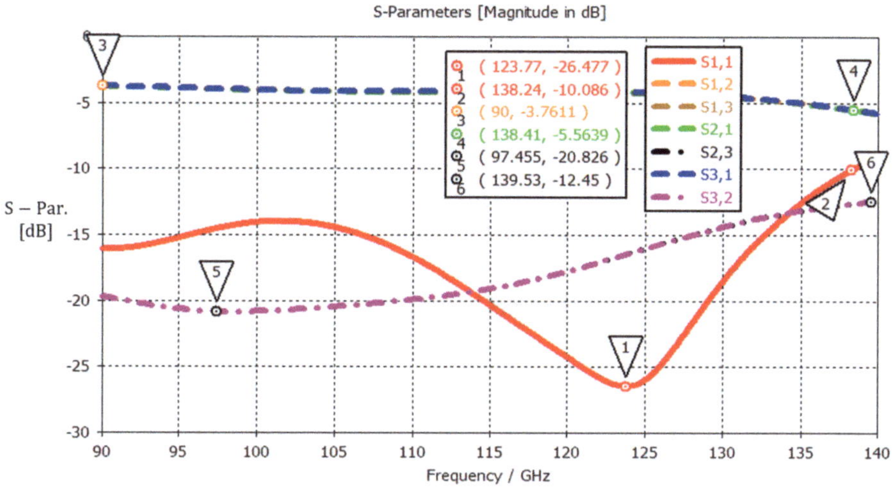

Fig. 5.7 The simulated S-parameters of the designed THz Wilkinson 1 to 2 equal power divider

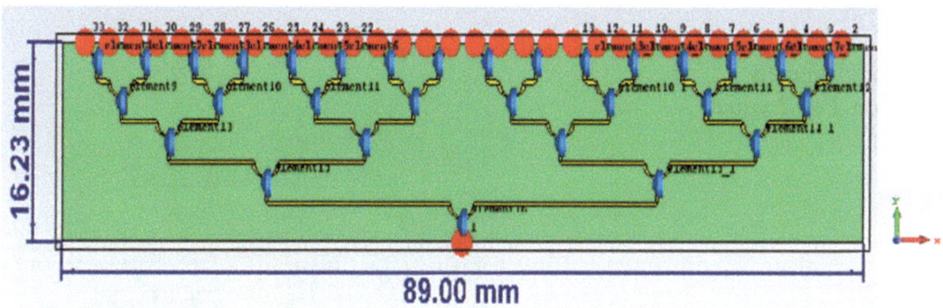

Fig. 5.8 The designed THz Wilkinson 1 to 32 equal power divider

the insertion losses (S_{21} and S_{31}) of this Wilkinson divider are 3.76–5.56 dB for 90–138.41, while the isolation (S_{32} and S_{23}) is better than −12.4 dB for 90–139.53 GHz and 123.77 GHz is the resonance frequency.

(b) *THz Wilkinson 1 to 32 Equal Power Divider*

The designed THz Wilkinson 1 to 32 equal power divider was based on the same laminate, laminate parameters, and the designed simulator as the designed THz Wilkinson 1 to 2 equal power divider. This divider includes 31 identically cascaded THz Wilkinson 1x2 equal power dividers. Figure 5.8 shows the designed THz Wilkinson 1 to 32 equal power divider.

Figure 5.9 shows the simulated S-parameters of the designed THz Wilkinson 1 to 32 equal power divider. From Fig. 5.9, it has been shown that BW($S_{11} \leq -10$dB) of this Wilkinson power divider is 29 GHz. Furthermore, it has been shown that this Wilkinson divider's insertion losses (S21, S31..., S33,1) are 21.34–27.18 dB for 101.1–119.86, while 119.56 GHz is the resonance frequency.

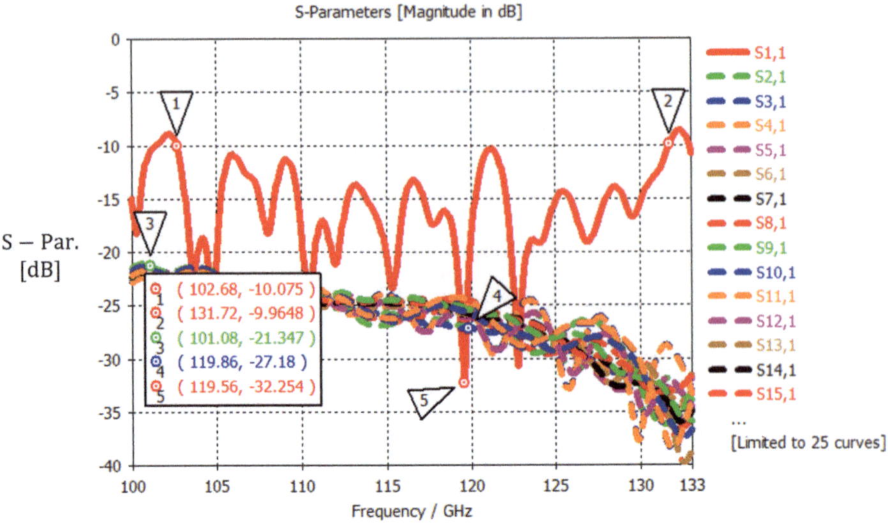

Fig. 5.9 The simulated S-parameters of the designed THz Wilkinson 1 to 32 equal power divider

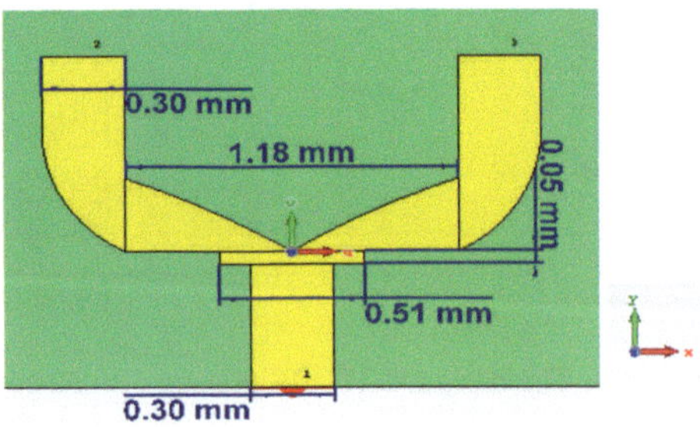

Fig. 5.10 The designed THz T-junction 1 to 2 equal power divider

5.6.2 *THz T-Junction Power Divider*

(a) *THz T-Junction 1 to 2 Equal Power Divider*

The designed THz T-junction 1 to 2 equal power divider is based on rolled copper Rogers RO3003 laminate, which has the following parameters of a copper thickness (t), substrate height (h), loss tangent (tanδ), and dielectric constant ε_r of 17.5 μm, 127 μm, (0.0019, 3.02) @ 100 GHz, respectively, for the resonance frequency of 113.28 GHz. When the Lq, W_{50}, and $W_{35.35}$ were set to 51 μm, 296 μm, and 514 μm, respectively. The CST MWS was the designed simulator. Figure 5.10 shows the designed THz T-junction 1 to 2 equal power divider.

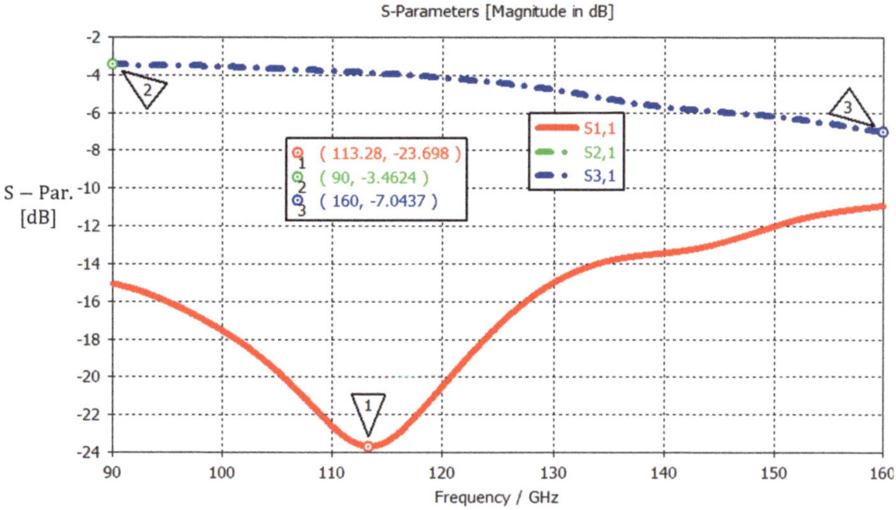

Fig. 5.11 The simulated S-parameters of the designed THz T-junction 1 to 2 equal power divider

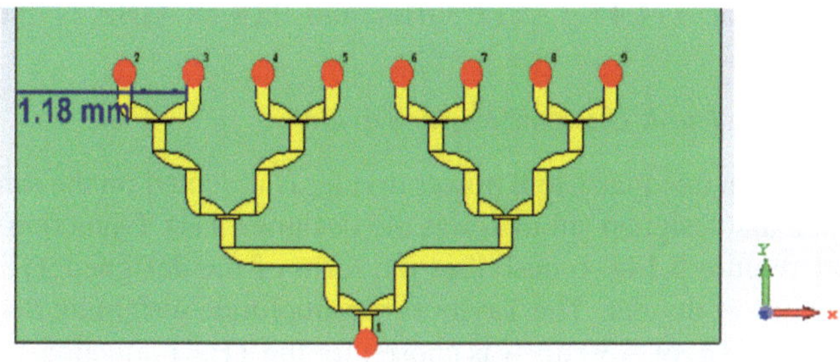

Fig. 5.12 The designed THz T-junction 1 to 8 equal power divider

Figure 5.11 shows the simulated S-parameters of the designed THz T-junction 1 to 2 equal power divider. From Fig. 5.11, it has been shown that BW($S_{11} \leq -10$dB) of this T-junction power divider is >70 GHz. Furthermore, it has been shown that this T-junction divider's insertion losses (S_{21} and S_{31}) are 3.46–7.04 dB for 90–160, and 113.28 GHz is the resonance frequency.

(b) *THz T-Junction 1 to 8 Equal Power Divider*

The designed THz T-junction 1 to 8 equal power divider was based on the same laminate parameters and the designed simulator as the designed THz T-junction 1 to 2 equal power divider. This divider includes seven identically cascaded THz T-junction 1x2 equal power dividers. Figure 5.12 shows the designed THz T-junction 1 to 8 equal power divider.

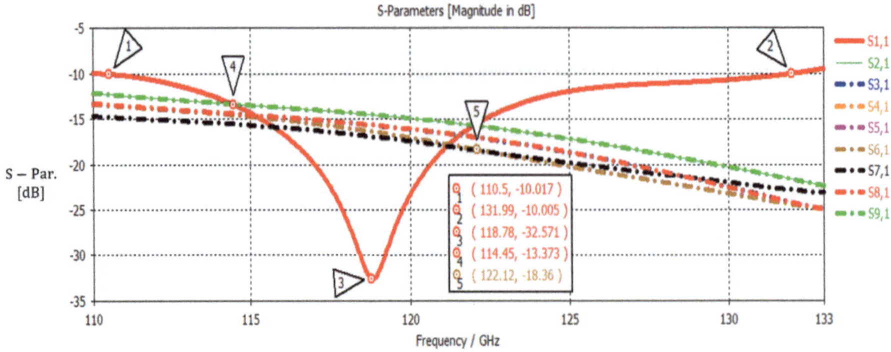

Fig. 5.13 The simulated S-parameters of the designed THz T-junction 1 to 8 equal power divider

Figure 5.13 shows the simulated S-parameters of the designed THz T-junction 1 to 8 equal power divider. From Fig. 5.13, it has been shown that BW($S_{11} \leq -10dB$) of this T-junction power divider is 21.5 GHz. Furthermore, it has been shown that this T-junction divider's insertion losses (S_{21}, S_{31},... and S_{91}) are 13.37–18.36 dB for 114.45–122.12 GHz, and 118.78 GHz is the resonance frequency.

(c) *THz T-Junction 1 to 8 Unequal Power Divider*

This THz T-junction unequal 1 to 8 power divider is mounted on the same laminate, parameters, and the designed simulator as the designed THz T-junction 1 to 2 equal power divider. Figure 5.14a, b describes the form of the designed THz T-junction 1 to 8 unequal power divider. The Taylor n-bar amplitude weighting distribution [8] within the desired SLL of −30 dB was chosen for the THz T-junction 1 to 8 unequal power divider outputs. The power levels in black color at THz T-junction 1 to 8 unequal power divider (Fig. 5.14) are the necessary powers following the needed weighing factors. In contrast, the power levels written in red are normalized to the required general input power at the THz T-junction 1 to 8 unequal power divider. The dimensions of the substrate were 15 × 30 mm^2. The THz T-junction 1 to 8 unequal power divider dimensions were optimized with the parameter sweep at the CST MWS simulator. The impedances, widths, and $\lambda/4$ length of each divider are uploaded in Table 5.3.

Figure 5.15 describes the simulation result of the S_{11}, insertion losses (S_{21}, S_{31}, . . ., S_{91}) in (dB) of the designed THz T-junction 1 to 8 unequal power divider, depending on the frequency. It has been shown that the BW($S_{11} \leq -10dB$) of this divider is: 13.56 GHz(11.78%), while the resonance frequency is: $f_0 = 116.9$ GHz, and the insertion losses ($S_{21}, S_{31}, .., S_{91}$) are between 10.33 – 17.09 dB at 108.57–121.87 GHz.

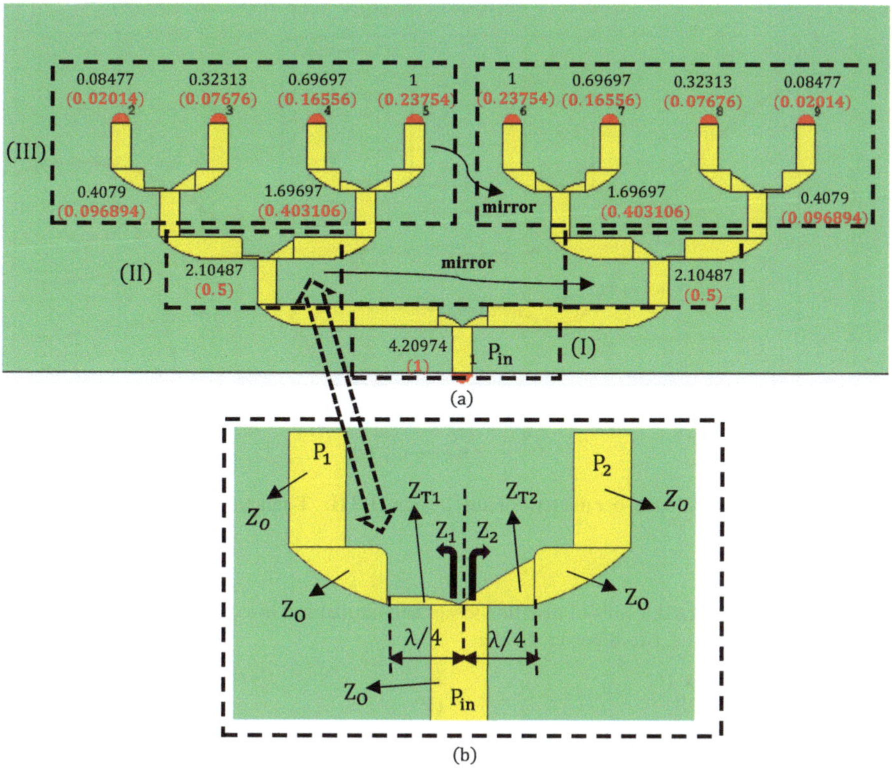

Fig. 5.14 The structure of the THz T-junction 1 to 8 unequal power divider, (**a**) The designed THz T-junction 1 to 8 unequal power divider, (**b**) The form of the THz T-junction 1 to 2 unequal power divider

Table 5.3 THz T-junction unequal 1 to 8 power divider dimensions

Section	The impedance of the microstrip line (Ω)	Width of the microstrip line (μm)	λ/4 length (μm)
I	$Z_{T1} = 70.71$ $Z_{T2} = 70.71$	$w_{Z_{T1}} = 154$ $w_{Z_{T2}} = 154$	$L_{qZ_{T1}} = 290$ $L_{qZ_{T2}} = 290$
II	$Z_{T1} = 113.58$ $Z_{T2} = 55.67$	$w_{Z_{T1}} = 40$ $w_{Z_{T2}} = 246$	$L_{qZ_{T1}} = 298$ $L_{qZ_{T2}} = 286$
III	$Z_{T1} = 109.69$ $Z_{T2} = 56.17$ $Z_{T1} = 78.02$ $Z_{T2} = 65.13$	$w_{Z_{T1}} = 43$ $w_{Z_{T2}} = 242$ $w_{Z_{T1}} = 123$ $w_{Z_{T2}} = 182$	$L_{qZ_{T1}} = 298$ $L_{qZ_{T2}} = 286$ $L_{qZ_{T1}} = 292$ $L_{qZ_{T2}} = 290$

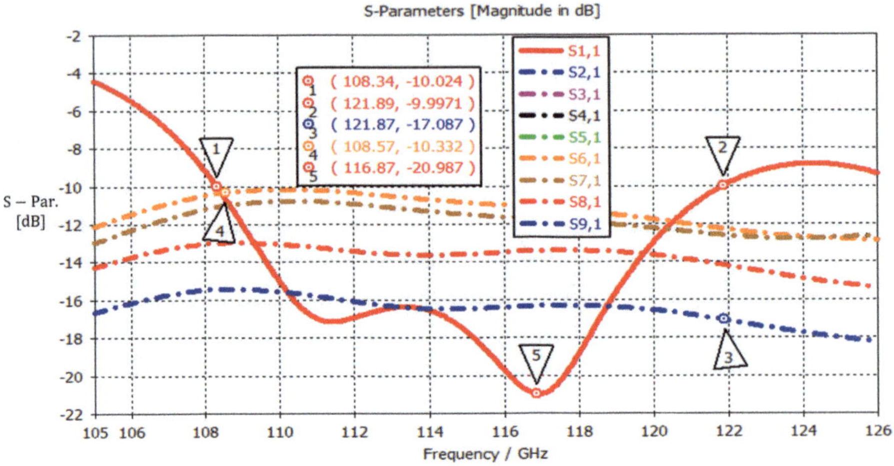

Fig. 5.15 The simulated S-parameters of the designed THz T-junction 1 to 8 unequal power divider

Table 5.4 The simulated results of insertion losses and normalized insertion losses of the designed THz T-junction unequal 1 to 8 power divider

Port number	Sx,1 (dB)	Sx,1 (V)	Normalized Sx,1
2, 9	$S_{21} = S_{91} = -16.33$	$S_{21} = S_{91} = 0.02328$	$S_{21} = S_{91} = 0.3013$
3, 8	$S_{31} = S_{81} = -13.41$	$S_{31} = S_{81} = 0.04560$	$S_{31} = S_{81} = 0.5901$
4, 7	$S_{41} = S_{71} = -11.74$	$S_{41} = S_{71} = 0.06699$	$S_{41} = S_{71} = 0.8670$
5, 6	$S_{51} = S_{61} = -11.12$	$S_{51} = S_{61} = 0.07727$	$S_{51} = S_{61} = 1$

Table 5.4 shows the simulation results of the insertion losses' and the normalized insertion losses at the resonance frequency of 116.9 GHz of the T-junction unequal 1 to 8 power divider.

Figure 5.16a shows the equalization of the simulation results of the S_{11} in (dB) of the designed THz T-junction 1 to 8 equal power divider. This comparison was carried out within the CST MWS simulator based on the FIT solver and the Ansys HFSS based on the FEM solver. It has been shown that a good agreement between the simulation results is attained regarding the designed THz T-junction 1 to 8 equal power divider that supports the suggested design. While Fig. 5.16b shows the comparison of the simulation results of S_{11} in (dB) of the designed T-junction 1 to 8 unequal power divider. This comparison was carried out within the CST MWS and the Ansys HFSS simulators. It has been shown that a good agreement between the simulation results is attained regarding the designed THz T-junction 1 to 8 unequal power divider that supports the suggested design.

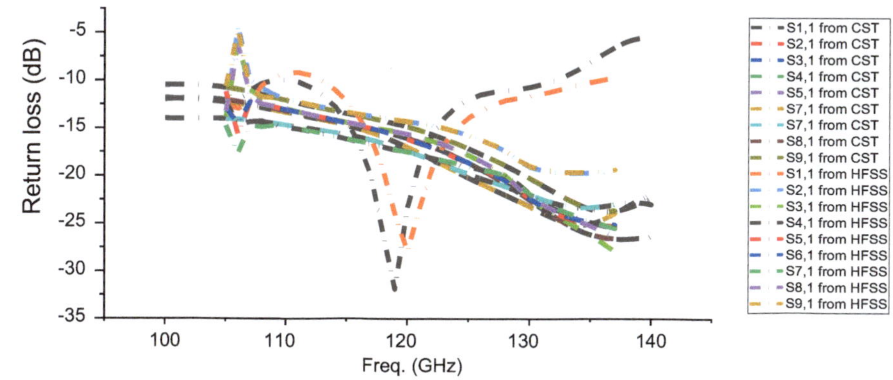

(a)

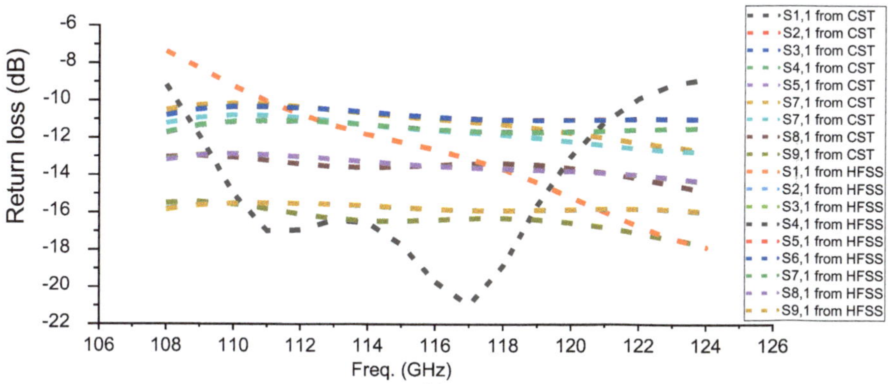

(b)

Fig. 5.16 Comparison of the simulation results of the S_{11} of the designed dividers between the CST MWS and the Ansys HFSS simulators, (**a**) THz T-junction 1 to 8 equal power divider, (**b**) THz T-junction 1 to 8 unequal power divider

5.6.3 THz SIW

(a) *THz Wideband SIW for 108 GHz*

The designed THz wideband SIW for 108 GHz is based on rolled copper Rogers RO3003 laminate, which has the following parameters of a copper thickness (t), substrate height (h), loss tangent (tanδ) and dielectric constant ε_r of 17.5 μm, 127 μm, (0.0019, 3.02) @ 100 GHz, respectively, for the resonance frequency of 108.1 GHz. The CST MWS was the designed simulator. This THz SIW includes 30 Vias. After optimization with the CST MWS, the THz SIW has the following dimensions: the Vias has an external diameter of 250 μm and an internal diameter of 110 μm. Each via hole was coated with copper at 20 μm on each side of the via hole. The gap between the via holes is 50 μm. The width of the output microstrip line was 340 μm.

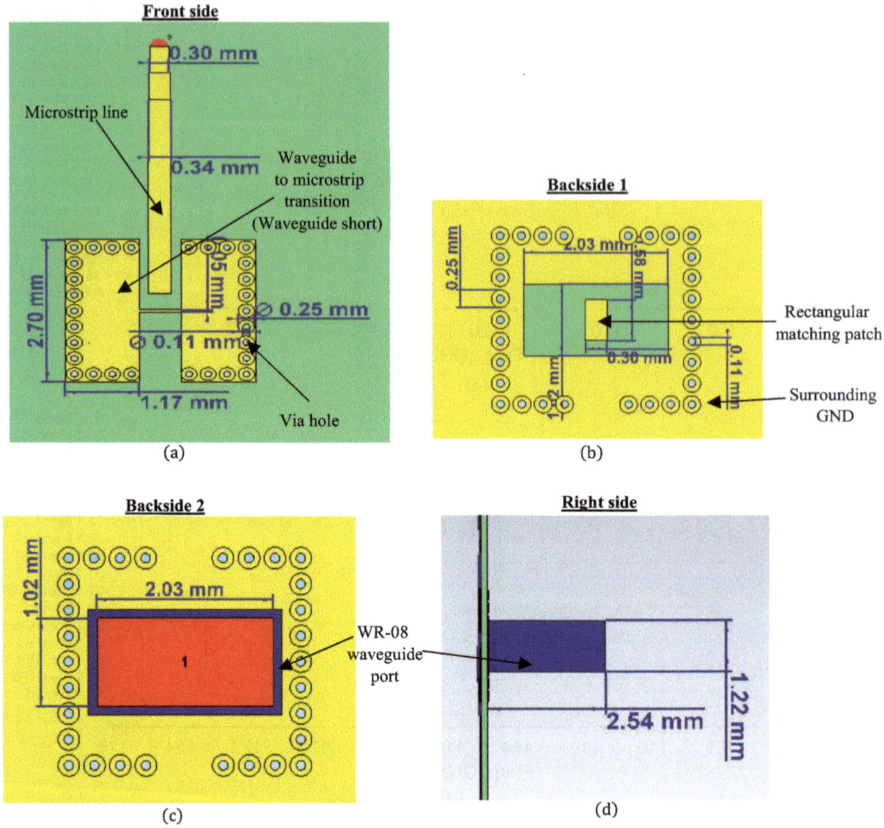

Fig. 5.17 Structure of the completely designed THz wideband SIW for 108 GHz, (**a**) Front side, (**b**) Backside 1, (**c**) Backside 2, (**d**) Right side

The half-size dimensions of the waveguide short were 2700 × 1170 μm^2. While the rectangular matching patch was 580 × 300 μm^2. The WR-08 waveguide length and width dimensions are 1020 × 2030 μm^2. Moreover, this waveguide covers the frequency range of 90–140 GHz. Figure 5.17 shows the designed THz wideband SIW for 108 GHz.

Figure 5.18 describes the simulation result of the return loss (S_{11}) and the insertion losses (S_{21}) of the designed THz wideband SIW for 108 GHz. From Fig. 5.18, the impedance BW ($S_{11} < -10$ dB) of the THz SIW for 108 GHz is 12.06 GHz, while the S_{21} is 2–2.4 dB for 105.7–117.7 GHz, while 108.1 GHz is the resonance frequency.

Figure 5.19 describes the simulation result comparison of the S_{11} and the S_{21} of the designed THz SIW for 108 GHz made within CST MWS FIT and CST MWS FEM solvers. From Fig. 5.19, the impedance BW($S_{11} < -10$ dB) and S_{21} is 12.06 GHz, 1.64–2.6 dB for 105.6–117.6 GHz, 16.02 GHz, and 1.64–2.6 dB for

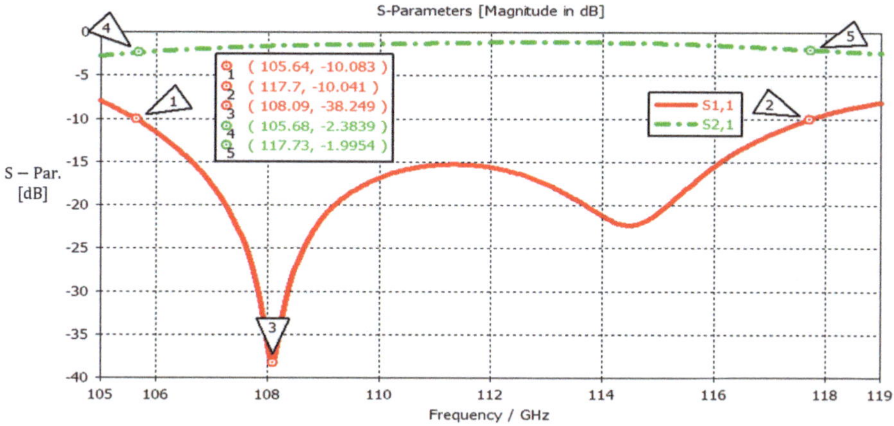

Fig. 5.18 Simulation results of S_{11} and S_{21} of the designed THz wideband SIW for 108 GHz

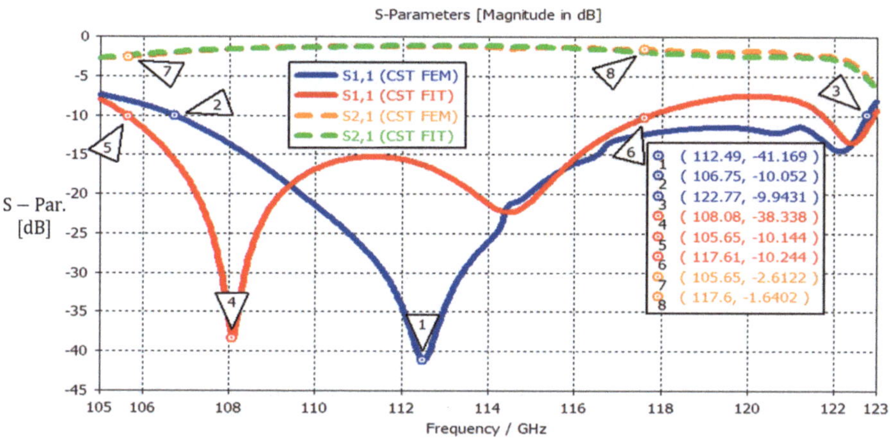

Fig. 5.19 Comparison of the simulation results of S_{11} and S_{21} of the designed THz wideband SIW for 108 GHz within CST MWS FIT and CST MWS FEM solvers

105.6–117.6 GHz within CST MWS FIT solver and CST MWS FEM solver, respectively. So, a good agreement was achieved for the designed THz SIW for 108 GHz.

(b) *THz SIW with Dual Outputs for 109 GHz*

The designed THz SIW with dual outputs for 109 GHz is based on the same laminate and laminate parameters as the previously designed THz wideband SIW. The CST MWS was the designed simulator. This THz SIW includes 30 Vias. After optimization with the CST MWS, the THz SIW for 115 GHz has the following dimensions: the Vias has an external diameter of 250 μm and an internal diameter of 110 μm.

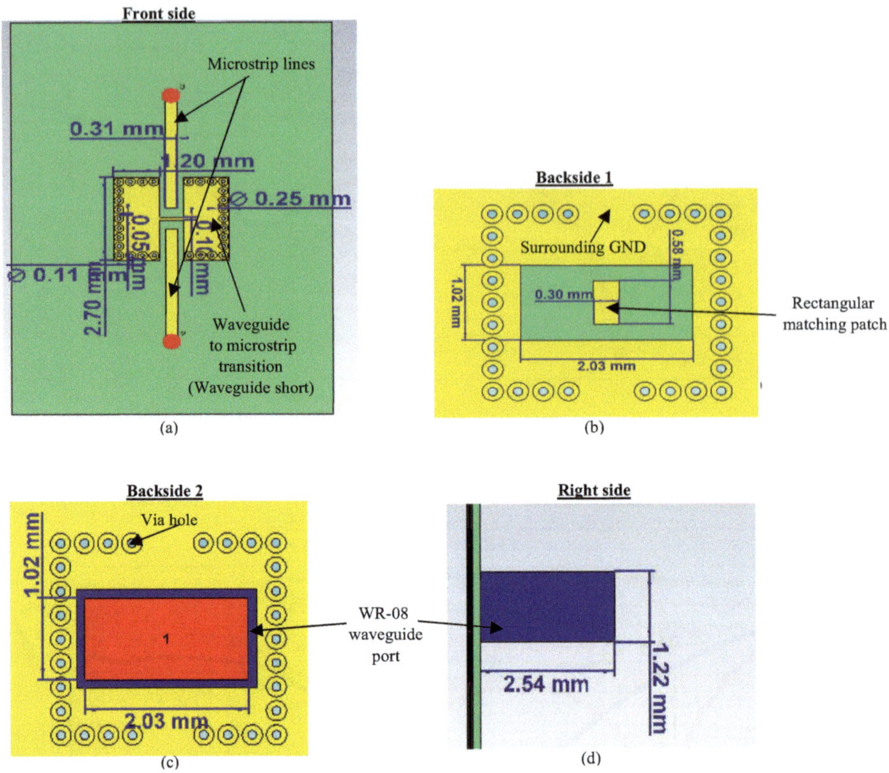

Fig. 5.20 Structure of the complete designed THz SIW with dual outputs for 109 GHz, (**a**) Front side, (**b**) Backside 1, (**c**) Backside 2, (**d**) Right side

Each via hole was coated with copper at 20 μm on each side of the via hole. The gap between the via holes is 50 μm. The half-size dimensions of the waveguide short were 2700 × 1200 μm^2. At the same time, the rectangular matching patch was 580 × 300 μm^2. The width of the output microstrip lines was 310 μm. In contrast, the WR-08 waveguide length and width dimensions are 1020 × 2030 μm^2. Moreover, this waveguide covers the frequency range of 90–140 GHz. Figure 5.20 shows the designed THz SIW with dual outputs for 109 GHz.

Figure 5.21 describes the simulation result of the return loss (S_{11}) and the insertion losses (S_{21} and S_{31}) of the designed THz SIW with dual outputs for 109 GHz. From Fig. 5.21, the impedance BW ($S_{11} < -10$ dB) of the THz SIW with dual outputs for 109 GHz is 5.8 GHz, while the S_{21} and S_{31} are 4.7–4.76 dB for 106.6–112.4 GHz, while 109.05 GHz is the resonance frequency.

Figure 5.22 describes the simulation result comparison of the S_{11}, S_{21}, and S_{31} designed THz SIW with dual outputs for 109 GHz made within CST MWS FIT and Ansys HFSS FEM solvers. From Fig. 5.22, the impedance BW($S_{11} < -10$ dB), S_{21},

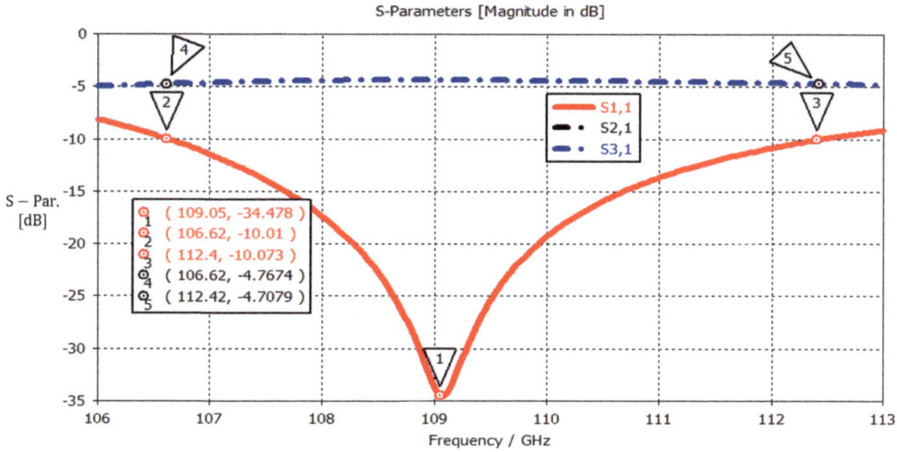

Fig. 5.21 Simulation results of S_{11}, S_{21}, and S_{31} of the designed THz SIW with dual outputs for 109 GHz

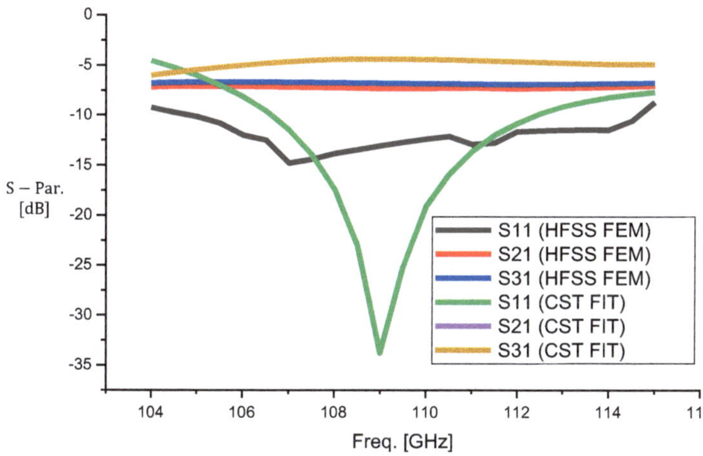

Fig. 5.22 Simulation results in comparison of S_{11}, S_{21}, and S_{31} of the designed THz SIW with dual outputs for 109 GHz within the CST MWS and Ansys HFSS

and S_{31} are 5.5 GHz, 4.39–5.99 dB for 105–115 GHz, 9.5 GHz, and 6.66–7.35 dB for 105–115 GHz within CST MWS FIT solver and Ansys HFSS FEM solver, respectively. So, a good agreement was achieved for the designed THz SIW with dual outputs for 109 GHz.

The performance summary of these designed and simulate THz passive components presented above in this chapter is presented in Table 5.5.

Table 5.5 The performance summary of the designed and simulate THz passive components in this chapter

THz passive technology	Freq. nom. (GHz)	BW (GHz)	Isolation (dB)	Insertion loss (dB)
Wilkinson 1 to 2 equal power divider	123.8[a]	>48.24[a]	>−12.4[a]	(3.76–5.56)[a]
Wilkinson 1 to 32 equal power divider	119.56[a]	29[a]	n/a	(21.34–27.18)[a]
T-junction 1 to 2 equal power divider	113.28[a]	>70[a]	n/a	(3.46–7.04)[a]
T-junction 1 to 8 equal power divider	118.78[b]	21.5[b]	n/a	(13.38–18.36)[b]
T-junction 1 to 8 unequal power divider	116.9[b]	13.56[b]	n/a	(10.33–17.09)[b]
Wideband SIW	108.1[b]	12.06[b]	n/a	(2–2.4)[b]
SIW with dual outputs	109[b]	5.8[b]	n/a	(4.7–4.76)[b]

Notes
[a]Simulated
[b]Simulated and compared to a different type of antenna solver/simulator

5.6.4 Discussion

An antenna within a high gain is necessary for 6G wireless communication networks to lighten the atmosphere's high path loss because resonance within water vapor is one of the significant solicitudes in the THz band [2–5]. So, an array of antennas will increase the microstrip antenna single radiator's gain. Thus, a feeding network that requires appropriate minimum losses power dividers, such as the T-junction and Wilkinson power splitter, is necessary [6, 7]. In this chapter, we show the design of the Wilkinson power divider and T-junction power divider within multiple configurations for THz frequency. Most dividers were equal power dividers, while one design was an unequal power divider needed for designing a microstrip array antenna with low SLLs. The performance summary is shown in Table 5.5, and from this table, we see that the insertion loss is proportional to the number of outputs. In theory, each multiplication of the outputs will increase the loss by 3 dB more.

At frequencies above 110 GHz, an SMA connector is unsuitable because of its small size [13], so we need to use a SIW. SIW is the integrated oblong shape waveguide built-in regular PCBs substrate that places within a couple of isolated brassy via-hole layouts and works under the same principles as standard oblong air-cavity waveguides [14]. This chapter shows the design of a couple of THz SIW. The performance summary is shown in Table 5.5, and from this table we see that the insertion loss is proportional to the number of outputs as the power divider. From this table, we can say that each multiplication of the outputs will increase the loss by 3 dB more.

5.7 Open Research Challenges

Based on this chapter, THz passive components for 6G have several open research challenges: (1) More research and developments on manufacturing 100-Ω resistors at THz bands for Wilkinson power divider. To our best knowledge, these resistors are not standard in the current mmWave/THz technology. (2) Additional analysis and evolutions on low loss and broadband THz power dividers, such as a T-junction and Wilkinson power splitters. (3) Further investigation and improvements on low loss and wideband THz SIW.

5.8 Summary

This chapter discusses THz passive components for 6G communication networks such as THz Wilkinson power divider, THz T-junction power divider, and THz SIW with their design considerations. Seven models were designed, and the simulation results are presented in this chapter. The first model was a THz Wilkinson 1 to 2 equal power divider, and the simulation results of resonance frequency, impedance BW, isolation, and insertion loss were 123.8 GHz, >48.24 GHz, −12.4 dB, and 3.76–5.56 dB, respectively. The second model was a THz Wilkinson 1 to 32 equal power divider, and the simulation results of resonance frequency, impedance BW, and insertion loss were 119.56 GHz, 29 GHz, and 21.34–27.18 dB, respectively. The third model was a THz T-junction 1 to 2 equal power divider, and the simulation results of resonance frequency, impedance BW, and insertion loss were 113.28 GHz, >70 GHz, and 3.46–7.04 dB, respectively. The fourth model was a THz T-junction 1 to 8 equal power divider, and the simulation results of resonance frequency, impedance BW, and insertion loss were 118.78 GHz, 21.5 GHz, and 13.38–18.36 dB, respectively. The fifth model was a THz T-junction 1 to 8 unequal power divider, and the simulation results of resonance frequency, impedance BW, and insertion loss were 116.9 GHz, 13.56 GHz, and 10.33–17.09 dB, respectively. The sixth model was a wideband THz SIW, and the simulation results of resonance frequency, impedance BW, and insertion loss were 108.1 GHz, 12.06 GHz, and 2–2.4 dB, respectively. Finally, the seventh model was a THz SIW with dual outputs, and the simulation results of resonance frequency, impedance BW, and insertion loss were 109 GHz, 5.8 GHz, and 4.7–4.76 dB, respectively.

The design and simulation were done within the CST MWS based on the FIT solver. At the same time, the comparison was made within CST MWS based on the FEM solver or by Ansys HFSS based on the FEM solver, and an acceptable agreement was achieved, which validates the proposed designs so that these THz passive components can be used for 6G communication networks after experimental verification.

References

1. B. Zhang, W. Chen, Y. Wu, K. Ding, and R. Li, "Review of 3D printed millimeter-wave and terahertz passive devices," *International Journal of Antennas and Propagation*, vol. 2017, pp. 1297931/1–10, June 2017.
2. M. H. Rahaman, A. Bandyopadhyay, S. Pal, and K. P. Ray, "Reviewing the scope of THz communication and a technology roadmap for implementation," *IETE Technical Review*, vol. 38, no. 5, pp. 465–478, Jun 2020.
3. G. Gui, H. Huang, Y. Song, and H. Sari, "Deep learning for an effective non-orthogonal multiple access scheme," *IEEE Transactions on Vehicular Technology*, vol. 67, no. 9, pp. 8440–8450, Sept. 2018.
4. Y. Wang, M. Liu, J. Yang, and G. Gui, "Data-driven deep learning for automatic modulation recognition in cognitive radios," *IEEE Transactions on Vehicular Technology*, vol. 68, no. 4, pp. 4074–4077, Apr. 2019.
5. C. Han and Y. Chen, "Propagation modeling for wireless communications in the terahertz band," *IEEE Communications Magazine*, vol. 56, no. 1, pp. 96–101, June 2018.
6. E.J. Wilkinson, "An n-way power divider," *IEEE Transactions on Microwave Theory and Techniques*, vol. 8, pp. 116–118, Jan. 1960.
7. D. M. Pozar, *Microwave Engineering*. John Wiley & Sons, Inc.4th Edition, 2012.
8. C. A. Balanis, *Antenna Synthesis, and Continuous Sources*. In the book: *Antenna Theory Analysis and Design*. John Wiley & Sons, 4th Edition, NJ, pp. 385–430, 2016.
9. U. Nissanov, "THz equal and unequal 1 to 8 T-junction power dividers," *Sensors International*, vol. 26, no. 100113, pp. 1–8, June 2021.
10. M. Liu, Zi-Rui Feng, Q. Wu, "Design of a millimeter-wave conformal low sidelobe microstrip antenna array on A cone surface," *Proceedings of IEEE China-Japan Joint Microwave Conference*, Shanghai, China, 10–12 Sep. 2008, pp. 121–124.
11. Y. Yohandri and J. T. S. Sumantyo, "Development of low sidelobe level array antenna for synthetic aperture radar sensor," *IOP Publishing International Conference on Mathematics and Natural Sciences (IConMNS 2017)*, Bali, Indonesia, 6–7 Sep. 2017, pp. 1–7.
12. Y. A. Rayisiwi and T. Hariyadi, "Design of A 1:12 power divider at 5 GHz for ground surveillance radar application," *IOP Publishing International Symposium on Materials and Electrical Engineering (ISMEE2017)*, July 2018, pp. 1–7.
13. H.W. Banning, "IEEE Standard for Precision Coaxial Connectors (DC to 110 GHz)," pp. 1–119, Sep. 2007. https://doi.org/10.1109/IEEESTD.2007.4317507
14. K. Wu and Y. Cassivi, "The substrate integrated circuits – a new concept for high-frequency electronics and optoelectronics," *Proceedings of IEEE 6th International Conference on Telecommunications in Modern Satellite, Cable and Broadcasting Service*, 1–3 Oct. 2013, pp. 1–9.
15. G. Q. Luo, T. Y. Wang, and X. H. Zhang, "Review of low-profile substrate integrated waveguide cavity-backed antennas," *International Journal of Antennas and Propagation*, vol. 2013, pp. 746920/1–8, Oct. 2013.
16. Y. Arfat, S. P. Singh, S. Khan, and S. Arya, "Modelling and design of substrate integrated waveguide using two parallel rows of rectangular conducting slots," *International Journal of Current Engineering and Technology*, vol. 3, no. 5, pp. 2101–2103, Dec. 2013.
17. V. Zhurbenko, *Passive Microwave Components, and Antennas*. InTechOpen, Croatia, pp. 226–246, 2010.
18. T. Djerafi, A. Doghri, and K. Wu, *Substrate Integrated Waveguide Antennas*. In the book: *Handbook of Antenna Technologies*. Springer Science, Singapore, pp. 1586–1652, 2015.
19. K. Seo, *Planar Microstrip-To-Waveguide Transition in Millimeter-Wave Band*. In the book *Advancement In Microstrip Antennas with Recent Applications*. InTechOpen, Croatia, pp. 249–277, 2013.

20. J. E. Rayas-Sanchez and V. Gutierrez-Ayala, "A general EM-based design procedure for single-layer substrate integrated waveguide interconnects with microstrip transitions," *Proceedings of IEEE International Microwave Symposium (MTT-S) Digest*, Atlanta, USA, 15–20 June 2008, pp. 983–986.
21. K. J. Russell, "Microwave power combining techniques," *IEEE*, vol. 27, no. 5, pp. 472–478, May 1979.
22. S. Tu, J. Liu, T. Wang, Z. Yang, and K. Wang, "Design of a 94 GHz millimeter-wave four-way power combiner based on circular waveguide structure," *Electronics*, vol. 10, no. 15, pp. 1795/1–12, July 2021.
23. H. Seidfaraji, M. I. Fuks, C. Christodoulou, and E. Schamiloglu, "Efficient power combiner for THz radiation," *AIP Advances*, vol. 6, pp. 085220/1–10, Aug. 2016.
24. U. Nissanov, G. Singh, and N. Kumar, "High gain microstrip array antenna with SIW and FSS for beyond 5G at THz band," *OPTIK*: *International Journal for Light and Electron Optics*, vol. 236c, pp. 166568/1–13, Mar. 2021.
25. U. Nissanov, G. Singh, E. Gelbart, and N. Kumar, "Highly directive microstrip array antenna with FSS for future generation cellular communication at THz band," *Wireless Personal Communications*, vol. 118, no. 1, pp. 599–617, Jan. 2021.
26. F. D. L. Peters, B. Boukari, S. O. Tatu, and T. A. Denidni, "77 GHz millimeter wave antenna array with Wilkinson divider feeding network," *Progress In Electromagnetics Research Letters*, vol. 9, pp. 193–199, Jan. 2009.
27. M. N. Uddin and S. Choi, "Non-uniformly powered and spaced corporate feeding power divider for high-gain beam with low SLL in millimeter-wave antenna array," *Sensors*, vol. 20, no. 17, pp. 4753/1–14, Aug. 2020.
28. J. Li, L. Li, L. Lu, H. Shi, H. Huo, and A. Zhang, "Four-way waveguide power divider design for W-band applications," *International Journal of RF and Microwave Computer-Aided Engineering*, vol. 28, no. 5, pp. e21242/1–5, Jan. 2018.
29. Y. Lee, S. M. Sim, R. L. Kang, I. L. Garro, Y. Wang, Y. H. Jang, and J. M. Kim, "Two-way waveguide power divider using 3D printing and electroless plating," *Proceedings of IEEE 48th European Microwave Conference (EuMC)*, Madrid, Spain, 23–27 Sep. 2018, pp. 219–222.
30. S. H. Shehab, N. C. Karmakar, and J. Walker, "Substrate-integrated-waveguide power dividers: An overview of the current technology," *IEEE Antennas and Propagation Magazine*, vol. 62, no. 4, pp. 27–38, Aug. 2020.
31. Z. Zhen-Yu and W. Ke, "Broadband half-mode substrate integrated waveguide (HMSIW) Wilkinson power divider," *Proceedings of IEEE International Microwave Symposium (MTT-S)* Digest, Atlanta, USA, 15–20 June 2008, pp. 879–882.
32. O. D. Gürbüz, K. Topalli, M. Ünlü, Ş. Demir, and Tayfun Akin, "A 35 GHz coplanar waveguide power divider," *Proceedings of IEEE 10th Mediterranean Microwave Symposium*, Guzelyurt, N. Cyprus, 25–27 Aug. 2010, pp. 22–24.
33. K. Al-Shamaileh, N. Dib, S. Abushamleh, "A compact coplanar waveguide Wilkinson power divider based on signal traces and adjacent grounds width modulation," *Microwave and Optical Technology Letters*, vol. 60, no. 9. pp. 2224–2227, Sep. 2018.
34. W. Gomaa, R. L. Smith, H. Esmaeilsabzali, and T. E. Darcie, "Terahertz power divider using symmetric CPS transmission line on a thin membrane," *IEEE Access*, vol. 8, pp. 214425–214433, Nov. 2020.
35. A. Gouda, C. D. López, V. Desmaris, D. Meledin, A. B. Pavolotsky, and V. Belitsky, "Millimeter-wave wideband waveguide power divider with improved isolation between output ports," *IEEE Transactions on Terahertz Science and Technology*, vol. 11, no. 4, pp. 408–416, May 2021.
36. F. Barrera, A. Siligaris, B. Blampey, J. L. G. Jimenez, "A D-band 4-ways power splitter/combiner implemented on a 28nm bulk CMOS process," *Proceedings of IEEE 14th European Microwave Integrated Circuits Conference (EuMIC)*, Paris, France, 30 Sep. 2019–01 Oct. 2019, pp. 868–871.

37. A. Treizebre, S. Laurette, Y. Xu, R. G. Bosisio, and B. Bocquet, "THz power divider circuits on planar Goubau lines (PGLs)," *Progress In Electromagnetics Research C*, vol. 26, pp. 219–228, Dec. 2011.
38. K. S. Reichel, R. Mendis, and D. M. Mittleman, "A broadband terahertz waveguide T-junction variable power splitter," *Scientific Reports*, vol. 6, pp. 28925/1–6, June 2016.
39. K. B. Kumar and T. Shanmuganantham, "A 3-port Y-junction power splitter for 60GHz applications," *Proceedings of IEEE 4th International Conference on Devices, Circuits, and Systems (ICDCS)*, Coimbatore, India, 16–17 March 2018, pp. 1–5.
40. R. Aloui, Z. Houaneb, and H. Zairi, "Substrate integrated waveguide circular antenna for terahertz application," *Progress In Electromagnetics Research C*, vol. 96, pp. 229–242, Sep. 2019.
41. K. Dhwaj, Y. Zhao, R. Al Hadi, X. Li, F. M. C. Chang, and T. Itoh, "A 0.55 THz on-chip substrate integrated waveguide antenna," *Proceedings of IEEE 43rd International Conference on Infrared, Millimeter, and Terahertz Waves (IRMMW-THz)*, Nagoya, Japan, 09–14 Sep. 2018, pp. 1–2.
42. A. A. Althuwayb, M. Alibakhshikenari, B. S. Virdee, H. Benetatos, F. Falcone, and E. Limiti, "Antenna on chip (AoC) design using metasurface and SIW technologies for THz wireless applications," *Electronics*, vol. 10, no. 9, pp. 1120/1–8, May 2021.

Chapter 6
Terahertz Microstrip Antenna with Frequency Selective Surface for 6G Communication

6.1 Introduction

Following Chap. 5, which presents THz passive components, this chapter will discuss in detail two other THz passive components, the THz unit-cell (Unit-cell) microstrip planar antenna radiator and THz Unit-cell frequency selective surface (FSS) [1]. The design base of each THz microstrip planar array antenna is the design of the THz scalable structure Unit-cell planar antenna radiator [2]. Similarly, the design base of each THz array of FSSs is the design of the THz scalable structure Unit-cell planar FSS [3].

6.2 Terahertz Unit-Cell Radiator Design Considerations

In microstrip antenna arrays, the design of the fundamental radiating element that makes up the antenna array is a design of the Unit-cell that can be replicated according to the antenna's desired gain requirements. So, first, as presented in Chap. 4, for THz microstrip antenna design considerations and to mitigate conductor and substrate losses, surface waves losses, and radiation losses [4–5], the design consideration of the THz Unit-cell antenna radiator should be the same by choosing laminate and laminate parameters following Eqs.(4.32)–(4.36) [6–7]. The second stage is designing and optimizing the THz Unit-cell radiator with maximum gain, impedance BW, and desired resonance frequency, which suits the demands [8]. These design considerations are shown in the following flowchart diagram Fig. 6.1.

U. Nissanov, G. Singh, *Antenna Technology for Terahertz Wireless Communication*,
https://doi.org/10.1007/978-3-031-35900-2_6

Fig. 6.1 Flowchart diagram of designing THz Unit-cell antenna radiator

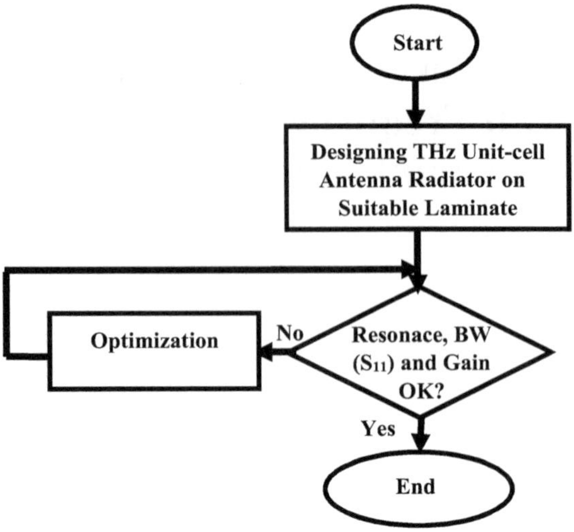

6.3 Terahertz Unit-Cell Frequency Selective Surface Design Considerations

The FSSs are an array of configurations consisting of a majority of slimmer conducting wires. Frequently printed on a dielectric substrate for support, i.e., invented man-fabricated components, which do not subsist in the environment, that set up bandstop filter and bandpass filter features while stimulated by an EM signal at a random corner to the plateau of the FSSs and modeled to absorb, transmit, or reflect EM waves established on the frequency of the transmit waves [9–10].

A couple of fundamental classes are (a) the conducting wires pattern and (b) the slots pattern, supported by a dielectric laminate to create the FSSs, and they act as non-active EM wave filters. While a bandpass FSS configuration complements the bandstop FSS configuration, a perfectly conducting plane is created when both arrays are on top [9–10]. Thus, at the THz band, the FSSs are also called metasurface. Fig. 6.2 exhibits the incident plane electric wave (E^i) to the lossless microstrip substrate with bandstop FSS, i.e., periodic wires type of FSS array, traveling perpendicular, i.e., normal incidence, to the lossless microstrip substrate with bandstop FSSs.

The incident plane electric wave (E^i) will partially transmit (E^t) in the ahead way, and partially reflect (E^r) in a specular way. Under resonance conditions and without grating lobes, the amplitude of the electric reflected signal may equal to (E^r), while $E^t = 0$. The reflection coefficient (Γ) is defined by [9–10]:

$$\Gamma = E^r / E^i \tag{6.1}$$

while the transmission coefficient (T) is defined by [9–10]:

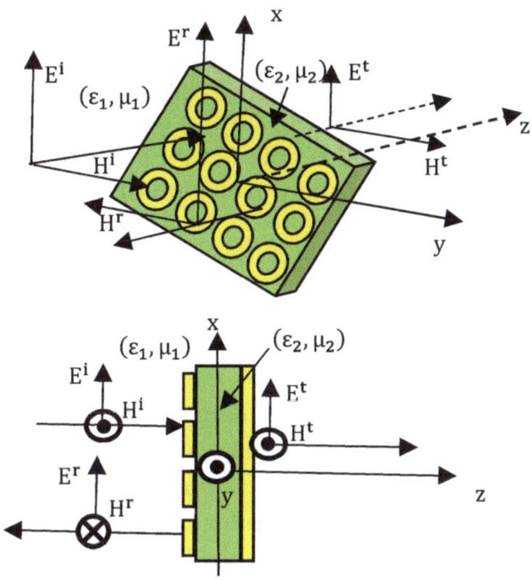

Fig. 6.2 Wave reflection and transmission at normal incidence to lossless microstrip substrate with bandstop FSS

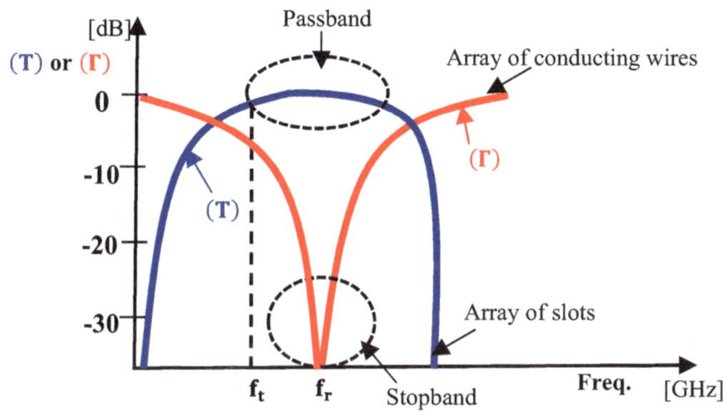

Fig. 6.3 Typical frequency response of the FSS

$$T = E^t/E^i \tag{6.2}$$

For periodic slots type array, the medium is excited by a current magnetic field $(\vec{H})$, opposite to periodic conducting wires type array, which needs an electric current field $(\vec{E})$ to excite the matter. Fig. 6.3 exhibits a typical transmission coefficient ($T = S_{21}$) response of a grid of wires and reflection coefficient ($\Gamma = S_{11}$) response of a grid of slots, wherever the layer reflects the incident polarized waves at multiple frequencies, i.e., bandstop or reflection band, and wherever the layer is diaphanous to those waves at other frequencies, i.e., bandpass or transmission band. where f_r is the resonant frequency/center frequency of the bandstop, and f_t is the starting of the bandpass. The −10 dB in the ($\Gamma = S_{11}$) and −0.5 dB level in the

($T = S_{21}$) levels define the BWs For the array of conducting wires cases. The resonance is because superior FSSs induced lower currents beside the bandpass. The layer acts as a brassy layer at a frequency of resonance. This brassy layer is mainly reflective and shows a bandpass at a resonance frequency resulting from solid EM waves in the slots. Any FSSs magnitude (α) need to be approximately more negligible from the wavelength of the dielectric laminate $\left(\lambda_{\varepsilon_r} = c/\left(f_0\sqrt{\varepsilon_r}\right)\right)$ [9–10]:

$$\Gamma = S_{11} \leq -10\ \text{dB},\ \ T = S_{21} \leq -0.5\ \text{dB},\ \ \alpha < \lambda_{\varepsilon_r} < \lambda_o/\sqrt{\varepsilon_r} < c/(f_0\sqrt{\varepsilon_r}) \quad (6.3)$$

FSSs can reduce interference of the microstrip antenna with other devices and reduce the scattering of the incident wave. Nevertheless, in the THz regime, the existence of the resonating sizes of FSSs causes the creation of the SLLs in the reflected and transmitted waves, which is the critical feature of the FSSs, so design attention is a necessity to take due to the reflection response, and transmission of repeated FSSs is dependent on the FSSs shape, incidence EM waves angle. Using FSSs to deflect the antenna beam introduces some problems. The first is, due to the closeness of the FSSs to the radiating elements, so the return loss (S_{11}) of the antenna is affected, while the frequency of the antenna's resonance is changed. So using FSSs to deflect the beam for specific narrowband applications would not be comfortable. The second is that as more FSSs are added to increase the range of deflecting the antenna's beam, the insertion loss of the antenna increases [11].

An array of the Unit-cells FSSs may be used to increase the gain of the microstrip antenna, which may act as a reflector and may be applied after each antenna to reflect the EM fields in the direction of a primary main beam. The array of Unit-cells FSSs may act as a reflector if its bandstop is the microstrip antenna frequency of the resonance band, and the distance betwixt array of Unit-cells FSSs and microstrip antenna should be approximately $\lambda_0/2$. In this technique, the EM rays form a unidirectional beam after striking the reflector, effectively increasing the antenna's gain [12].

Multilayered FSSs must be used to achieve a wideband response in the wanted frequency range because of the discontinuities equipped by the layers. While the Unit-cells of an FSS are practically excited with the EM fields, they may be replaced with a resonance matching circuit. The electric equivalent circuit model for Unit-cell bandstop FSSs is a serial link of capacitance (C) and inductance (L) [12–13]. In contrast, the electric equivalent circuit model for Unit-cell bandpass FSSs configuration is a shunt, i.e., parallel, a connection between capacitance (C) and inductance (L). The f_r is the frequency of the resonance and can be computed by the following formula [3]:

$$f_r = \frac{1}{2\pi\sqrt{L \cdot C}} \quad (6.4)$$

L and C appropriately represent the equivalent inductance and the capacitance of a Unit-cell FSS.

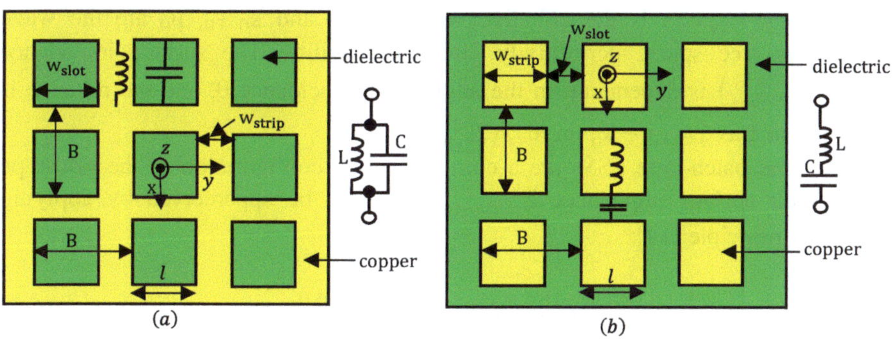

Fig. 6.4 Typical types of the FSSs, (**a**) Grid-type FSSs (Bandpass), and (**b**) Patch-type FSSs (Bandstop)

6.3.1 Equivalent Electric Circuits Models for Bandpass and Bandstop FSSs

The bandpass and bandstop FSSs need to be classified into two categories to find analytical formulations of the surface impedance. (a) The first is called a grid-type configuration, while the second (b) is called a patch-type configuration, shown in Fig. 6.4, appropriately [11–13]. Therefore, Fig. 6.4a, b appropriately exhibits the grid and patch-type FSSs.

From Fig. 6.4a, the recurring existence of B in the two x- and y-ways when the width of the grid w_{strip} is extra lesser than B ($w_{strip} < < B$). For a random incidence angle, the impedance layer of the grid-type FSSs of the TM and TE polarized plateau is given by [11–13]:

$$Z^{TE}_{grid-type\ FSSs} = j\frac{\eta_{eff}}{2}\alpha \tag{6.5}$$

$$Z^{TM}_{grid-type\ FSSs} = j\frac{\eta_{eff}}{2}\alpha \cdot \xi \tag{6.6}$$

The $\eta_{eff} = \sqrt{\mu_0\mu_{eff}/(\varepsilon_0\varepsilon_{eff})}$ is the intrinsic impedance of the efficacious dielectric laminate that the grid-type FSSs mount. If the grid-type FSSs are positioned in a specific laminate with EM variables ε_r and μ_r, then $\varepsilon_{eff} = \varepsilon_r$and $\mu_{eff} = \mu_r$; in case the grid-type FSSs are at the bond of a couple of laminates with various EM variables ($\varepsilon_{r1}, \ \mu_{r1}$) and ($\varepsilon_{r2}, \ \mu_{r2}$), then the proportional efficacious permeability and permittivity may be meant as $\mu_{eff} = (\mu_{r1} + \mu_{r2})/2$ and $\varepsilon_{eff} = (\varepsilon_{r1} + \varepsilon_{r2})/2$. α is specified like the grid-type FSSs variable in [12] for an electrically crowded array of grid-type FSSs (namely, $w_{strip} < < B$, $k_{eff}B < < 2\pi$) and equals to [11–13]:

$$\alpha = \frac{k_{eff} \cdot B}{\pi} \ln\left[\frac{1}{\sin(\pi \cdot w_{strip}/2B)}\right] \tag{6.7}$$

where $k_{eff} = k_0\sqrt{\varepsilon_{eff}}$ is the effective wavenumber and k_0 ε_0, μ_0 are the wave number in free space, permittivity, and permeability. The surface impedance $\left(Z^{TM}_{grid-type\ FSSs}\right)$ is dependent on the angle of the incidence θ, and the reliance is inclusive in ξ as $\xi = \left(1 - k_0^2 \sin^2\theta\right)/\left(2k_{eff}^2\right)$.

Since the patch-type FSSs are a complementary configuration of the grid-type FSSs, their surface impedance $Z^{k}_{patch-type\ FSSs}$ can be apperceived by applying Babinet's principle [14]:

$$Z^{TE}_{patch-type\ FSSs} = -j\frac{\eta_{eff}}{2\alpha\xi} \tag{6.8}$$

$$Z^{TM}_{patch-type\ FSSs} = -j\frac{\eta_{eff}}{2\alpha} \tag{6.9}$$

The per-unit-length inductance (L_i) of the strip in grid-type FSSs is [11–13]:

$$L_i = \mu_0\mu_{eff}\ln\left[\frac{1}{\sin\left(\pi \cdot w_{strip}/2B\right)}\right] \tag{6.10}$$

The per-unit-length capacitance (C_i) of the slot-type FSSs is [11–13]:

$$C_i = \varepsilon_0\varepsilon_{eff}\ln\left[\frac{1}{\sin(\pi \cdot w_{slot}/2B)}\right] \tag{6.11}$$

Thus, the surface impedance of the bandpass FSSs may be seen as a parallel connection of inductance (L) and capacitance (C) and can compute by [11–13]:

$$Z^{TE}_{pass-band\ FSSs} = \left(j\omega L_{pass-band\ FSSs}\right) \parallel \left(\frac{1}{j\omega C_{pass-band\ FSSs}}\right) \tag{6.12}$$

$$Z^{TM}_{pass-band\ FSSs} = \left(j\omega L_{pass-band\ FSSs}\right) \parallel \left(\frac{1}{j\omega C_{pass-band\ FSSs}}\right)\xi \tag{6.13}$$

where the $L_{pass-band\ FSSs}$ is the per-unit-length inductance and $C_{pass-band\ FSSs}$ is the per-unit-length capacitance that can be computed by [11–13]:

$$L_{pass-band\ FSSs} = l \cdot L_i \tag{6.14}$$

$$C_{pass-band\ FSSs} = \frac{(l - 2w_{slot})}{2}C_i \tag{6.15}$$

Thus the surface impedance of the bandstop FSSs can be seen as a serial connection of inductance (L) and capacitance (C) and can compute by [11–13]:

$$Z^{TE}_{stop-band\ FSSs} = \frac{j\omega L_{stop-band\ FSSs} + \left(1/j\omega C_{stop-band\ FSSs}\right)}{\xi} \tag{6.16}$$

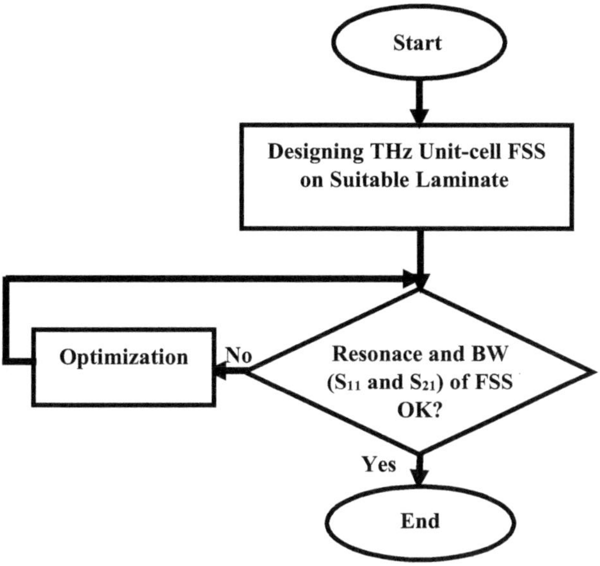

Fig. 6.5 Flowchart diagram of designing THz Unit-cell FSS

$$Z^{TM}_{stop-band\ FSSs} = j\omega L_{stop-band\ FSSs} + \left(1/j\omega C_{stop-band\ FSSs}\right) \quad (6.17)$$

where the $L_{stop-band\ FSSs}$ is the per-unit-length inductance and $C_{stop-band\ FSSs}$ is the per-unit-length capacitance that can be computed by [11–13]:

$$L_{stop-band\ FSSs} = l \cdot L_i \quad (6.18)$$

$$C_{pass-band\ FSSs} = l \cdot C_i \quad (6.19)$$

To conclude this paragraph, the design considerations of Unit-cell FSS are shown in the following flowchart diagram Fig. 6.5. First, the design consideration of THz Unit-cell FSS should choose the laminate and laminate parameters following eqs. (4.32)–(4.36) [6–7]. The second stage is designing and optimizing the THz Unit-cell FSS with desired resonance frequency and impedance BW (S_{11} and S_{21}), which is suitable for the demands [8].

6.4 Problem Formulation

Developing a high-gain, wider BW, and high-radiation efficiency antenna is a critical and essential apparatus for THz 6G wireless communication networks to amplify the signal's propagation distance beyond 100 m. In addition, wider BW can facilitate data rates beyond 20 Gb/s, which is needed for 6G, while high-radiation efficiency will maximize the propagation distance of the transmitted THz signals due to the lack of a high-power THz signal source. However, proportional to the mmWave band, the THz signals will involve extreme attenuations because of resonance within water vapor and, to meagerly, oxygen molecules at these higher THz frequencies [15–16].

The most efficient and suitable antenna is the microstrip planar antenna for THz communication [17–18], where a single Unit-cell THz planar microstrip patch antenna cannot achieve a gain of more than 5–6 dB [19], approximately, so a microstrip array technique is needed. Another proper technique to increase a microstrip antenna's gain is beam steering phased array microstrip antenna or MIMO microstrip antenna [20–21]. The basic design of the microstrip array antenna, beam steering microstrip antenna, and MIMO microstrip antenna is the design of the Unit-cell radiator, which is usually a single radiator and, after a proper design and optimization, can be integrated and multiplied because it is scalable, in the above technique to achieve a higher gain and directivity microstrip antenna [22–23].

An additional method to enhance the gain of a Unit-cell THz microstrip planar antenna is using an array of Unit-cells THz planar metamaterials (MtMs) or an array of Unit-cells THz planar FSSs. Furthermore, after proper design and optimization, it can be merged and multiplied because it is scalable, with a Unit-cell THz microstrip planar antenna to achieve a higher gain and directivity microstrip antenna as needed for THz communication networks [24–25].

6.5 Related Work

Alibakhshikenari et al. [26] demonstrated a THz on-chip 15 Unit-cells circular and rectangular aperture-fed microstrip array antenna based on silicon and polycarbonate laminates. The maximum gain and impedance BW measurement results at the fabricated THz on-chip antenna were 11.71 dB @ 300 GHz and 26 GHz (290–316 GHz) for $S_{11} \leq -10$ dB, respectively. Next, Bae et al. [27] demonstrated a THz dual-polarized L-probe differential-fed Unit-cell broadband circular patch antenna. The measurement results at the fabricated THz Unit-cell antenna of the maximum gain and impedance BW were 9.8 dB @ 125 GHz and 25 GHz (100–125 GHz) for $S_{11} \leq -10$ dB, respectively. Finally, Ma et al. [28] demonstrated a THz Unit-cell rectangular patch antenna fed by SIW. The maximum gain and impedance BW measurement results at the fabricated THz Unit-cell antenna were 7.26 dB @ 135 GHz and 29.1 GHz (125.3–154.4 GHz) for $S_{11} \leq -10$ dB, respectively.

However, Hlaing et al. [29] demonstrated a THz Unit-cell bandstop FSS utilizing carbon-based polyimide. The impedance BW and maximum return loss (RL) at the fabricated THz Unit-cell FSS were 1000 GHz (500–1500 GHz) and 40.5 dB @ 1000 GHz, respectively. Rashid et al. [30] demonstrated a THz Unit-cells array of bandpass FSSs for 306 GHz based on Rogers RO4350B microstrip laminate. The insertion loss (IL) S_{21} and BW at the fabricated THz Unit-cell FSS was 6 dB @ 264 GHz and 2 GHz (305–307 GHz), respectively. Finally, Ghavidel et al. [31] demonstrated a THz Unit-cells array bandpass double-layer FSSs using an

Table 6.1 THz passive components classification and performance according to related work in Chap. 6

Refs.	THz passive technology	Freq. nom. (GHz)	Measured BW (GHz)	Measured maximum gain (dB)	
[26]	On-chip circular and rectangular 15 unit-cells antenna array radiators	300	26 (290–316)	11.71	
[27]	Circular unit-cell antenna radiator	110	25 (100–125)	9.8	
[28]	Rectangular unit-cell antenna radiator	135	29.1 (125.3–154.4)	9.8	
Refs.	**THz passive technology**	**Freq. Nom. (GHz)**	**Measured BW (GHz)**	**Measured maximum return loss (dB)**	**Measured insertion loss (dB)**
[29]	Bandstop unit-cell FSS carbon-based polyimide	1000	1000 (500–1500)	40.5	n/a
[30]	Bandpass unit-cells FSSs array	306	2 (305–307)	n/a	6
[31]	Bandpass double layer unit-cells FSSs array	240	9 (235.5–244.5)	11	n/a

RT/duroid R5880 microstrip substrate. The impedance BW and maximum RL at the fabricated THz Unit-cells array of FSSs were 9 GHz (235.5–244.5 GHz) and 11 dB @ 240 GHz, respectively. The performance summary of these THz passive components discussed above is presented in Table 6.1 according to the related work.

6.6 Simulation Results and Discussion

6.6.1 THz Unit-Cell Planar Rectangular Microstrip Antenna for 112.5 GHz

The designed THz Unit-cell planar rectangular microstrip antenna, as shown in Fig. 6.6, is established on the rolled copper Rogers RO3003 with the tan (δ), ε_r, copper thickness (t), laminate height (h), (0.0019, 3.02) @ 100 GHz, 8.75 μm, 127 μm, appropriately. The resonance frequency of 112.5 GHz was the design goal.

The Unit-cell THz planar rectangular microstrip antenna sizes after calculation with Eqs. (4.16)–(4.24) and the optimized sizes following with parameter sweep at the CST MWS to get the best gain, BW, and dimensions of the ground plane of this designed THz Unit-cell planar rectangular microstrip antenna are presented at the following Table 6.2.

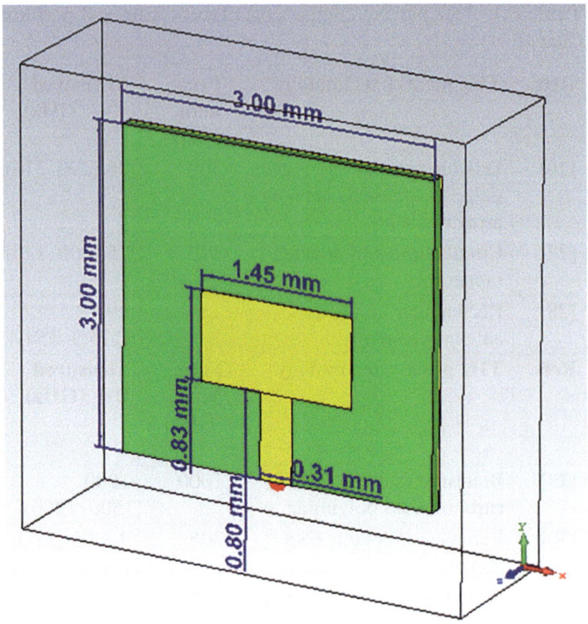

Fig. 6.6 Configuration of the designed THz Unit-cell planar rectangular microstrip antenna

Table 6.2 Dimensions of the designed THz Unit-cell planar rectangular microstrip antenna for 112.5 GHz

Parameter	Size according to equations	Size after optimization at CST MWS
W_x (the width of unit-cell)	0.94 mm	1.45 mm
L_x (length of unit-cell)	0.7 mm	0.83 mm
L_f (length of the feed line)	2.6 mm (M = 0)	0.8 mm
W_{50} (the width of the feeding line	0.297 mm	0.308 mm
L_g x W_g (GND dimensions)	1.462 × 1.702 mm^2	3 x 3 mm^2

From Table 6.2, it has been shown that the actual sizes of the THz Unit-cell planar rectangular microstrip antenna are different from the sizes that were apperceived according to Eqs. (4.16)–(4.24) because the equations give only approximate sizes. Moreover, for finding the actual dimensions of the THz Unit-cell planar rectangular microstrip antenna, because of the high frequency (112.5 GHz) losses and due to the fringing effects, we must optimize these dimensions with the simulation by the CST MWS simulator because these sizes need to be changed for our wanted BW of 6.04 GHz and resonance frequency of 112.5 GHz. The simulation result of the S_{11} of the designed THz Unit-cell planar rectangular microstrip antenna is presented in Fig. 6.7. From this Fig.; it has been shown that the BW of this THz Unit-cell planar

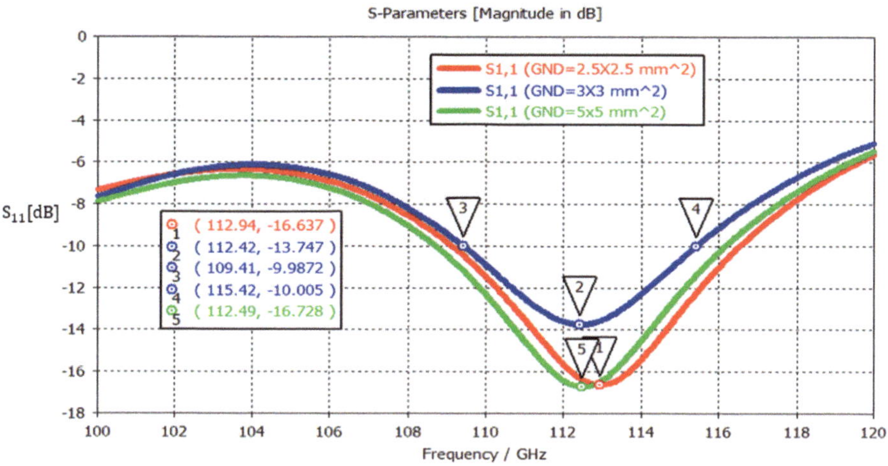

Fig. 6.7 Simulation results of the S_{11} for the THz Unit-cell planar rectangular microstrip antenna

rectangular microstrip antenna for ground sizes of 3 × 3 mm^2 was: BW($S_{11} \leq -10$ dB) = 6.04 GHz, while the frequency of the resonance was 112.94 GHz (GND = 2.5 × 2.5 mm^2), 112.42 GHz (GND = 3 × 3 mm^2), 112.49 GHz (GND = 5 × 5 mm^2). So in this design, the chosen ground was 3 × 3 mm^2 because this ground is relatively more minor than 5 × 5 mm^2 and close to our wanted resonance frequency of 112.5 GHz.

Figures 6.8a–c show the simulation results of the 3D farfield radiation pattern, gain, and directivity at the E-plane and H-plane of the designed THz Unit-cell planar rectangular microstrip antenna at a frequency of 112.5 GHz at the polar plots, respectively. From Fig. 6.8a, it has been shown that the directivity is 6.45 dBi, and the total efficiency is −0.43 dB for 112.5 GHz. From Fig. 6.8b, it has been shown that the gain is 5.75 dB in the E-plane with the direction of the main lobe at 39°, while the –3 dB angular width is 56.8° and the side lobe level (SLL) is −11.4 dB. The gain is −1.86 dB in the H-plane with the direction of the main lobe at −10°, while the –3 dB angular width is 78.6° and the SLL is −3.4 dB. From Fig. 6.8c, it has been shown that the directivity is 6 dBi in the E-plane with the direction of the main lobe at 39°, while the –3 dB angular width is 56.8° and the SLL is −11.4 dB. The directivity is −1.62 dBi in the H-plane with the direction of the main lobe at −10°, while the –3 dB angular width is 78.6°, and the SLL is −3.4 dB.

Simulation results from comparing the S_{11} and gain of the THz Unit-cell planar rectangular microstrip antenna made by the CST MWS FIT vs. CST MWS FEM solvers are attached in Fig. 6.9a, b. From Fig. 6.9a, b, the BW and the gain of the THz Unit-cell planar rectangular microstrip antenna are 6 GHz, 5.31–6.78 dB, 6.24 GHz, and 5–6.67 dB from the CST MWS FIT and the CST MWS FEM solver, respectively. So, an excellent agreement was achieved from this solver comparison.

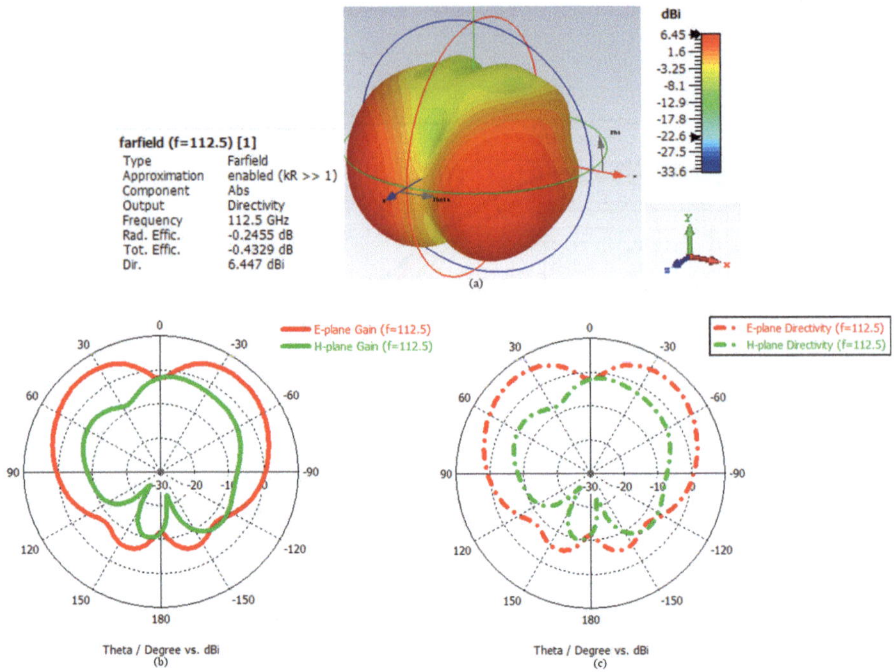

Fig. 6.8 Simulation results of the 3D farfield radiation pattern, E-plane, and H-plane of THz Unit-cell planar rectangular microstrip antenna (**a**) 3D farfield (**b**) Gain, and (**c**) Directivity

6.6.2 THz Unit-Cell Planar Crown Shape Microstrip Antenna for 127.6 GHz

The designed THz Unit-cell planar crown shape microstrip antenna, as shown in Fig. 6.10, is established on the rolled copper Rogers RO3003 with the tan (δ), ε_r, copper thickness (t), laminate height (h), (0.0019, 3.02) @ 100 GHz, 17.5 μm, 127 μm, appropriately. The resonance frequency of 127.5 GHz was the design goal.

The simulation result of the S_{11} of the designed THz Unit-cell planar crown shape microstrip antenna is presented in Fig. 6.11. From this Fig., the BW ($S_{11} < -10$ dB) is 5 GHz, while the minimum S_{11} is −30.47, and the resonance frequency is 127.6 GHz.

Figure 6.12a–c show the simulation results of the 3D farfield radiation pattern, gain, and directivity at the E-plane and H-plane of the designed THz Unit-cell planar crown shape microstrip antenna at a frequency of 127.5 GHz at the polar plots, respectively. From Fig. 6.12a, it has been shown that the gain is 5.43 dB, and the total efficiency is −0.34 dB for 127.5 GHz. From Fig. 6.12b, it has been shown that the gain is 5.47 dB in the E-plane with the direction of the main lobe at 48°, while

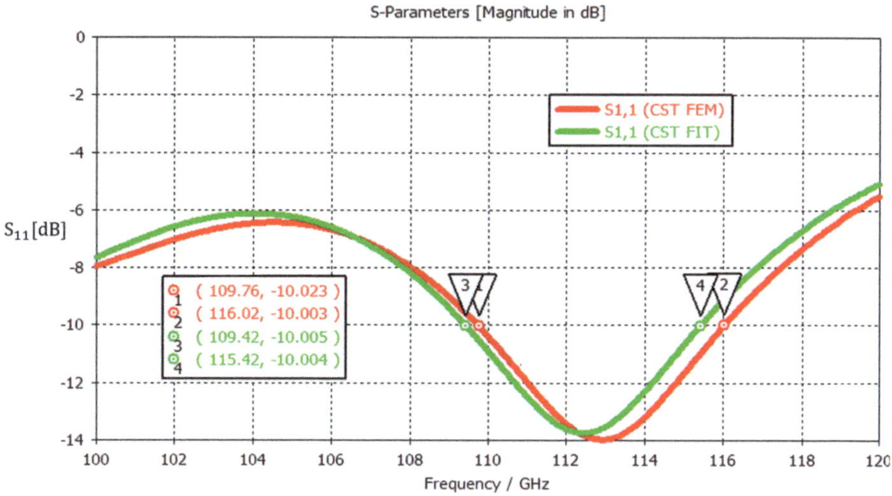

(a)

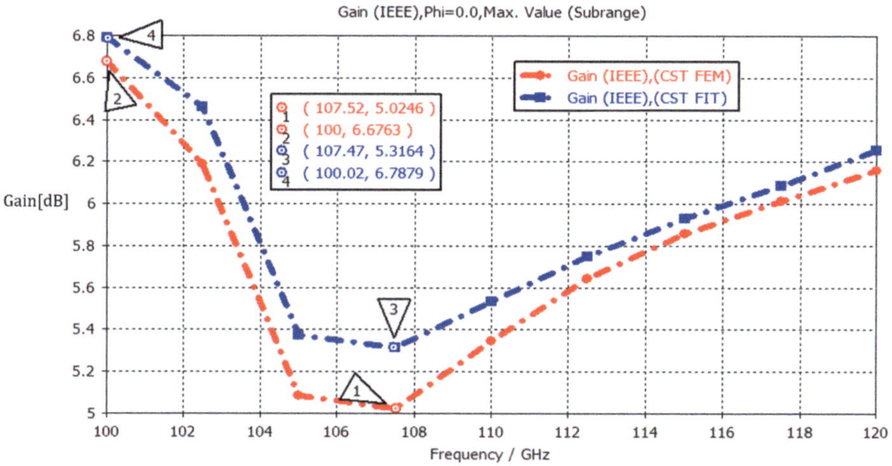

(b)

Fig. 6.9 Simulation results from the comparison made by the CST MWS FIT and CST MWS FEM solvers of THz Unit-cell planar rectangular microstrip antenna, (**a**) S_{11}, (**b**) Gain

the –3 dB angular width is 48° and the SLL is −3.8 dB. The gain is 3.72 dB in the H-plane with the direction of the main lobe at 16°, while the –3 dB angular width is 54.2° and the SLL is −5.8 dB. From Fig. 6.12c, it has been shown that the directivity is 5.8 dBi in the E-plane with the direction of the main lobe at 48°, while the –3 dB angular width is 46.2° and the SLL is −3.8 dB. The directivity is 4.05 dBi in the H-plane with the direction of the main lobe at 16°, while the –3 dB angular width is 54.2°, and the SLL is −5.8 dB.

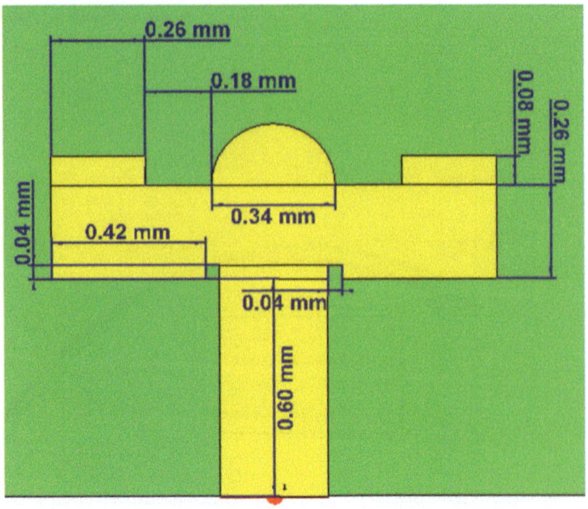

Fig. 6.10 Configuration of the designed THz Unit-cell planar crown shape microstrip antenna

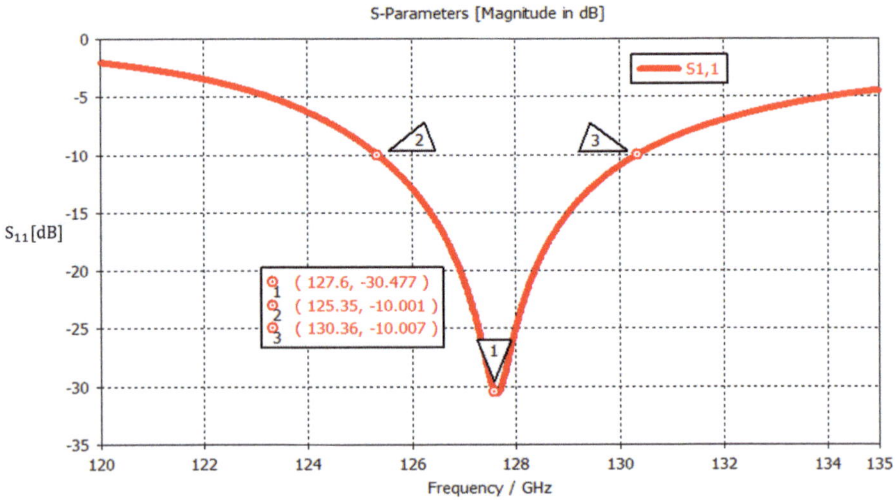

Fig. 6.11 Simulation results of the S_{11} for the THz Unit-cell planar crown shape microstrip antenna for 127.6 GHz

Simulation results from comparing the S_{11} and gain of the THz Unit-cell planar rectangular microstrip antenna made by the CST MWS FIT vs. CST MWS FEM solvers are attached in Fig. 6.13a, b. From Fig. 6.13a, b, the BW and the gain of the THz Unit-cell planar rectangular microstrip antenna are 5 GHz, 4.21–5.6 dB, 5 GHz, and 4.23–5.09 dB from the CST MWS FIT and the CST MWS FEM solver, respectively. So, an excellent agreement was achieved from this solver comparison.

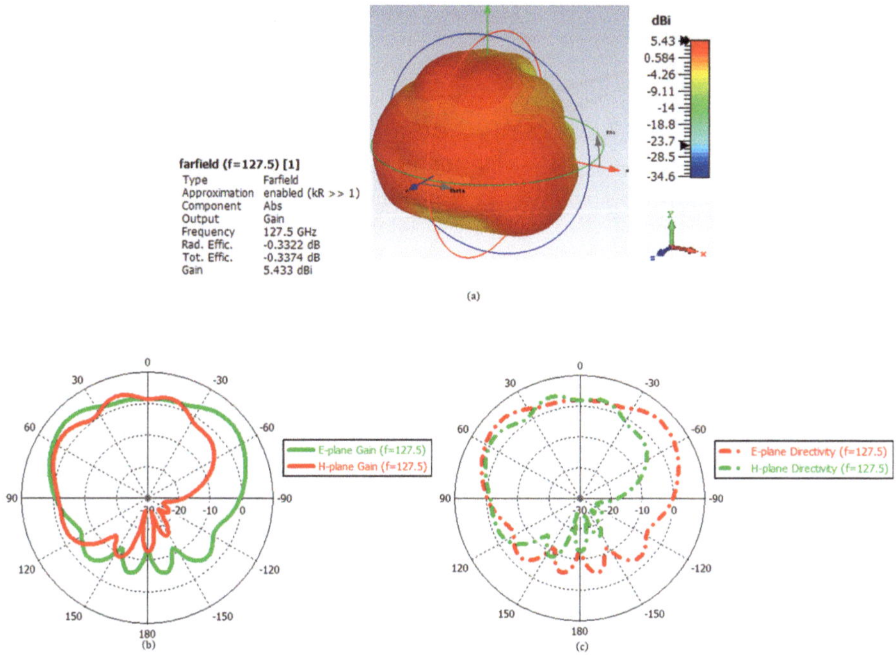

Fig. 6.12 Simulation results of the 3D farfield radiation pattern, E-plane, and H-plane of THz Unit-cell planar crown shape microstrip antenna, (**a**) 3D farfield, (**b**) Gain, (**c**) Directivity

6.6.3 THz Unit-Cell Bandstop Double Circular Ring FSS

The modeled THz Unit-cell bandstop double circular ring FSS was made with a double-printed circular ring copper with the standard electrodeposition (ED) copper foil Rogers RO3003 microstrip laminate with the parameters of substrate height (h), copper thickness (t), dielectric constant (ε_r), and loss tangent (tanδ) of 127 μm, 8.75 μm, (3.16, 0.0019) @ 100 GHz, respectively, with the sizes of $1 \times 1 \times 0.127\ mm^3$. The following Fig. 6.14 shows the designed THz Unit-cell bandstop double circular ring FSS.

The simulation results of the reflection coefficient S_{11} and the transmission coefficient S_{21} by variation of the diameter of the circular ring w_2 of the modeled THz Unit-cell bandstop double circular ring FSS are shown in Figs. 6.15a, b. From Fig. 6.15a, it may have been shown that increasing w_2 decreased the resonant frequency f_r of the THz Unit-cell bandstop FSS. While at the same time, increasing w_2 decreased the minimum S_{11} from −38.64 dB to − 47.35dB. Furthermore, from Fig. 6.15b, it may have been shown that increasing w_2 increased the S_{21} at all frequency ranges of 106.4–131.16 GHz, where $S_{21} \leq -0.5$ dB, so the w_2 was set to 0.762 mm because a peak S_{21} was achieved at −0.048 dB for f_r. The f_r is equal to

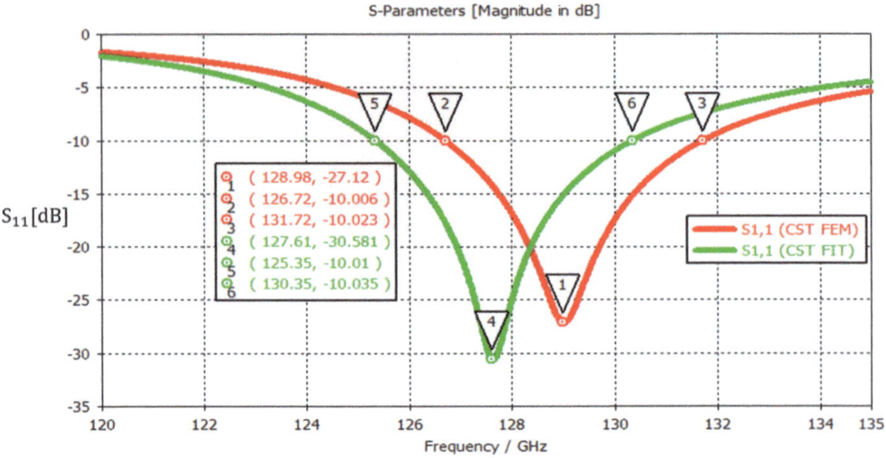

(a)

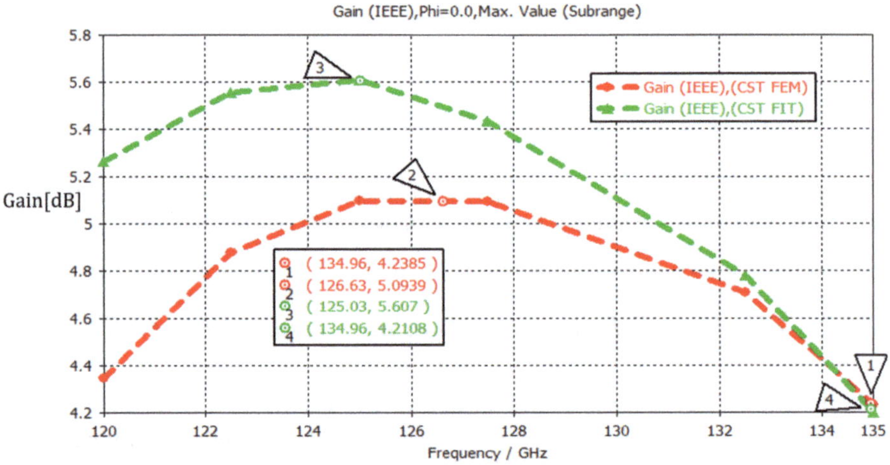

(b)

Fig. 6.13 Simulation results from the comparison made by the CST MWS FIT and CST MWS FEM solvers of THz Unit-cell planar crown shape microstrip antenna, (**a**) $S_{11,}$ and (**b**) Gain

120 GHz, meaning that THz Unit-cell bandstop FSS is reflective and presents a bandpass at a frequency range from 106.4 GHz to 131.16 GHz and is mostly reflective at the resonance frequency of 120 GHz. Under resonance conditions, the amplitude of the electric reflected signal from the THz Unit-cell bandstop double circular ring FSS might partially be reflected.

The simulated results of the S_{11} and the S_{21} of the THz Unit-cell bandstop double circular ring FSS for 120 GHz are shown in Figs. 6.16a–c. From these figures it has been shown that the transmission coefficient $S_{21} < -0.49$ dB for the frequency range of 106.53–131.07 GHz, while it has been shown that the reflection coefficient $S_{11} \leq -10$ dB for the frequency range of 106.56–131.18 GHz and the BW ($S_{11} \leq -10$ dB) of the THz Unit-cell, bandstop FSS is equal to 24.54 GHz.

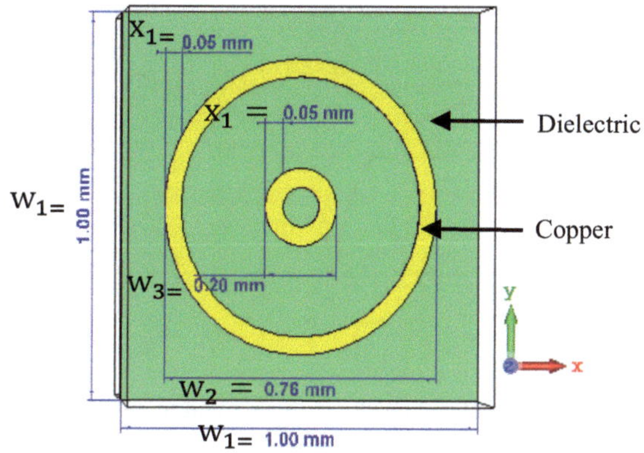

Fig. 6.14 Structure of the designed THz Unit-cell bandstop double circular ring FSS

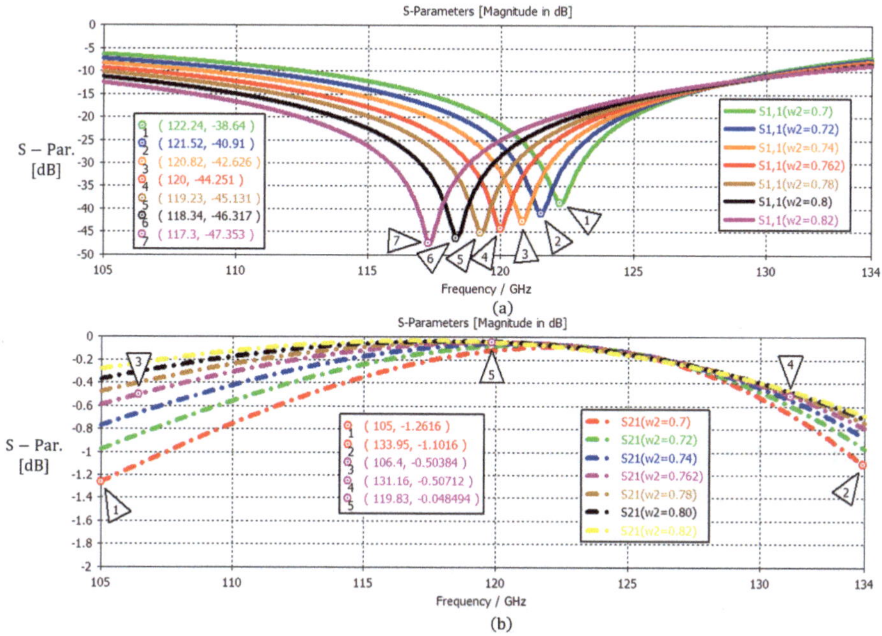

Fig. 6.15 Simulation results of the S_{11} and the S_{21} by variation of the diameter of the circular ring w_2 for designed THz Unit-cell bandstop double circular ring FSS, (**a**) S_{11}, (**b**) S_{21}

The simulated angular stability results of the S_{11} and the S_{21} and the resonance frequency fr of the THz Unit-cell bandstop FSS were investigated by varying the EM plane wave incidence angle from $\Theta = 0°$ (normal incidence) to 30°, with a ± 10° step, at the CST MWS simulator, as shown in Fig. 6.17.

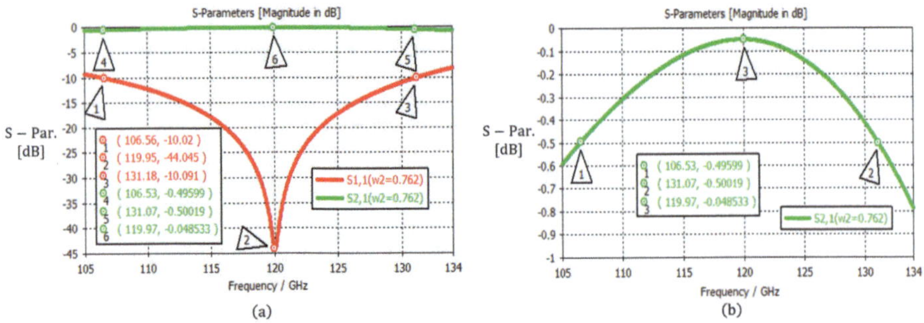

Fig. 6.16 Simulated results of the S_{11} and the S_{21} of the modeled THz Unit-cell bandstop double circular ring FSS, (**a**) S_{11}, and (**b**) S_{21}

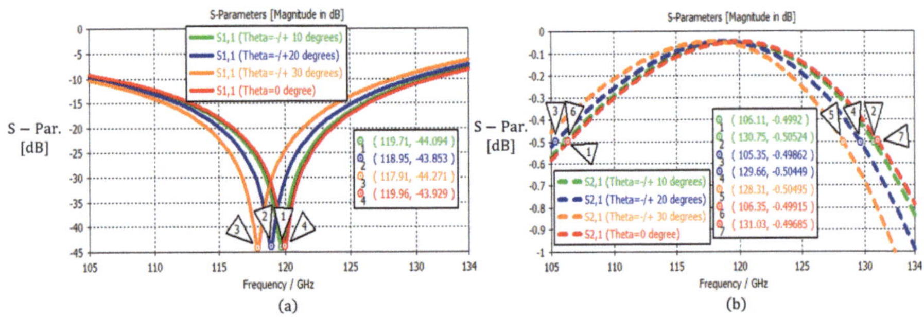

Fig. 6.17 Simulated angular stability results of the S11 and the S21 and the resonance frequency (f_r) of the modeled THz Unit-cell bandstop double circular ring FSS, (**a**) S11, (**b**) S21

6.6.4 THz Unit-Cell Bandstop Discontinuous Circular Ring FSS

The modeled THz Unit-cell bandstop discontinuous circular ring FSS was made with a discontinuous circular ring copper with the rolled copper foil Rogers RO3003 microstrip laminate with the parameters of substrate height (h), copper thickness (t), dielectric constant (ε_r), and loss tangent (tanδ) of 127 μm, 8.75 μm, (3.02, 0.0019) @ 100 GHz, respectively, with the sizes of $1.6 \times 1.6 \times 0.127\ mm^3$. The following Fig. 6.18 shows the designed THz Unit-cell bandstop discontinuous circular ring FSS.

The simulated results of the S_{11} and the S_{21} of the THz Unit-cell bandstop discontinuous circular ring FSS are shown in Fig. 6.19. From this figure it has been shown that the transmission coefficient $\mathbf{S_{21} < -0.53\ dB}$ for the frequency range of 113.13–124.31 GHz, while it has been shown that the reflection coefficient $\mathbf{S_{11} \leq -10\ dB}$ for the frequency range of 113.08–123.3 GHz and the BW($\mathbf{S_{11} \leq -10\ dB}$) of the THz Unit-cell bandstop discontinuous circular ring FSS equals 10.22 GHz.

Fig. 6.18 Structure of the designed THz Unit-cell bandstop discontinuous circular ring FSS

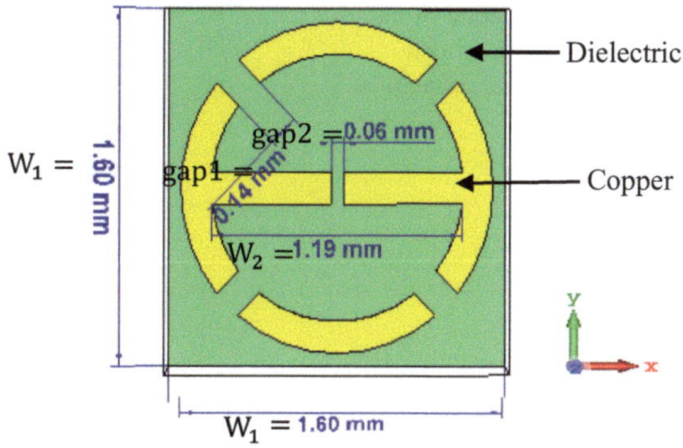

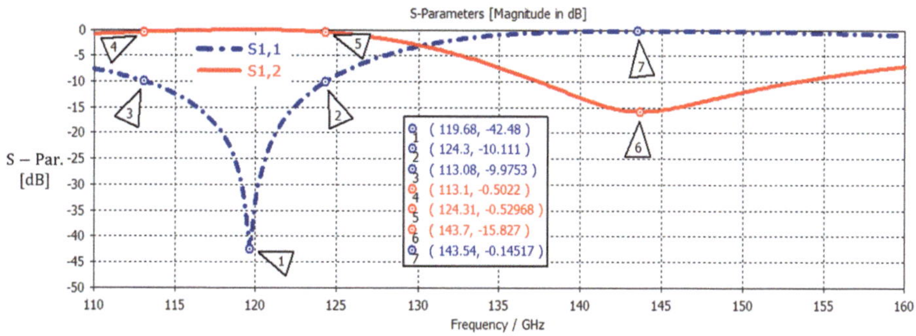

Fig. 6.19 Simulated results of the S_{11} and the S_{21} of the modeled THz Unit-cell bandstop discontinuous circular ring FSS

6.6.5 THz Unit-Cell Bandstop Double Rectangular Ring FSS

The modeled THz Unit-cell bandstop double rectangular ring FSS was made with a double rectangular ring copper with the rolled copper foil Rogers RO3003 microstrip laminate with the parameters of substrate height (h), copper thickness (t), dielectric constant (ε_r) and loss tangent (tanδ) of 127 μm, 8.75 μm, (3.02, 0.0019) @ 100 GHz, respectively, with the sizes of 1.1 × 1.1 × 0.127 mm^3. The following Fig. 6.20 is shown the designed THz Unit-cell bandstop double rectangular ring FSS.

The simulated results of the S_{11} and the S_{21} of the THz Unit-cell bandstop double rectangular ring FSS are shown in Fig. 6.21. From this Fig., it has been shown that the transmission coefficient $S_{21} < -0.28$ dB for the frequency range of 108.31–115.73 GHz, while it has been shown that the reflection coefficient $S_{11} \leq -10$ dB for the frequency range of 105.89–118.7 GHz and the BW ($S_{11} \leq -10$ dB) of the THz Unit-cell bandstop double rectangular ring FSS equals 12.81 GHz.

Fig. 6.20 Structure of the designed THz Unit-cell bandstop double rectangular ring FSS

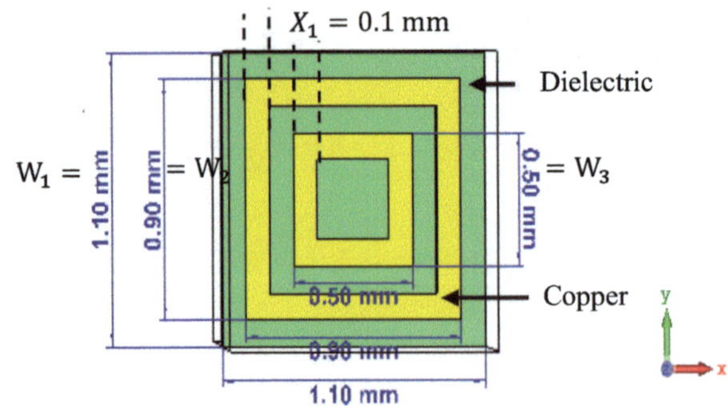

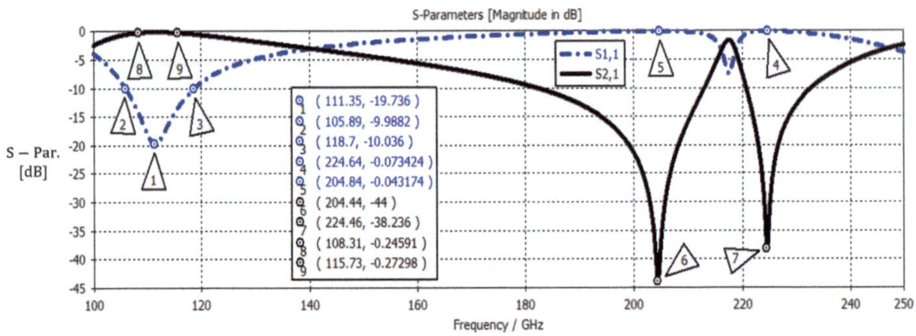

Fig. 6.21 Simulated results of the S_{11} and the S_{21} of the modeled THz Unit-cell bandstop double rectangular ring FSS

6.6.6 *THz Unit-Cell Bandstop Double Discontinuous Rectangular Ring and Rectangular Dipoles FSSs*

The modeled THz Unit-cell bandstop double discontinuous rectangular ring and rectangular dipoles FSSs were made with the rolled copper foil Rogers RO3003 microstrip laminate with the parameters of substrate height (h), copper thickness (t), dielectric constant (ε_r), and loss tangent (tanδ) of 127 μm, 8.75 μm, (3.02, 0.0019) @ 100 GHz, respectively, with sizes of 1.25 × 1.25 × 0.127 mm^3. Figure 6.22 (a)–(b) shows the designed THz Unit-cell bandstop double discontinuous rectangular ring and rectangular dipoles FSSs mounted on the same laminate on opposite sides.

The simulated results of the S_{11} and the S_{21} of the THz Unit-cell bandstop double discontinuous rectangular ring and rectangular dipoles FSSs are shown in Fig. 6.23. From this figure it has been shown that the transmission coefficient $S_{21} < -1.15$ dB for the frequency range of 104.74–111.78 GHz, while it has been shown that the reflection coefficient $S_{11} \leq -10$ dB for the frequency range of 104.79–111.77 GHz and the BW($S_{11} \leq -10$ dB) of the THz Unit-cell bandstop double discontinuous rectangular ring and rectangular dipoles FSSs equals 6.38 GHz.

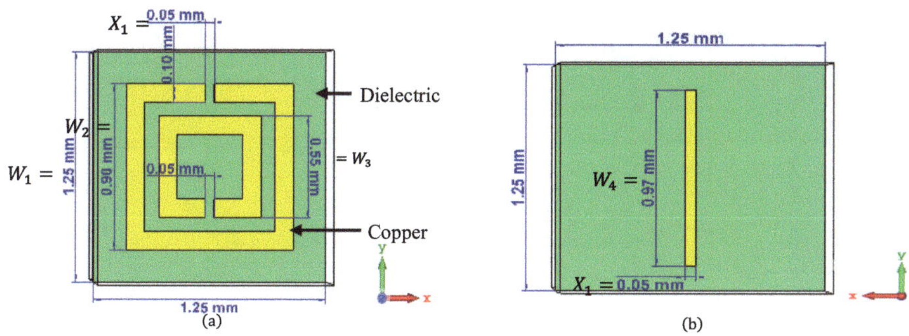

Fig. 6.22 Structure of the designed THz Unit-cell bandstop double discontinuous rectangular ring and rectangular dipoles FSSs, (**a**) Internal FSS, and (**b**) External FSS

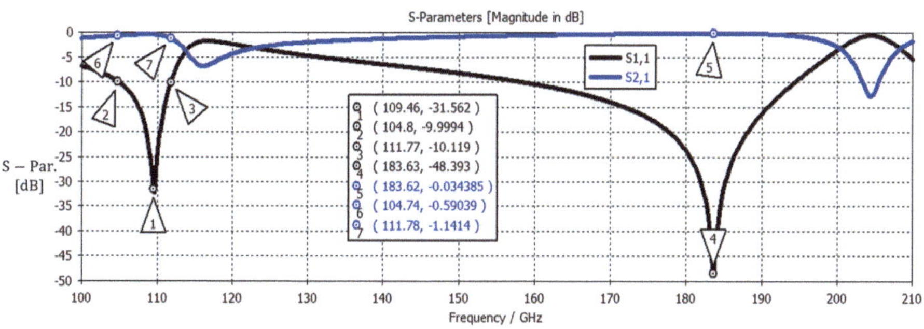

Fig. 6.23 Simulated results of the S_{11} and the S_{21} of the modeled THz Unit-cell bandstop double discontinuous rectangular ring and rectangular dipoles FSSs

6.6.7 *THz Unit-Cell Bandstop and Bandpass Steering Wheel Shape FSSs*

The modeled THz Unit-cell bandstop steering wheel shape and bandpass complementary steering wheel shape FSS were made with the Taconic TLY-5 microstrip laminate with the parameters of substrate height (h), copper thickness (t), dielectric constant (ε_r), and loss tangent (tanδ) of 127 μm, 17.5 μm, 2.2, 0.0009, respectively, with the sizes of $1 \times 1 \times 0.127$ mm^3. Figure 6.24 (a)–(b) shows the designed THz Unit-cell bandstop steering wheel shape and bandpass complementary steering wheel shape FSS, respectively.

The simulated results of the S_{11} and the S_{21} of the THz Unit-cell bandstop steering wheel shape FSS is shown in Fig. 6.25. From this figure, it has been shown that the transmission coefficient $S_{21} < -0.48$ dB for the frequency range of 103.87–153.4 GHz, while it has been shown that the reflection coefficient $S_{11} \leq -10$ dB for the frequency range of 104.11–153.46 GHz and the BW ($S_{11} \leq -10$ dB) of the THz Unit-cell bandstop steering wheel shape FSS equals 49.35 GHz.

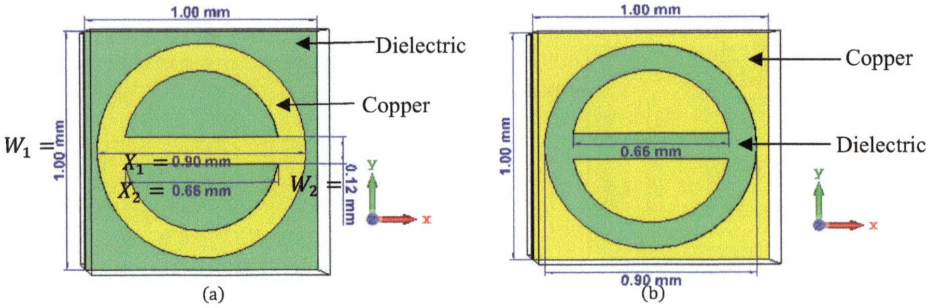

Fig. 6.24 Structure of the designed THz Unit-cell bandstop and bandpass steering wheel shape FSS, (**a**) Bandstop, and (**b**) Bandpass

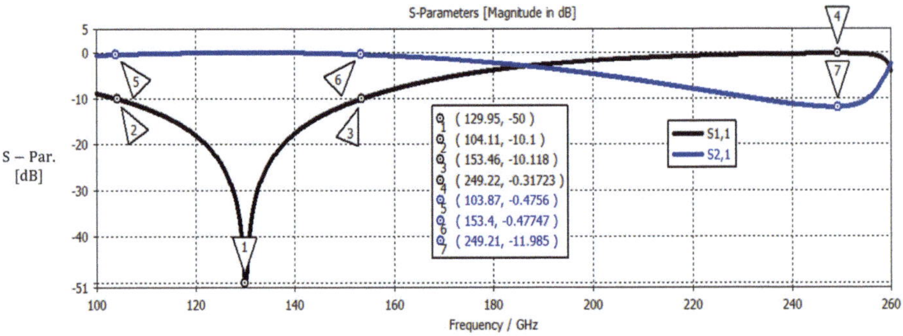

Fig. 6.25 Simulated results of the S_{11} and the S_{21} of the modeled THz Unit-cell bandstop steering wheel shape FSS

The simulated results of the S_{11} and the S_{21} of the THz Unit-cell bandpass complementary steering wheel shape FSS are shown in Fig. 6.26. From this figure, it has been shown that the transmission coefficient $\mathbf{S_{21}} < -\mathbf{0.47\ dB}$ for the frequency range of 131.96–175.21 GHz, while it has been shown that the reflection coefficient $\mathbf{S_{11}} \leq -\mathbf{10\ dB}$ for the frequency range of 131.91–175.06 GHz and the BW($\mathbf{S_{11}} \leq -\mathbf{10\ dB}$) of the THz Unit-cell bandpass complementary steering wheel shape FSS equals 43.15 GHz.

6.6.8 *THz Unit-Cell Bandstop Cross-Shape FSS*

The modeled THz Unit-cell bandstop cross-shape FSS were made with the Taconic TLY-5 microstrip laminate with the parameters of substrate height (h), copper thickness (t), dielectric constant (ε_r), and loss tangent (tanδ) of 127 μm, 17.5 μm, 2.2, 0.0009, respectively, with the sizes of 1.13 × 1.13 × 0.127 mm^3. Figure 6.27 shows the designed THz Unit-cell bandstop cross-shape FSS.

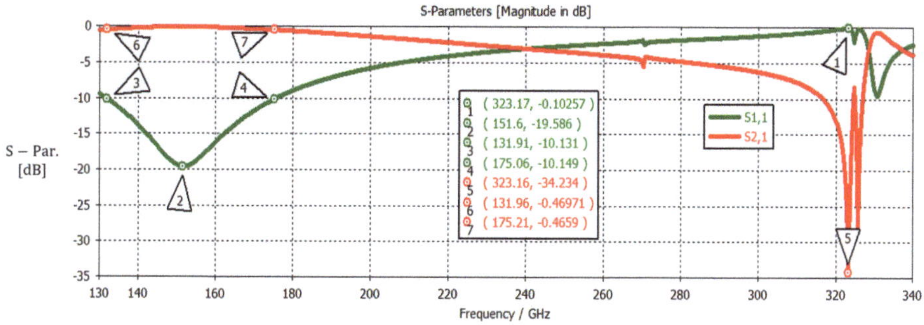

Fig. 6.26 Simulated results of the S_{11} and the S_{21} of the modeled THz Unit-cell bandpass complementary steering wheel shape FSS

Fig. 6.27 Structure of the designed THz Unit-cell bandstop cross-shape FSS

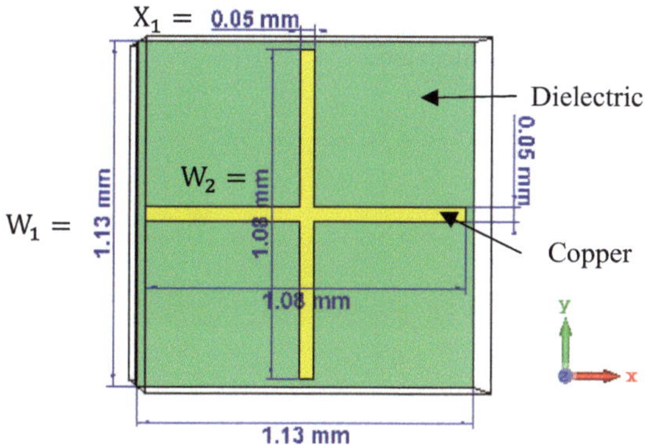

The simulated results of the S_{11} and the S_{21} of the THz Unit-cell bandstop cross-shape FSS are shown in Fig. 6.28. From this figure, it has been shown that the transmission coefficient $S_{21} < -0.55$ dB for the frequency range of 96.72–112 GHz, while it has been shown that the reflection coefficient $S_{11} \leq -10$ dB for the frequency range of 96.8–111.75 GHz and the BW($S_{11} \leq -10$ dB) of the THz Unit-cell bandstop cross-shape FSS equals 14.95 GHz.

6.6.9 THz Unit-Cell Bandstop Jerusalem Cross-Shape FSS

The modeled THz Unit-cell bandstop Jerusalem cross-shape FSS was made with the Isola Astra MT77 microstrip laminate with the parameters of substrate height (h), copper thickness (t), dielectric constant (ε_r), and loss tangent (tanδ) of 127 μm, 17.5 μm, 3, 0.0017, respectively, with the sizes of 0.65 × 0.65 × 0.127 mm^3. Figure 6.29 shows the designed THz Unit-cell bandstop cross-shape FSS.

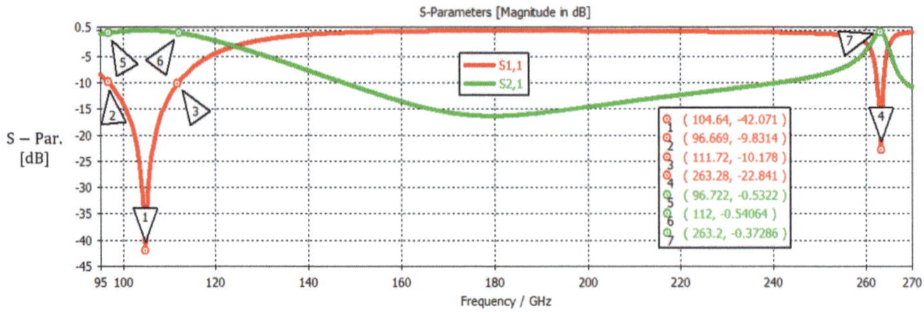

Fig. 6.28 Simulated results of the S_{11} and the S_{21} of the modeled THz Unit-cell bandstop cross-shape FSS

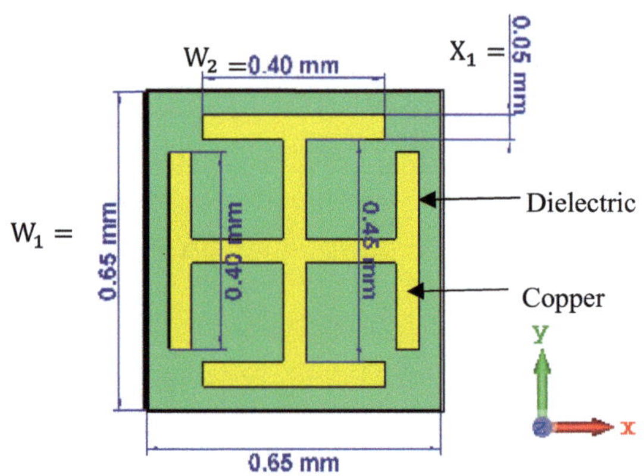

Fig. 6.29 Structure of the designed THz Unit-cell bandstop Jerusalem cross-shape FSS

The simulated results of the S_{11} and the S_{21} of the THz Unit-cell bandstop Jerusalem cross-shape FSS are shown in Fig. 6.30. From this figure, it has been shown that the transmission coefficient $S_{21} < -0.63$ dB for the frequency range of 93.07–115.03 GHz, while it has been shown that the reflection coefficient $S_{11} \leq -10$ dB for the frequency range of 93.1–114.61 GHz and the BW($S_{11} \leq -10$ dB) of the THz Unit-cell bandstop Jerusalem cross-shape FSS equals 21.51 GHz.

6.6.10 THz Unit-Cell Bandstop and Bandpass Tilted Cross-Shape FSSs

The modeled THz Unit-cell bandstop steering wheel shape and bandpass complementary steering wheel shape FSS were made with the Isola Astra MT77 microstrip laminate with the parameters of substrate height (h), copper thickness (t), dielectric constant (ε_r), and loss tangent (tanδ) of 127 μm, 17.5 μm, (2.82, 0.001) @ 125 GHz,

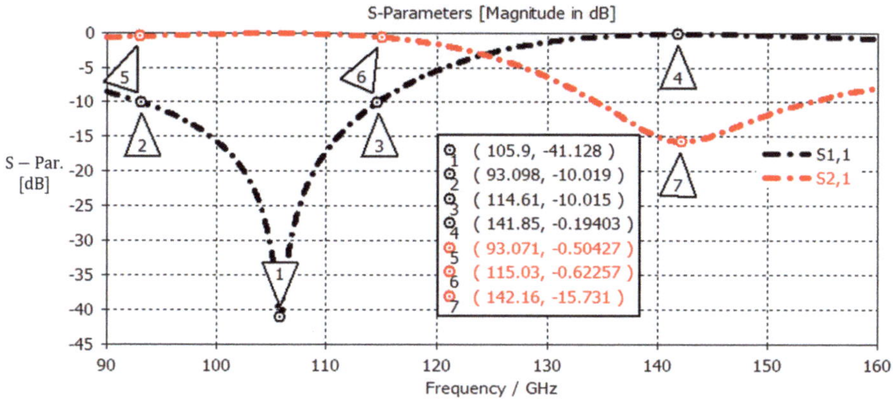

Fig. 6.30 Simulated results of the S_{11} and the S_{21} of the modeled THz Unit-cell bandstop Jerusalem cross-shape FSS

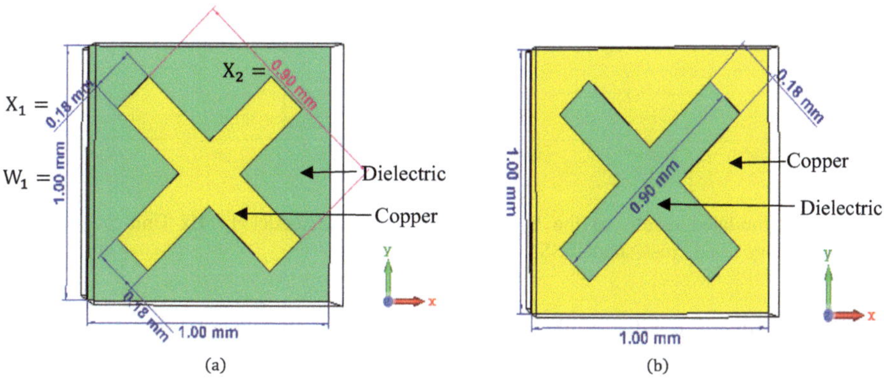

Fig. 6.31 Structure of the designed THz Unit-cell bandstop and bandpass tilted cross-shape FSS, (**a**) Bandstop, and (**b**) Bandpass

respectively, with the sizes of $1 \times 1 \times 0.127$ mm^3. Figure 6.31a, b shows the designed THz Unit-cell bandstop tilted cross-shape and bandpass complementary tilted cross-shape FSS, respectively.

The simulated results of the S_{11} and the S_{21} of the THz Unit-cell bandstop tilted cross-shape FSS are shown in Fig. 6.32. From this Fig., it has been shown that the transmission coefficient $S_{21} < -0.48$ dB for the frequency range of 135.7–167.06 GHz, while it has been shown that the reflection coefficient $S_{11} \leq -10$ dB for the frequency range of 135.48–167.2 GHz and the BW ($S_{11} \leq -10$ dB) of the THz Unit-cell bandstop steering wheel shape FSS equals 31.72 GHz.

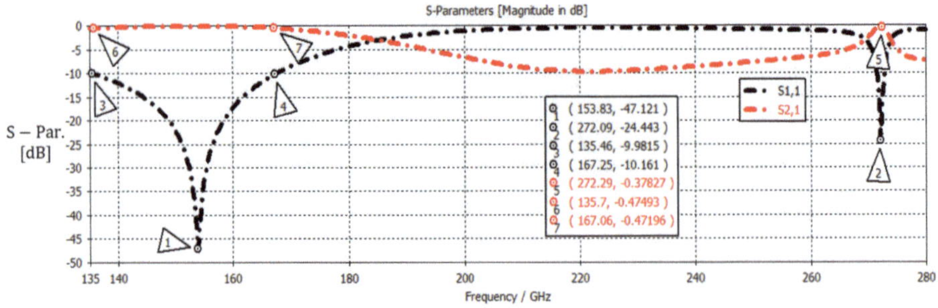

Fig. 6.32 Simulated results of the S_{11} and the S_{21} of the modeled THz Unit-cell bandstop tilted cross-shape FSS

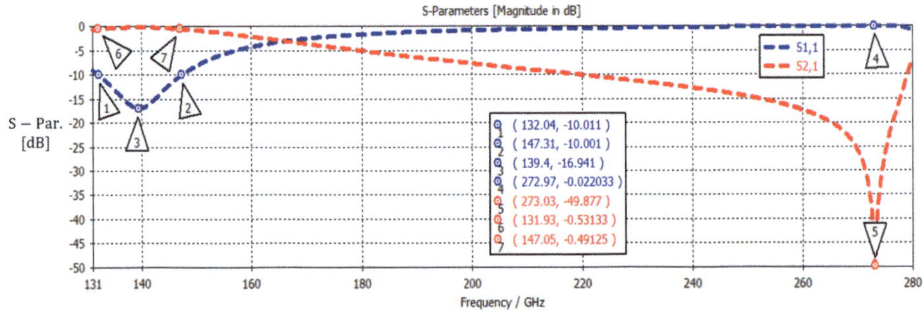

Fig. 6.33 Simulated results of the S_{11} and the S_{21} of the modeled THz Unit-cell bandpass complementary tilted cross-shape FSS

The simulated results of the S_{11} and the S_{21} of the THz Unit-cell bandpass complementary tilted cross-shape FSS are shown in Fig. 6.33. From this figure, it has been shown that the transmission coefficient $\mathbf{S_{21} < -0.54\ dB}$ for the frequency range of 131.93–147.05 GHz, while it has been shown that the reflection coefficient $\mathbf{S_{11} \leq -10\ dB}$ for the frequency range of 132.04–147.31 GHz and the BW ($\mathbf{S_{11} \leq -10\ dB}$) of the THz Unit-cell bandpass complementary steering wheel shape FSS equals 15.27 GHz.

The performance summary of these designed and simulate THz passive components presented above in Chap. 6 is presented in Table 6.3.

6.6.11 Discussion

In microstrip antenna arrays, the design of the fundamental radiating element that makes up the antenna array is a design of the Unit-cell that can be replicated according to the antenna's desired gain requirements. Therefore, the Unit-cell radiator should have maximum gain, impedance bandwidth, and desired resonance frequency, which suits the demands [8]. In comparison, the FSSs are an array of

Table 6.3 The performance summary of the designed and simulate THz passive components in Chap. 6

THz passive technology	Freq. nom. (GHz)	BW (GHz)	Minimum S_{11} (dB)	Minimum S_{21} (dB)
Unit-cell bandstop double circular ring shape FSS	120	25.54	−44.04	−0.05 max.
Unit-cell bandstop discontinuous circular ring shape FSS	119.68	10.22	−42.48	−15.83
Unit-cell bandstop double rectangular ring shape FSS	111.35	12.81	−19.77	−44
Unit-cell bandstop double discontinuous rectangular ring shape and rectangular dipole FSS	109.46	6.38	−48.4	−13
Unit-cell bandstop and bandpass steering wheel shape FSSs	129.95	49.35	−50	−34.23
Unit-cell bandstop cross-shape FSS	104.64	14.95	−42.07	−17
Unit-cell bandstop Jerusalem cross-shape FSS	105.9	21.51	−41.13	−15.73
Unit-cell bandstop and bandpass tilted cross-shape FSSs	153.83	31.72	−47.12	−49.88
THz passive technology	**Freq. Nom. (GHz)**	**BW (GHz)**	**Maximum gain (dB)**	
Unit-cells planar rectangular microstrip antenna	112.42[a]	6.04[a]	6.78[a]	
Unit-cell crown shape planar microstrip antenna	127.6[a]	5[a]	5.67[a]	

[a]Simulated and compared to a different type of antenna solver/simulator

configurations consisting of a majority of slimmer conducting wires that set up bandstop filter and bandpass filter features while stimulated by an electromagnetic signal at a random corner to the plateau of the FSSs and modeled to absorb, transmit, or reflect electromagnetic waves established on the frequency of the transmit waves [9–10]. In this chapter, we show the design and simulation of the Unit-cell microstrip antenna radiators. The performance summary is shown in Table 6.3. This table shows that the gain of the Unit-cell microstrip antenna radiator is about 5–7 dB, while the impedance BW of this Unit-cell antenna radiator is about 3.9–5%. Furthermore, we show the design and simulation of unit-cell bandstop and bandpass FSSs at 104–154 GHz resonance frequencies. The maximum impedance BW of the unit-cell FSS is 49.35 GHz (38.1%), while the minimum S_{11} and S_{21} of the unit-cell FSS are −50 dB and −49.88 dB, respectively.

6.7 Open Research Challenges

Based on this chapter, the THz Unit-cell microstrip antenna radiator and THz Unit-cell FSS for 6G have several open research challenges:

1. Further investigation and improvements on high gain and wideband THz Unit-cell microstrip antenna radiators.
2. Additional analysis and evolutions on low loss and broadband THz Unit-cell bandstop and bandpass FSSs with minimum S_{11} and S_{21} low than −50 dB.

6.8 Summary

Chapter 6 discusses the THz Unit-cell microstrip antenna radiator and THz Unit-cell FSS for 6G with their design considerations. Ten models were designed, and the simulation results are presented in this chapter. The first model was a Unit-cell planar rectangular microstrip antenna radiator for a resonance frequency of 112.42 GHz, and the achieved gain and impedance BW were 6.78 dB and 6.04 GHz. The second model was a Unit-cell crown shape planar microstrip antenna radiator for a resonance frequency of 127.6 GHz, and the achieved gain and impedance BW were 5.67 dB and 5 GHz. The third model was a Unit-cell bandstop double circular ring shape FSS for a resonance frequency of 120 GHz. The achieved impedance BW, S_{11}, and S_{21} were 25.54 GHz, −44.04 dB, and −0.05 dB, respectively. The fourth model was a Unit-cell bandstop discontinuous circular ring shape FSS for a resonance frequency of 119.68 GHz. The achieved impedance BW, S_{11}, and S_{21} were 10.22 GHz, −42.48 dB, and −15.83 dB, respectively. The fifth model was a Unit-cell bandstop double rectangular ring shape FSS for a resonance frequency of 111.35 GHz. The achieved impedance BW, S_{11}, and S_{21} were 12.81 GHz, −19.77 dB, and −44 dB, respectively. The sixth model was a Unit-cell bandstop double discontinuous rectangular ring shape and rectangular dipole FSS for a resonance frequency of 109.46 GHz. The achieved impedance BW, S_{11}, and S_{21} were 6.38 GHz, −48.4 dB, and −13 dB, respectively. The seventh model was a Unit-cell bandstop and bandpass rectangular steering wheel shape FSSs for a resonance frequency of 129.95 GHz. The achieved impedance BW, S_{11}, and S_{21} were 49.35 GHz, −50 dB, and −34.23 dB, respectively. The eighth model was a Unit-cell bandstop cross-shape FSS for a resonance frequency of 104.64 GHz. The achieved impedance BW, S_{11}, and S_{21} were 14.95 GHz, −42.07 dB, and −17 dB, respectively. The ninth model was a Unit-cell bandstop Jerusalem cross-shape FSS for a resonance frequency of 105.9 GHz. The achieved impedance BW, S_{11}, and S_{21} were 21.51 GHz, −41.13 dB, and −15.73 dB, respectively. Finally, the tenth model was a bandstop and bandpass tilted cross-shape FSSs for a resonance frequency of 153.83 GHz. The achieved impedance BW, S_{11}, and S_{21} were 31.72 GHz, −47.12 dB, and −49.88 dB, respectively. The suggested THz models were designed and simulated with the FIT solver within the CST MWS. In contrast, the first and second models were compared with the FEM solver within the CST MWS simulator, so a good agreement was achieved, verifying the simulation results.

References

1. B. Zhang, W. Chen, Y. Wu, K. Ding, and R. Li, "Review of 3D printed millimeter-wave and terahertz passive devices," *International Journal of Antennas and Propagation*, vol. 2017, pp. 1297931/1–10, June 2017.
2. H. Jalili and O. Momeni, "A 0.46-THz 25-element scalable and wideband radiator array with optimized lens integration in 65-nm CMOS," *IEEE Journal of Solid-State Circuits*, vol. 55, no. 9, pp. 2387–2400, Sep. 2020.
3. R. S. Anwar, L. Mao, and H. Ning, "Frequency selective surfaces: A review," *Applied Sciences*, vol. 8, no. 9, pp. 1689/1–46, Sep. 2018.
4. G. Kouemou, *Radar Technology*. IntechOpen, Croatia, pp. 297–317, 2010.
5. D. Liu, B. Gaucher, U. Pfeiffer, and J. Grzyb, *Advanced Millimeter-Wave Technologies Antennas, Packaging, and Circuits*. John Wiley and Sons: UK, 2009.
6. L. K. S. Granados, "Antennas for Millimeter-Wave Applications," Ph.D. Electrical and Electronic Engineering Thesis, Santa María Tonantzintla University, Puebla, Mexico, 2016.
7. U. Nissanov, G. Singh, and A. Akinola, "Sixth-generation (6G) microstrip antenna with high-gain," *International Journal on Communications Antenna and Propagation*, vol. 11, no. 4, pp. 279–287, Aug. 2021.
8. U. Nissanov and G. Singh, "Beamforming D-band phased array microstrip antennas," *Sensors International*, vol. 3, pp. 100196/1–14, July 2022.
9. F. Capolino, *Theory and Phenomena of Metamaterials*. CRC Press: Boca Raton, 2009.
10. B. A. Munk, *Frequency-Selective Surfaces - Theory and Design*. John Wiley & Sons: New York, 2000.
11. F. Costa, A. Monorchio, and G. Manara, "An overview of equivalent circuit modeling techniques of frequency selective surfaces and metasurfaces," *Applied Computational Electromagnetics Society Journal*, vol. 29, no. 12, pp. 960– 976, Dec. 2014.
12. T. Gao, F. Huang, Y. Chen, W. Zhu, X. Ju, and X. Wang, "Resonant coupling effects in a double-layer THz bandpass filter," *Applied Sciences*, vol. 10, no. 15, pp. 5030/1–7, July 2020.
13. Y. Xu and M. He, "Design of multilayer frequency-selective surfaces by equivalent circuit method and basic building blocks," *International Journal of Antennas and Propagation*, vol. 2019, pp. 9582564/1–13, Aug. 2019.
14. M. Babinet, "Memoires d'optique méteorologique," *Academy Science*, vol. 4, pp. 638–648, 1837.
15. K. R. Jha and G. Singh, "Terahertz Planar Antennas for Next Generation Communication," Springer International Publishing: Switzerland, 2014.
16. K. Tekbiyik, A. R. Ekti, G. K. Kurt, and A. Gorcin, "Terahertz band communication systems: Challenges, novelties, and standardization efforts," *Physical Communication*, vol. 35, pp. 100700/1–18, May 2019.
17. I. F. Akyildiz, C. Han, and S. Nie, "Combating the distance problem in the millimeter-wave and terahertz frequency bands," *IEEE Communications Magazine*, vol. 56, no. 6, pp. 102–108, June 2018.
18. H. J. Visser, *Array, and Phased Array Antenna Basics*. John Wiley & Sons, UK, 2005.
19. A. A. Althuwayb, "MTM-and SIW-inspired bowtie antenna loaded with AMC for 5G mm-wave applications," *International Journal of Antennas and Propagation*, vol. 2021, pp. 6658819/1–7, Jan. 2021.
20. R. J. Mailloux, *Phased Array Antenna Handbook*. Artech House, 2nd Edition, USA, 2005.
21. S. Drabowitch, A. Papiernik, H. D. Griffiths, J. Encinas, and B. L. Smith, *Modem Antennas*. Springer, 2nd Edition, Netherlands, 2005.
22. T. B. Ogilvie, A Novel Unit-cell Antenna for Highly Integrated Phased Arrays in the SHF Band," M. Sc. Thesis, Electrical and Electronic Engineering, California Polytechnic State University, San Luis Obispo, June 2013.

23. S. Syedakbar, M. Dharani, "Design of unit-cell structure for MIMO antenna," *International Journal of Engineering Research & Technology*, vol. 5, no. 13, pp. 1–3, May 2017.
24. S. S. Al-Bawri, M. T. Islam, M. S. Islam, M. J. Singh, and H. Alsaif, "Massive metamaterial system-loaded MIMO antenna array for 5G base stations," *Scientific Reports*, vol. 12, pp. 14311/1-16, Aug. 2022.
25. U. Nissanov, G. Singh, E. Gelbart, and N. Kumar, "Highly directive microstrip array antenna with FSS for future generation cellular communication at THz band," *Wireless Personal Communications*, vol. 118, no. 1, pp. 599–617, Jan. 2021.
26. M. Alibakhshikenari, B. S. Virdee, M. Khalily, C. H. See, and R. A. Abd-Alhameed, "High-gain on-chip antenna design on silicon layer with aperture excitation for terahertz applications," *IEEE Antennas and Wireless Propagation Letters*, vol. 19, no. 9, pp. 1576–1580, Sep. 2020.
27. H. H. Bae, T. H. Jang, H. Y. Kim, and C. S. Park, "Broadband 120-GHz L-probe differential feed dual-polarized patch antenna with soft-surface," *IEEE Transactions on Antennas and Propagation*, vol. 69, no. 10, pp. 6185 – 6195, Oct. 2021.
28. C. Ma, S. Ma, L. Dai, Q. Zhang, H. Wang, and H. Yu, "Wideband and high-gain D-band antennas for next-generation short-distance wireless communication chips," *IEEE Transactions on Antennas and Propagation*, vol. 69, no. 7, pp. 3700 – 3708, July 2021.
29. M. Z. Hlaing, V. Karthikeyan, W. Wu, B. J. Chen, A. K. Ng, C. H. Chan, M. M. De Souza, and V. A. L. Roy, "3D microstructured frequency selective surface based on carbonized polyimide films for terahertz applications," *Advanced Optical Materials*, vol. 10, pp. 2102178/1–11, March 2022.
30. A. K. Rashid and Q. Zhang, "Low-cost terahertz three-dimensional frequency selective structure: efficient analysis and characterization," *IEEE Transactions on Terahertz Science and Technology*, vol. 10, no. 1, pp. 1–8, Jan. 2020.
31. A. Ghavidel, M. Kokkonen, and S. Myllymäki, "A double-layer FSS filter for sub-THz applications," *Scientific Reports*, vol. 11, pp. 19773/1–8, Oct. 2021.

Chapter 7
Gain and Bandwidth Enhancement Techniques for Terahertz Planar Antenna for 6G Communication

7.1 Introduction

The frequency ranges 0.1–1 THz (sub-THz) could expedite state-of-the-art technology to realize 6G wireless communication with data rate speeds up to 1 Tb/s. However, this frequency band exposure increased rapidly in free-space path loss (FSPL) and atmospheric absorption by water vapor molecules in the air. Therefore, there are frequency bands where the attenuation is less than 0.1 dB/m, allowing short-range mobile wireless communication up to a couple of 100 m. For example, at frequencies below 1 THz, the frequency windows are around 100 GHz and 300 GHz. 6G wireless communication's main component is developing high-gain, wider bandwidth (bandwidth), and high-radiation efficiency antennas because of the limited mmWave/THz power source for compensating the FSPL and atmospheric absorption [1, 2]. So this chapter presents gain and bandwidth enhancement techniques for the THz planar microstrip antenna to support signal propagation at least 100 m and a data rate of at least 20 Gb/s.

7.2 Terahertz Planar Microstrip Array Antenna Technologies

A single Unit-cell microstrip patch antenna might need to be more efficient. From here, we must use the microstrip antenna array technique to boost the efficiency and gain of the microstrip antenna. In the feed network of the microstrip array antenna, we can feed the elements using a single line or with the help of many lines. Depending on their way of feeding, the array can be classified as (a) Corporate/parallel fed, (b) Series fed, and (c) Hybrid fed [3–7]. It is well-known that the maximal gain of a Unit-cell oblong microstrip antenna can be reached up to 8 dB, so to increase the gain, we can use the array technique. Furthermore, as a rule of

U. Nissanov, G. Singh, *Antenna Technology for Terahertz Wireless Communication*,
https://doi.org/10.1007/978-3-031-35900-2_7

Fig. 7.1 The basic configuration of an N-elements linear array

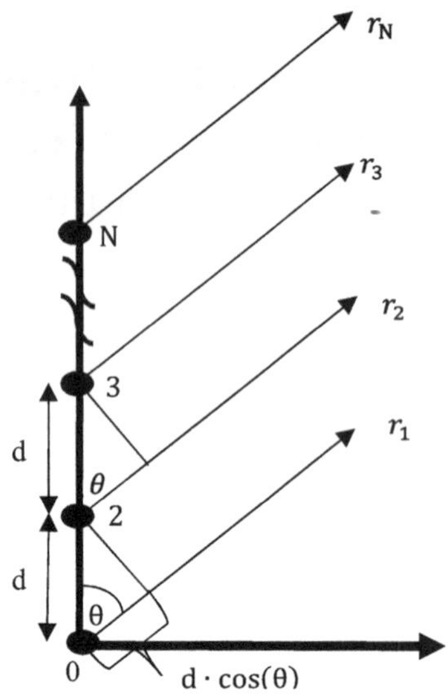

thumb, it is known that each multiplication by 2 of the number of radiators at corporate/parallel/hybrid will enhance the antenna gain by 3 dB to the overall antenna gain by proper design [8].

We have a single isotropic radiating element, which has a radiated field that is relative to [9]:

$$\frac{e^{j(k \cdot r)}}{4\pi \cdot r} \tag{7.1}$$

So, the radiation intensity related to this is permanent. Therefore, we can write the non-normalized array factor (AF) of such a state as [9]:

$$AF = 1 \tag{7.2}$$

Now consider the array shown in Fig. 7.1, which illustrates the basic configuration of an N-elements linear array. The transmitting signal pattern is a plane wave at angle θ to the array plateau. Every radiator is transmitted within a signal of "one" magnitude. All the radiators are at the same interval (*d*) from each other, so a linear array is used at a total length of (L) [9]:

$$L = (N - 1)d \tag{7.3}$$

Thus, AF will be equal to [9]:

$$\begin{aligned} AF &= 1 + e^{j(kd\cos\theta)} + e^{j2(kd\cos\theta)} + \ldots + e^{j(N-1)(kd\cos\theta)} \\ &= \sum\nolimits_{n=1}^{N} e^{j(n-1)(kd\cos\theta)} \end{aligned} \tag{7.4}$$

which can be inscribed as [9]:

$$AF=\sum_{n=1}^{N} e^{j(n-1)\varphi}=1+e^{j\varphi}+e^{j2\varphi}+\ldots+e^{j(N-1)\varphi},\ \text{where}\ \varphi =(k\cdot d\cos\theta) \tag{7.5}$$

The indicated function is a function of φ and is similar to a Fourier series, where the AF combines a set of sinusoids at a duplicate of a "fundamental frequency" φ.

If the baseline is in the middle of the array, the AF will be equal to [9]:

$$AF=\left[\frac{\sin\left(\frac{N}{2}\right)\varphi}{\sin\left(\frac{1}{2}\right)\varphi}\right] \tag{7.6}$$

N is the linear array radiators' number of elements (N = 1, 2, 3, ... , ∞). The computation of the directivity of a broadside linear array is performed using the following steps:

The maximal directivity, which happens in the broadside direction of a uniformly excited equal space (d) linear array, can be computed as follows, and the normalized AF is [9]:

$$(AF)_n=\left[\frac{\sin\left(\frac{N}{2}k\cdot d\ \cos\ \theta\right)}{N\cdot\ \sin\left(\frac{1}{2}k\cdot d\ \cos\ \theta\right)}\right] \tag{7.7}$$

If $d \ll \lambda$, we can use the small angle estimation for the denominator [9]:

$$(AF)_n\approx\left[\frac{\sin\left(\frac{N}{2}k\cdot d\ \cos\ \theta\right)}{\frac{N}{2}k\cdot d\ \cos\ \theta}\right] \tag{7.8}$$

The normalized radiation intensity yielded by this AF is equaled to [9]:

$$U(\theta)=\left[(AF)_n\right]^2=\left[\frac{\sin\left(\frac{N}{2}k\cdot d\ \cos\ \theta\right)}{\frac{N}{2}k\cdot d\ \cos\ \theta}\right]^2 \tag{7.9}$$

The average radiation intensity is equaled to [9]:

$$U_0=P_{rad}/(4\pi)=\frac{1}{4\pi}\int_0^{2\pi}\int_0^{\pi}\left[\frac{\sin\left(\frac{N}{2}k\cdot d\ \cos\ \theta\right)}{\frac{N}{2}k\cdot d\ \cos\ \theta}\right]^2 \sin\theta d\theta d\Phi \tag{7.10}$$

The average radiation intensity can be estimated by making the following changes [9]:

$$Z = \frac{N}{2} k \cdot d \cos \theta \rightarrow dZ = -\frac{N}{2} k \cdot d \sin \theta d\theta \rightarrow \sin \theta d\theta = -\frac{2}{N \cdot k \cdot d} dZ \quad (7.11)$$

Furthermore, we get [9]:

$$U_0 = -\frac{1}{2} \frac{2}{N \cdot k \cdot d} \int_{N \cdot k \cdot d/2}^{-N \cdot k \cdot d/2} \left[\frac{\sin Z}{Z} \right]^2 dZ \quad (7.12)$$

If the array is long enough ($N \cdot k \cdot d$ very large), then we can estimate U_0 as [9]:

$$U_0 \approx -\frac{1}{2} \frac{2}{N \cdot k \cdot d} \int_{\infty}^{-\infty} \left[\frac{\sin Z}{Z} \right]^2 dZ = \frac{\pi}{N \cdot k \cdot d} \quad (7.13)$$

$$\text{Directivity} = \frac{U_{max}}{U_0} = \frac{1}{U_0} = \frac{N \cdot k \cdot d}{\pi} = 2 \frac{N \cdot d}{\lambda} \quad (7.14)$$

The indicated result supposes that the array length (L) is long enough ($L = Nd \gg \lambda$), and if $d = \lambda/2$:

$$\text{Directivity} = N \quad (7.15)$$

For isotropic radiators, the directivity equals N, the number of radiators at the linear array ($N = 1, 2, 3, \ldots, \infty$). It is known that the gain of an antenna is equal [9]:

$$G = \eta \cdot D \quad (7.16)$$

$$G[dB] = 10 \log G = 10 \log(\eta \cdot D) \quad (7.17)$$

G is the antenna's gain, η is the antenna efficiency, and D is the antenna's directivity.

7.2.1 Series Fed

The serial array antennas can be sorted as traveling-wave antennas or standing wave (resonant) antennas. The open terminal of the lead reflections for resonant antennas creates voltage-standing-wave (VSW) patterns. It suggests that the current through the lengthwise of the wire resets at the open terminal of it. For the traveling-wave antennas, the radiator is terminated with a matched impedance, and the current is generated by waves traveling without reflections, therefore manufacturing homogeneous forms in terms of voltage and current. The radiator could connect numerous radiators, and the derived radiation shape will affect the radiator's radiation pattern. Besides, the resonant array has a narrow bandwidth because of the phase coherence

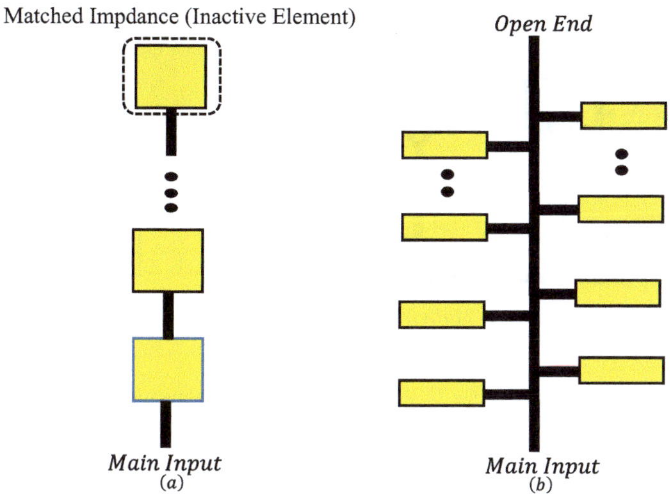

Fig. 7.2 The basic configurations of a serial-fed microstrip array antenna, (**a**) Traveling wave antenna, and (**b**) Standing wave antenna

requirement of the multiply bounced waves. Due to these disadvantages of the resonant array, a traveling-wave array feeding technique must be used. In a full series-fed array, the input power to the antenna should come from one end of the array. At this end feeding, the primary beam angle can be compassionate to squint, i.e., frequency change, because of the continuous phase change of the series-fed radiators. Integrating corporate-fed and series-fed techniques must be used to avoid this problem as frequency changes. The series fed is tight; hence the transmission line losses connected to this array are less than those of the parallel-fed array. The disadvantages of serial fed are the narrow bandwidth and high SLL [10, 11]. Figure 7.2 exhibits the basic configuration of the serially fed microstrip array antenna, while Fig. 7.2a exhibits the basic pattern of the traveling wave antenna, and Fig. 7.2b exhibits the standing wave antenna.

7.2.2 Corporate Fed

The parallel/corporate fed contains numerous feed lines, which help dominate the feed of any line. In addition, the parallel-fed mesh provides power divides by 2 N (i.e., N = 1, 2, 4, 16) by tapered lines or by applying quarter-wavelength transformers, as shown in Fig. 7.3. Thus, corporate-fed arrays are adaptable and generic. Furthermore, with the already stated technique, the designer has higher supervisees of any radiator feed (phase and amplitude), and it is ultimate for beam-shaped arrays, multi-beam arrays, or steering phased arrays [12, 13].

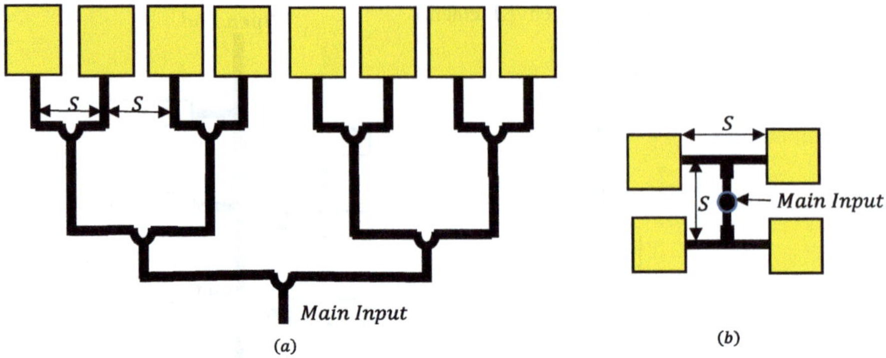

Fig. 7.3 The basic configurations of a corporate/parallel fed microstrip array antenna, (**a**) 2N elements corporate fed microstrip array, and (**b**) N × N elements corporate fed microstrip array

Figure 7.3a exhibits the basic configuration of 2N elements corporate fed microstrip antenna, while Fig. 7.3b exhibits the basic configuration of N × N elements corporate fed microstrip antenna. The main disadvantages of N × N elements corporate fed microstrip antenna are higher N number, high impedance lines (more than 110Ω) need to be used, and at the present fabrication limits, which is 50 μm is not possible. Therefore, the distance (S) between each branch should be [9]:

$$S \leq \lambda_0/2 \leq c/(2 \cdot f_0) \tag{7.18}$$

where S should be smaller than or equal to $\lambda_0/2$ to suppress grating lobes.

7.2.3 *Hybrid Fed*

The third way combines a serial and corporate fed, also called a hybrid fed. So, a broadband impedance bandwidth can be apperceived [14, 15]. Figure 7.4 exhibits the basic configuration of the hybrid-fed microstrip antenna. The distance (S) between each branch/column should be the same as the corporate-fed microstrip array antenna.

To summarize this clause, Table 7.1 exhibits the advantages and disadvantages of each microstrip array antenna technique [10–15].

7.2.4 *Side Lobe Levels Reduction by Taylor Amplitude Weighing/Tapering Technique*

One of the severe drawbacks of constant-fed serial arrays compared to corporate fed is the high SLLs, which cannot be smaller than the limits of approximately −13.45 dB [9, 16, 17]. In addition, extreme SLLs may decrease the transmitter

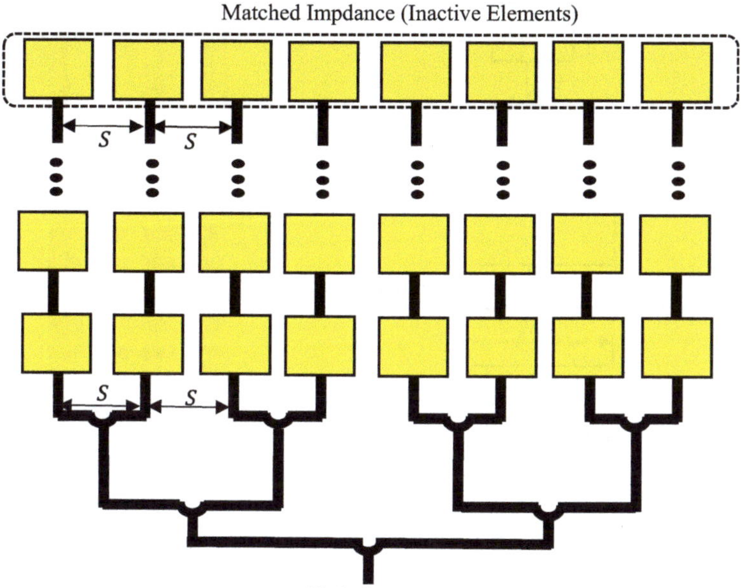

Fig. 7.4 The basic configurations of a hybrid-fed microstrip array antenna

Table 7.1 Classification of microstrip array antenna techniques

Microstrip array antenna techniques	Advantages	Disadvantages
Serial fed – traveling wave	Simple to design and fabrication, no need for power dividers, minimal reflections, minimal line losses, by amplitude tapering low SLL	Low bandwidth, gain up to ≈16 dB
Serial fed – standing (resonance) wave	Simple to design and fabricate, with no need for power dividers, by amplitude tapering low SLL	Squint, low bandwidth, reflections, gain up to ≈16 dB
2N elements corporate/parallel fed	High bandwidth, gain more than 20 dB. By amplitude tapering low SLL, shaped-beam arrays, multi-beam arrays, or steering phased arrays	Insertion losses of the power dividers reduce the gain, quarter-wave transformer for impedance matching, 50 μm minimal widths of lines
N × N elements corporate/parallel fed	High bandwidth, gain more than 20 dB. By amplitude tapering low SLL. Shaped-beam arrays, multi-beam arrays, or steering phased arrays	Insertion losses of the power dividers reduce the gain, quarter-wave transformer for impedance matching, 50 μm minimal widths of lines
Hybrid fed	High bandwidth, gain more than 20 dB. Convenient line widths. By amplitude tapering low SLL. Shaped-beam arrays, multi-beam arrays, or steering phased arrays	Insertion losses of the power dividers reduce the gain, quarter-wave transformer for impedance matching, 50 μm minimal widths of lines

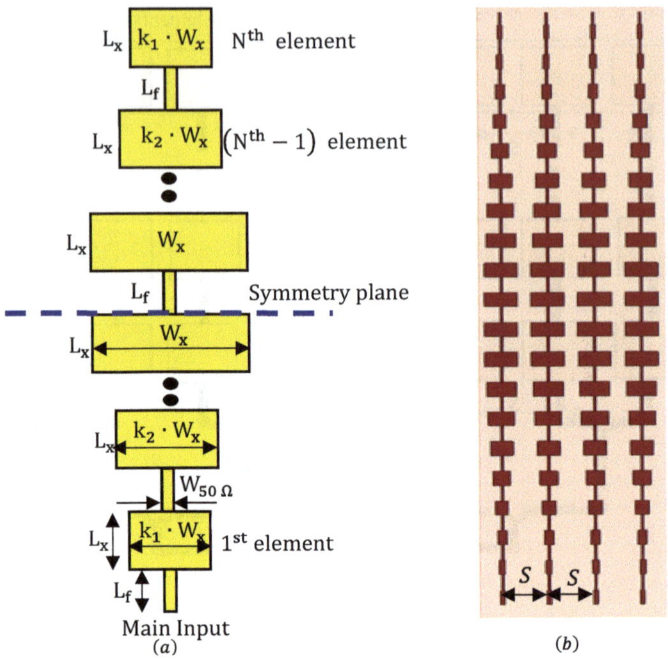

Fig. 7.5 The basic configuration of a microstrip array antenna with the Taylor amplitude weighing technique, (**a**) One column of the microstrip antenna with the Taylor weighing, and (**b**) A microstrip antenna array with the Taylor amplitude

efficiency when their power is confined, lowering the spread of the transmit signal. Therefore, the majority standard method is to insert an amplitude taper across the array to help reduce SLLs, by a minor lowering at the total array gain. This method duplicates any radiator width by different magnitude coefficients (k_n). Figure 7.5 exhibits the basic configuration of the microstrip antenna array with the Taylor magnitude weighing method, where Fig. 7.5a shows one column of the microstrip antenna with the Taylor amplitude weighing and Fig. 7.5b shows the microstrip antenna array with the four columns Taylor amplitude weighing/tapering.

The tapering at the middle of any row radiator is more significant than the rest, and other radiators in any order weigh less, reducing the SLLs, but the primary lobe is more comprehensive than a regular array. The middle is symmetrical in any column. The width of any radiator is marked as W_{patch}, and it is equal to $W_x \cdot k_n$. The longitude of any radiator is marked as L_x, while the longitude of the fed lead is marked as L_f and the width of the fed lead is marked as $W_{50\ \Omega}$. This width should be equivalent to the impedance of 50 Ω, where S should be smaller than or equal to a half free-space wavelength to suppress grating lobes [9]:

$$L_f = L_x = \lambda_0/2 \tag{7.19}$$

where Taylor n-bar matched magnitude factors are marked as k_n. Furthermore, it may be calculated with the MATLAB simulator's *taylorwin* (L, n-bar, SLL) function. This command returns an L-point Taylor factor number (n-bar) of almost permanent SLLs, near the primary lobe and maximal SLLs at (dB) proportional to the primary maximum lobe. We must divide each amplitude coefficient (k_n) by the maximal amplitude coefficient to achieve the normalized amplitude coefficients. Correspondingly, we must weigh the input signal at any row's radiators. To get the normalized power coefficient $(k_n)^2$, for each input port, there is a need to use the following equation [9]:

$$\text{Normalized power coefficient} = (k_n)^2 \tag{7.20}$$

The Taylor n-bar magnitude tapering technique with the wanted SLLs has been used in some of our designs in this thesis because this technique causes a radiation form, is kept at an even nominal level, and the rest lobes fade uniformity, which near the main lobe.

7.3 Placement Configuration of Frequency Selective Surfaces for Terahertz Antenna's Gain Enhancement

Figure 7.6 exhibits the primary configuration planar microstrip antenna, which included bandstop FSS.

The upper substrate (superstrate) included bandstop FSSs to reduce the surface waves and as a reflector for gain enhancement. The gaps between the microstrip array and the FSSs substrates were an air interval, and h_2 was the distance interval. Figure 7.7 exhibits the primary configuration planar microstrip antenna combined with bandpass and bandstop FSSs.

The upper substrate (superstrate) included bandpass FSSs to reduce the surface waves and substrate losses and as a reflector for gain enhancement. The lower substrate included bandstop FSS elements, acting as a ground plane to reflect the plane wave in phase. The gaps between the microstrip array and the FSS substrates were an air interval, and h_2 was the distance interval.

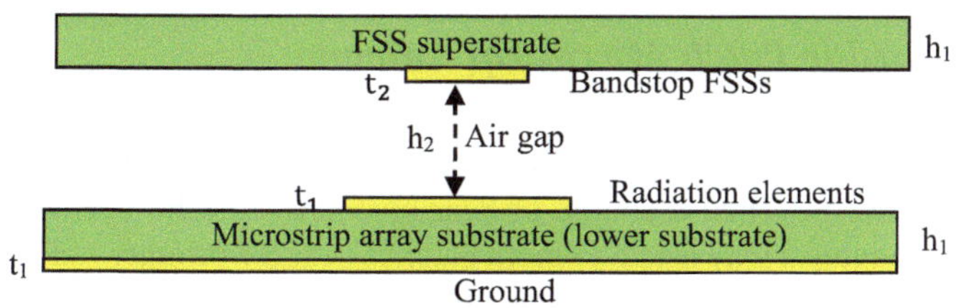

Fig. 7.6 The basic configuration of a microstrip antenna combined with bandstop FSSs

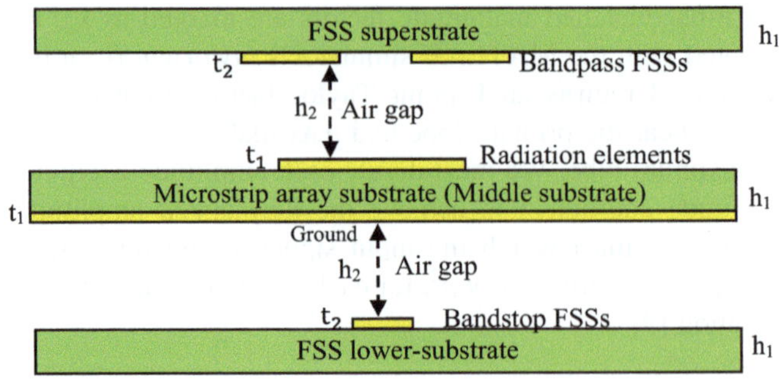

Fig. 7.7 The basic configuration of a microstrip antenna combined with bandpass and bandstop FSSs

7.4 Terahertz Planar Microstrip Array Antenna Bandwidth Enhancement Techniques

With the fast growth of wireless communications, broadband microstrip planar antennas must be used. Planar microstrip patch antennas are devices operating at resonant frequencies, and one of the main disadvantages of microstrip antennas is limited impedance bandwidth, typically around 2–5%. Usually, an antenna is a resonant configuration, and its impedance input changes significantly with frequency, although the feed impedance is unaltered. If the antenna can be matched relatively, to its feed, over a specific frequency sweep, that frequency sweep is determined as its bandwidth. Antenna bandwidth specifies the frequencies range that the antenna's performance matches to a particular criterion. The good-matched bandwidth is necessary to fully cover the necessary working frequency sweep to any defined standard, like $S_{11} \leq -10$ dB [9].

There are some valuable techniques to increase the impedance bandwidth of THz planar microstrip antennas, such as (1) Using another radiating element form alternatively of the accepted nonstandard patches patterns, (2) Managing microstrip substrate with low ε_r, (3) Using a proximity coupling or an aperture feeding, (4) Using a matching network, (5) Inserting parasitic patches, and (6) using a Log-periodic planar microstrip antenna [18–23].

7.4.1 *Parasitic Patch Microstrip Antenna*

Operation near some resonances is a feasible way of enhancing the bandwidth. This approach has well-excited a couple or near resonances in the operating frequency range. The straightest way is to add more patches near the primary radiating element. These parasitic patches (dummy elements) are excited utilizing the EM coupling

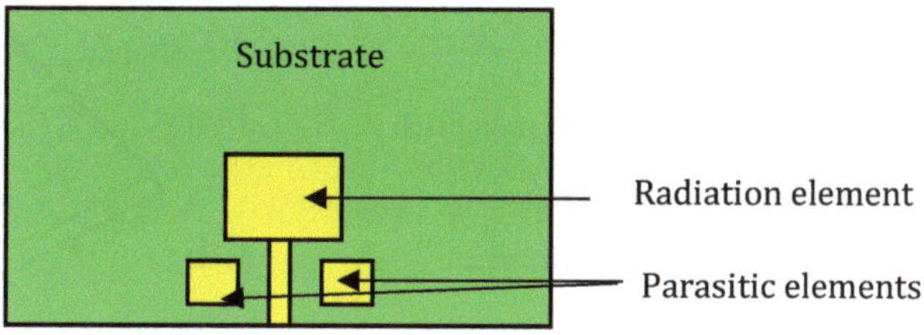

Fig. 7.8 The basic configuration of the co-planar parasitic patch technique

between the primary radiation patch [9, 24]. The parasitic patches may be loaded above in the identical plateau as the primary radiator. Figure 7.8 exhibits the basic formation of the co-planar parasitic patch technique.

The tiny gaps let the parasitic patches be significantly coupled to the radiation element. Moreover, the well-excited resonances are near to one another such that a wider impedance bandwidth was obtained that is generally wider from one radiator.

7.4.2 Using Slots at the Microstrip Antenna Radiators

A possible way to enhance the impedance bandwidth is to cut one slot or multiple slots of varied shapes on the radiator for a single-layered microstrip antenna. These parasitic slots will not increase the volume but will introduce additional resonances, increasing the impedance bandwidth [9, 25].

7.4.3 Proximity Coupled Feed Microstrip Antenna

The proximity-coupled fed technique [26] increases the microstrip array antenna impedance BW. This category of feeding is also called the EM coupling technique. Figure 7.9 exhibits the basic configuration of the E-shaped radiation patch at the proximity coupled feed technique, where Fig. 7.9a shows the cross-section and Fig. 7.9b shows the perspective view.

This design uses two dielectric substrates, so the feeding line comprises two substrates and an E-shaped radiator above the top laminate. The feed line terminates in an open end underneath the radiating patch. The tuning stub on the feed line is small enough to be used in array applications. It can be seen that there is no direct connection between the fed line and the E-shaped radiator. So, the fields are coupled to an E-shaped radiator via the disclosed termination of the stub. Via adjusting the disclosed termination of the fed line compared to the radiation patches, a suitable coupling (capacitive nature) between them can be achieved, causing an increase in

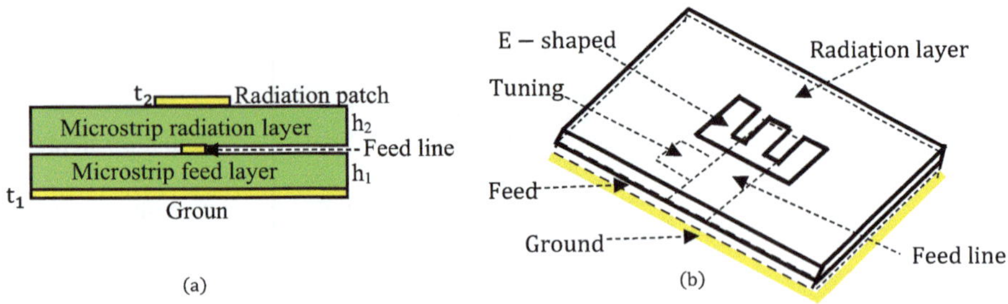

Fig. 7.9 The basic configuration of proximity coupled feed technique, (**a**) cross-section, and (**b**) perspective section

the antenna bandwidth. The primary superiority of this feeding method is that it provides higher bandwidth because of the vast enlargement in the thickness of the microstrip antenna and rejects spurious feed irradiation. The main drawback of this feed technique is that it is challenging to produce due to a couple of dielectric surfaces that demand appropriate alignment [26].

7.4.4 Log-Periodic Microstrip Antenna

A log-periodic antenna (LPA) is a wideband antenna with radiation characteristics, and impedance is invariably repetitive as a logarithmic function of the stimulation frequency. The spacing and length of the radiating elements of LPAs logarithmically expand from particular end to the other. The principle of LPAs demands that all magnitudes of the array radiators be scaled within a log-periodic way to obtain proper frequency scaling [27–29].

The quasi-log-periodic antenna was accomplished by editing different narrow bandwidth radiating elements. Each has its frequency band of operation and serial connection of the elements at different sizes, the basic impedance bandwidth is increased, roughly the sum of bandwidths of each radiator element, and the smallest and largest sizes of the configurations can be selected. A log-periodic microstrip antenna is a form of traveling-wave antenna. Its terminal portion works with the impedance-matching load to absorb radiated energy and reduce reflected energy. The terminal section is essential since it can significantly improve the antenna's radiation pattern in the working frequency band, but terminal matching load absorption can notably reduce the total efficiency of a microstrip antenna. The design principle for LPA needs scaling of magnitudes from point to point; thus, the rendition is cyclic within the logarithm of the frequency [9, 27–29]. Figure 7.10 exhibits the 24th Unit-cell element log-periodic microstrip array antenna formation.

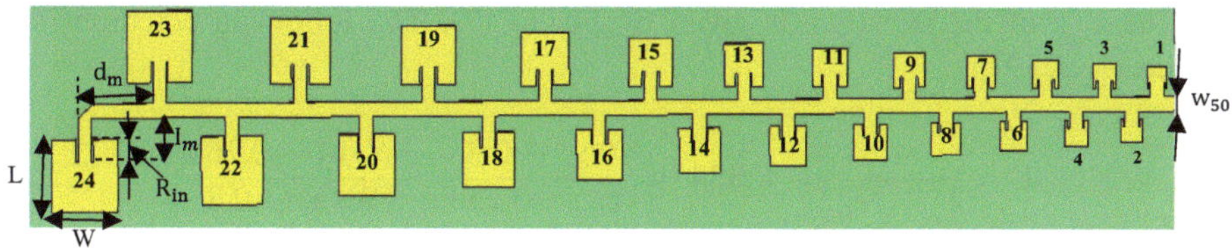

Fig. 7.10 The formation of the 24th Unit-cell element log-periodic microstrip array antenna

This technique may be exploited with an array of radiators antenna. The following equations are apperceived [27–29]:

$$\tau = \frac{L_{n+1}}{L_n} = \frac{W_{n+1}}{W_n} = \frac{I_{mn+1}}{I_{mn}} \tag{7.21}$$

$$\log(f_{min}) - \log(f_{max}) = (n-1)\log(1/\tau) \tag{7.22}$$

The patch width (W), the Inset (I_m), and the length (L) are connected to the scale coefficient by (τ). The number of the microstrip array patch elements is (n). Where f_{min} is the low frequency that the antenna needs to have $S_{11} \leq -10$ dB, and f_{max} is the high frequency, the antenna needs to have $S_{11} \leq -10$ dB. I_m is the feed line at a length of (λ/4), which causes phase shifting of 90° to the signal at the patch where it is connected. If we multiply every magnitude of the array per τ, it scales to itself, component (n) changes to component (n + 1), and component (n + 1) changes to (n + 2). The self-scaling character suggests the array will have identical radiation features at any frequencies that are correlated per τ factor [27–29]:

$$f_1, f_2 = f_1/\tau, f_3 = f_1/\tau^2, \qquad f_4 = f_1/\tau^3, \ldots, f_n = f_1/\tau^{n-1} \tag{7.23}$$

$$\ln\,(f_1/f_2) = \ln(\tau),\ \ln\,(f_1/f_3) = 2\cdot\ln(\tau),\ \ldots,\ \ln\,(f_1/f_n) = (n-1)\cdot\ln(\tau) \tag{7.24}$$

f_1 is the frequency of the resonance of the smallest radiator element, and f_2 is the frequency of the resonance of the subsequent smallest radiator element. Where d_m is the distance between element (n) and element (n + 1) and is determined following the upper frequency of the following antenna radiator, so the entry gazing to the subsequent upper frequency necessary to be an open-circuit previously the subsequent radiator (n + 1) is linked, at this pattern, the distance between a couple of radiators is approximately a half wavelength (λ/2), which causes phase shifting of 180° to the signal and varying logs periodically.

The microstrip lines on the surface can be described as connecting capacitors and inductors per unit length. Therefore, the microstrip line has a characteristic impedance, the square root of the inductance per unit length divided by the capacitance per unit length. In order to transmit maximum power to the radiating element/rectangular patch, it is necessary that the input impedance of the radiating element, which is dependent on the width of the feeding line of the radiating element, be in the same impedance as the impedance of the input signal (generator) to the antenna that is 50 Ω. So the widths of I_m and d_m should be at a 50 Ω impedance.

The $\varepsilon_{r_{eff}}$ of the substrate is given by Eq. (4.17). The following equations are apperceived [9]:

$$L = \frac{c}{2 \cdot f_r \sqrt{\varepsilon_{r_{eff}}}} - 0.824 \cdot h \cdot \frac{\varepsilon_{r_{eff}} + 0.3}{\varepsilon_{r_{eff}} - 0.258} \cdot \frac{(W/h) + 0.264}{(W/h) + 0.8} \tag{7.25}$$

$$W = L \tag{7.26}$$

The substrate permittivity is (ε_r) and the substrate thickness is (h). A quadratic radiator (L = W) should be applied to avoid high-order modes, disturbing the polarization and the radiation pattern. The microstrip antenna input impedance (Z_a) is given by [27–29]:

$$Z_a = 90 \cdot \frac{{\varepsilon_r}^2}{\varepsilon_r - 1} (L/W)^2 \tag{7.27}$$

The characteristic impedance (Z_T) of the transition section is given by [27–29]:

$$Z_T = \sqrt{(Z_0 \cdot Z_a)} \tag{7.28}$$

$$R_{in} = \frac{L}{\pi} \cos^{-1}\{\theta\}, \theta^\circ = \sqrt{(Z_0/Z_T)}, (\theta^\circ) = (rad) \tag{7.29}$$

where R_{in} is the microstrip line inset fed for locating the exact inset length to obtain 50 Ω, and Z_0 is the wanted characteristic impedance of the signal generator, which equals 50 Ω.

7.5 Problem Formulation

High transmission data rates, high reliability, low latency, and no interference operation are significant applications shifting from communications to documentary movies and healthcare positioning. The motivation for this requirement is critical modern evolution in communication networks (6G wireless networks) in a higher frequency band. This evolution makes the THz band the critical candidate for next-generation 6G wireless networks [30]. The sub-THz could be a potential facilitator as robust successor technology to comprehend 6G wireless communication speeds up to 1000 Gb/s [31]. In addition, the sub-THz frequency causes a significant increase in atmospheric absorption and free space power loss (FSPL).

Therefore, the high-gain antennas are a crucial component to compensate for the path loss in the environment (indoor or outdoor). The planar microstrip antenna arrays within various feeding networks can supply superior gain but have high loss and narrow bandwidth, exclusively at high THz frequencies, because substrate materials and feeding losses reasonably alter antennas' performance and reduce antenna radiation efficiency. These losses could be mitigated by adequately choosing the microstrip laminate sizes and electric performances [32, 33], enhancing the impedance bandwidth and gain of a THz planar microstrip array antenna, and proper techniques can be used as reported in [34, 35].

7.6 Related Work

Ding et al. [36] demonstrated a THz 16 × 16 slot corporate array antenna fed by a ridge gap waveguide (RGW). The corporate fed was chosen to enhance antenna gain and impedance bandwidth. The measured results of the peak realized gain, impedance bandwidth, and peak efficiency were 31.8 dB, 17.5 GHz (155–172.5 GHz), and 63%, respectively. Next, Aqlan et al. [37] demonstrated a THz metallic Fabry–Perot cavity (FPC) antenna with six layers for 300 GHz. The peak realized gain and impedance bandwidth measured results were 17.7 dB @ 289 GHz and 22 GHz (282–304 GHz). Finally, Warmowska et al. [38] demonstrated a THz 4 × 4 slot corporate patch array antenna with six dielectric and metallic layers for 350 GHz. The peak realized gain simulation result and measured impedance bandwidth results were 18.4 dB @ 335 GHz and 27 GHz. The corporate fed was chosen because it can simultaneously enhance antenna gain and impedance bandwidth.

Bae et al. [39] demonstrated a THz dual-polarized L-probe differential-fed patch antenna for 120 GHz. The peak realized gain, impedance bandwidth, and isolation results were 9.8 dB @ 125 GHz, 25 GHz (100–125 GHz), and 28.5 dB, respectively. Lu et al. [40] demonstrated a THz beam-steering leaky-wave InP-based antenna. The peak realized gain, impedance bandwidth, and steering angle measured 11 dB @ 290 GHz, 45 GHz (230–250 GHz and 275–300 GHz), and −46° to 42°, respectively. Finally, Xiao et al. [41] demonstrated a THz 8 × 8 slot substrate integrated cavity (SIC) array antenna based on the low-temperature co-fired ceramic (LTCC) process within Ferro A6M laminate with eight layers for 137 GHz. The corporate fed was chosen to enhance antenna gain and impedance bandwidth. The measured results of the peak realized gain, impedance bandwidth, and peak efficiency were 20.5 dB @ 136.5 GHz, 14.7 GHz (130.3–145 GHz), and 59.2%, respectively. The performance summary of these THz antennas discussed above is presented in Table 7.2 according to the related work.

7.7 Simulation Results and Discussion

7.7.1 THz Low SLLs Microstrip Array Antenna

A couple of THz low SLLs were designed with 28 × 32 radiators at hybrid fed for gain and bandwidth enhancement, while the second was with bandstop FSSs for gain enhancement besides the 28 × 32 radiators at hybrid fed. The designed THz low SLLS microstrip array antenna is based on rolled copper Rogers RO3003 laminate,

Table 7.2 THz antennas classification and performance according to related work in this chapter

References	THz antenna	Gain and bandwidth enhancement technique	Freq. nom. (GHz)	Measured bandwidth (GHz)	Measured maximum realized gain (dB)
[36]	16 × 16 slot array fed by metallic RGW	Corporate fed	163.7	17.5 (155–172.5)	31.8
[37]	Metallic FPC	n/a	300	22 (282–304)	17.7
[38]	4 × 4 slot metallic and dielectric layers	Corporate fed	350	27	18.4
[39]	L-probe differential fed patch	Nonstandard patch	120	25 (100–125)	9.8
[40]	Leaky wave InP-based beam steering	Nonstandard patch	290	45	11
[41]	8 × 8 slot SIC within Ferro A6M laminate based LTCC	Corporate fed	137	14.7 (130.3–145)	20.5

which has the following parameters of a copper thickness (t), substrate height (h), loss tangent (tanδ), and dielectric constant ε_r of 8.75 μm, 127 μm, (0.0019, 3.02) @ 100 GHz, respectively, for the resonance frequency beyond 115 GHz. To get a low SLLs = −30 dB microstrip array antenna, we need to use the Taylor n-bar amplitude tapering technique, as explained in this chapter and Chap. 5. Therefore, the following Tables 7.3 and 7.4 are presented, where Table 7.3 exhibits the Taylor normalized magnitude factors founded by *the taylorwin* (28, 50, −30) function at the MATLAB simulator. These normalized magnitude factors are needed to calculate the radiating patch's widths, while Table 7.4 exhibits the Taylor normalized magnitude factors founded by the *taylorwin (32, 50, −30)* function at the MATLAB simulator. These normalized magnitude factors were needed to calculate the coefficients for multiplying each input port (P1–P32). Each port was multiplied with these coefficients at the CST MWS simulator to add the whole sources jointly, i.e., in simultaneous mode.

The magnitude of the substrate for the first one designed was 80 × 60 mm^2, which equals $31.52\lambda_0 \times 23.64\lambda_0$ @ 118.2 GHz. Figure 7.11a presents the first THz low SLL microstrip array antenna configuration. While Fig. 7.11b presents zoomed partial structure of the radiation elements.

Figure 7.12 exhibits the configuration of the second THz low SLLs microstrip array antenna with the designed FSSs, where Fig. 7.12a shows the Microstrip array antenna substrate (lower substrate), Fig. 7.12b shows part of the FSS superstrate with FSSs at the superstrate, Fig. 7.12c, shows the complete second microstrip array antenna (up view), and Fig. 7.12d shows the complete microstrip array antenna (side

Table 7.3 Taylor normalized amplitude coefficients for SLLs = −30 dB

Patches numbers from the bottom	Amplitude coefficients (Taylor n-bar amplitude tapering)	Normalized amplitude coefficients
1,28	0.26067	0.1769
2,27	0.40658	0.2759
3,26	0.50788	0.3446
4,25	0.61586	0.4179
5,24	0.73098	0.4960
6,23	0.84954	0.5764
7,22	0.96663	0.6559
8,21	1.07809	0.7315
9,20	1.18099	0.8014
10,19	1.27281	0.8637
11,18	1.35066	0.9165
12,17	1.41148	0.9578
13,16	1.45289	0.9858
14,15	1.47373	1

Table 7.4 Taylor normalized power coefficients for SLLs = −30 dB

Port numbers	Amplitude coefficients (Taylor n-bar amplitude tapering)	Normalized amplitude coefficients	Normalized power coefficients
P1, P32	0.35697	0.24210	0.058612
P2, P31	0.38680	0.26233	0.068817
P3, P30	0.46825	0.31757	0.100850
P4, P29	0.56871	0.38570	0.148764
P5, P28	0.66501	0.45101	0.203410
P6, P27	0.76958	0.52193	0.272410
P7, P26	0.87142	0.59100	0.349281
P8, P25	0.97347	0.66020	0.435864
P9, P24	1.07106	0.72639	0.527642
P10, P23	1.16235	0.78830	0.621416
P11, P22	1.24606	0.84507	0.714143
P12, P21	1.31839	0.89413	0.799468
P13, P20	1.37964	0.93567	0.875478
P14, P19	1.42611	0.96719	0.935456
P15, P18	1.45846	0.98913	0.978378
P16, P17	1.47449	1	1

view). This antenna is established on two rolled copper Rogers RO3003 substrates with the same substrate parameters as the first THz low SLLs microstrip array antenna with an air gap of h2 = 1352 μm between both PCBs. This antenna included 41 UCell bandstop FSSs on the superstrate, and the 28 × 32 radiators on the lower substrate, with Taylor n-bar amplitude tapering of SLLs of −30 dB (Table 7.3). All

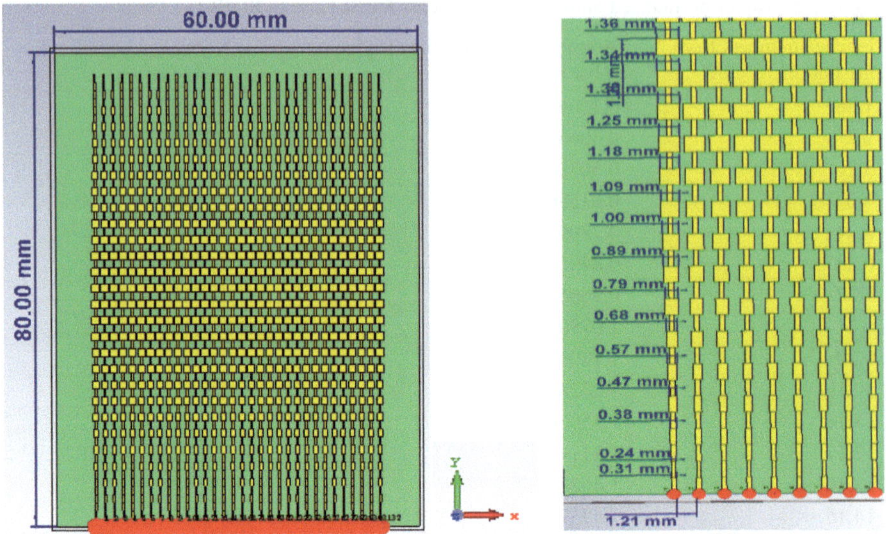

(a) (b)

Fig. 7.11 The configuration of the first THz low SLLs microstrip array antenna, (**a**) The complete structure of the first THz low SLLs microstrip array antenna, and (**b**) Zoomed partial structure of the radiation elements

magnitudes of this antenna, including the numbers of the UCell bandstop FSSs, and the air gap distance (h_2) were optimized to get this antenna's best gain and bandwidth with reasonable time simulation. The magnitudes of the substrates for the second THz low SLLs microstrip array antenna with the designed FSSs were 80×60 mm^2, which equals $33.04\lambda_0 \times 24.78\lambda_0$ @ 123.9 GHz.

The simulation result of the S_{11} of the designed first THz low SLLs microstrip antenna is presented in Fig. 7.13. From this figure, it has been shown that the bandwidth of this first THz low SLLs microstrip antenna is more than 7.86 GHz, while the fractional bandwidth is more than 6.67% ; furthermore, the frequency of the resonance of this offered antenna is 118.2 GHz, while the minimum S_{11} is equal to −33.01 dB.

Figures 7.14a–c show the simulation results of the 3D far-field radiation pattern, gain, and directivity at the E-plane and H-plane of the designed first THz low SLLs microstrip antenna at a frequency of 119 GHz at the polar plots, respectively. From Fig. 7.14a, it has been shown that the directivity is 23.9 dBi, and the total efficiency is −1.69 dB for 119 GHz. From Fig. 7.14b, it has been shown that the gain is 17.1 dB in the E-plane with the direction of the main lobe at 0°, while the –3 dB angular width is 3.1° and the SLL is −28 dB. The gain is 23.7 dB in the H-plane with the direction of the main lobe at 53°, while the –3 dB angular width is 7.7°, and the SLL is −3 dB. From Fig. 7.14c, it has been shown that the directivity is 18.5 dBi in the E-plane with the direction of the main lobe at 0°, while the –3 dB angular width is

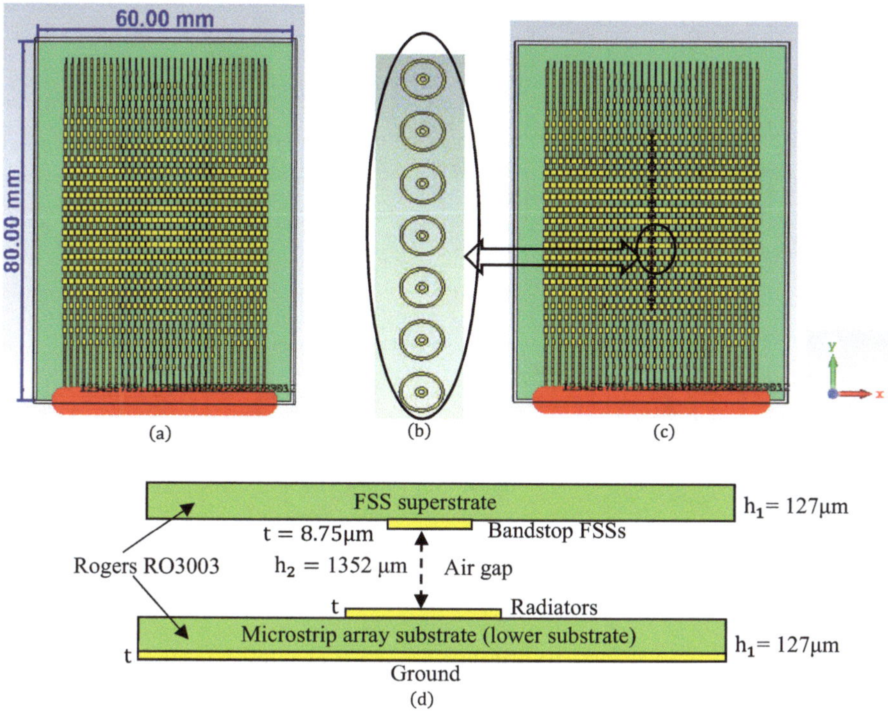

Fig. 7.12 The configuration of the second THz low SLLs microstrip array antenna with the designed bandstop FSSs, (**a**) Microstrip array antenna substrate (Lower substrate), (**b**) Part of FSS superstrate with FSSs at the superstrate, (**c**) The complete microstrip array antenna (Up view), and (**d**) The complete microstrip array antenna (Side view)

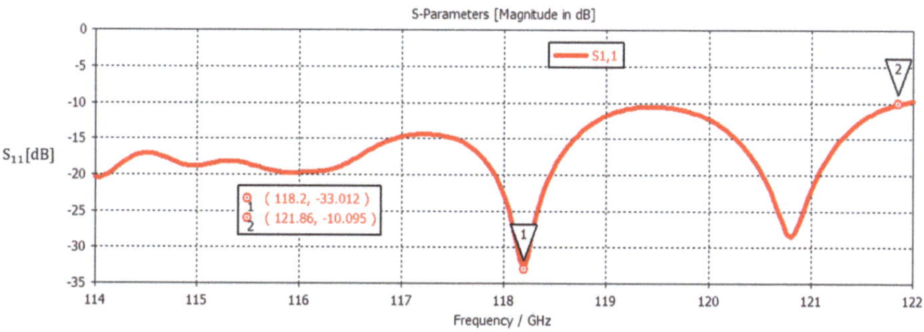

Fig. 7.13 Simulation result of S_{11} for the first THz low SLLs microstrip array antenna

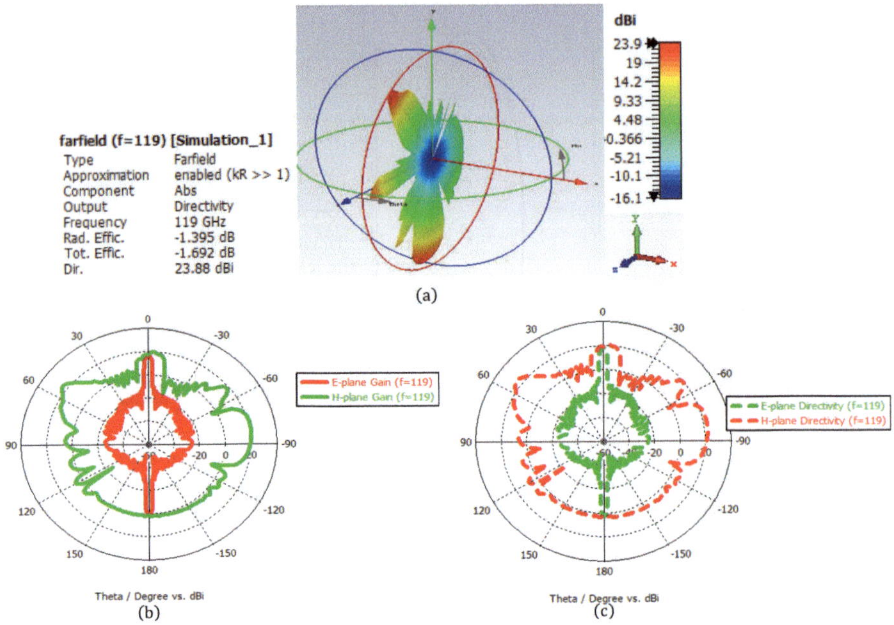

Fig. 7.14 Simulation results of the 3D far-field radiation pattern, E-plane, and H-plane of first THz low SLLs microstrip array antenna, (**a**) 3D far-field, (**b**) Gain, and (**c**) Directivity

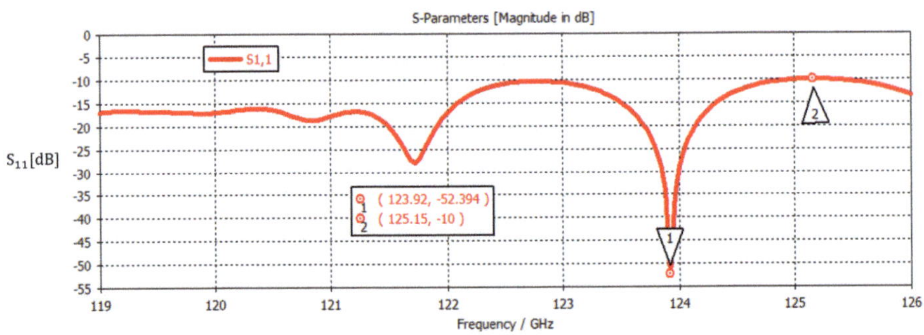

Fig. 7.15 Simulation result of S_{11} for the second THz low SLLs microstrip array antenna with the FSSs

3.1° and the SLL is −28 dB. The directivity is 25.1 dBi in the H-plane with the direction of the main lobe at 53°, while the –3 dB angular width is 7.7°, and the SLL is −3 dB.

The simulation result of the S_{11} of the designed second THz low SLLs microstrip antenna with the FSSs is presented in Fig. 7.15. From this figure, it has been shown that bandwidth($S_{11} \leq -10$dB) is more than 6.15 GHz, while the fractional bandwidth is more than 5%; moreover, the resonance frequency of this second antenna is 123.92 GHz.

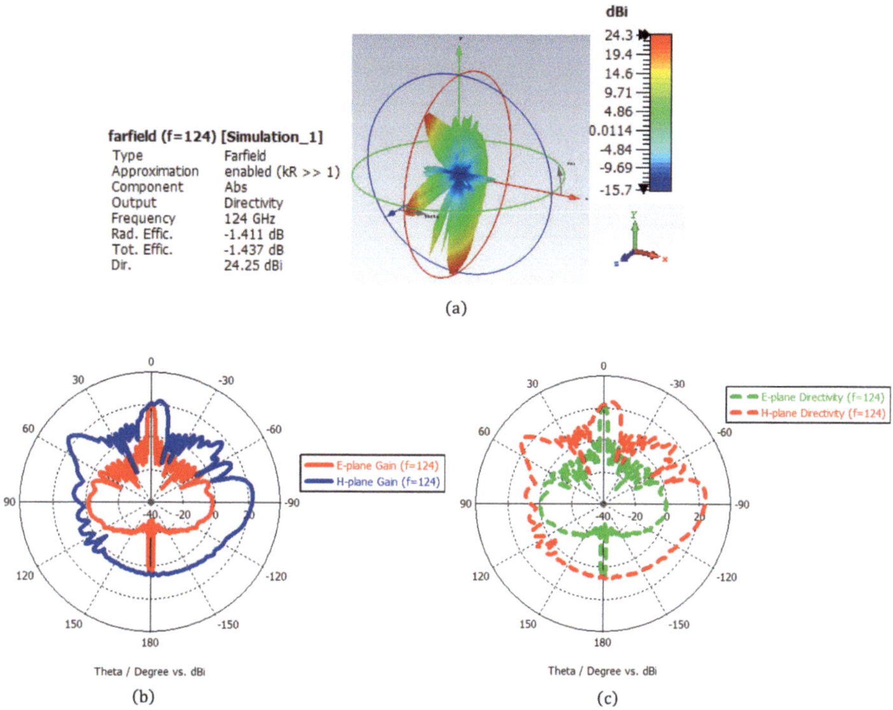

Fig. 7.16 Simulation results of the 3D far-field radiation pattern, E-plane, and H-plane of second THz low SLLs microstrip array antenna with FSSs (**a**) 3D far-field, (**b**) Gain, and (**c**) Directivity

Figures 7.16a–c show the simulation results of the 3D far-field radiation pattern, gain, and directivity at the E-plane and H-plane of the designed first THz low SLLs microstrip antenna at a frequency of 119 GHz at the polar plots, respectively. From Fig. 7.16a, it has been shown that the directivity is 24.3 dBi, and the total efficiency is −1.44 dB for 124 GHz. From Fig. 7.16b, it has been shown that the gain is 20.6 dB in E-plane with the direction of the main lobe at 0°, while the –3 dB angular width is 3° and the SLLs is −20.16 dB. The gain is 23.7 dB in the H-plane with the direction of the main lobe at 52°, while the –3 dB angular width is 7.4° and the SLLs is −1.1 dB. From Fig. 7.16c, it has been shown that the directivity is 22 dBi in the E-plane with the direction of the main lobe at 0°, while the –3 dB angular width is 3° and the SLL is −20.16 dB. The directivity is 25.1 dBi in the H-plane with the direction of the main lobe at 52°, while the –3 dB angular width is 7.4° and the SLL is −1.1 dB.

The simulation results from comparisons between offered first and second antennas are shown in Fig. 7.17a, b. From Fig. 7.17a, it has been shown that the minimum S_{11} decreased by 13.4 dB, approximately, by adding FSS to the antenna. From Fig. 7.17b, it has been shown that the antenna gain increased by more than 3.5 dB at 123–126 GHz because of FSS added to the antenna.

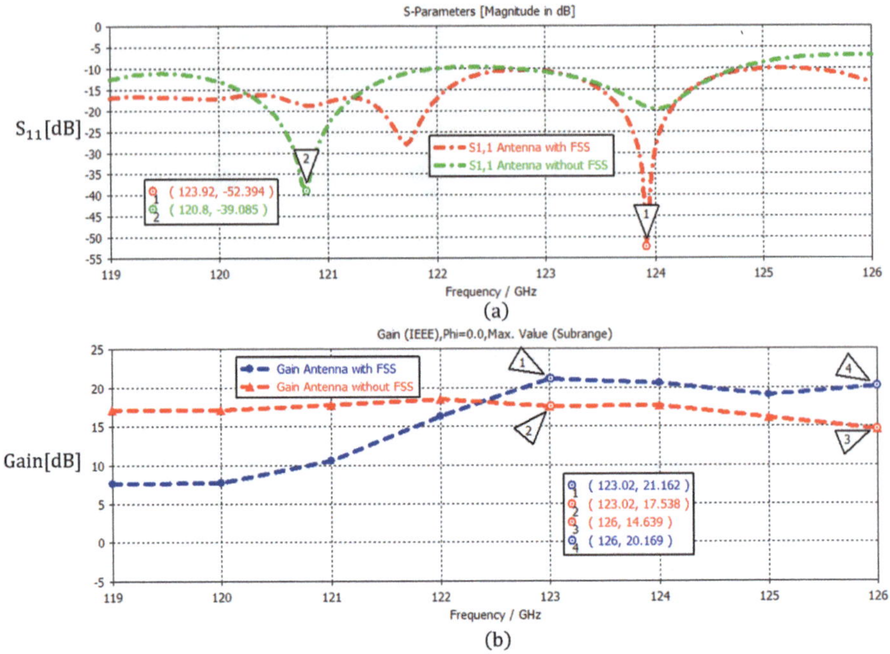

Fig. 7.17 The simulation results in comparison of the first and second THz low SLLs microstrip array antennas, (**a**) S_{11}, (**b**) Gain

The simulated results from comparisons between Ansys HFSS and CST MWS simulators for the offered first antenna are shown in Fig. 7.18. From these simulation results, it has been shown that the simulation results of the maximal gain and the bandwidth of the first antenna is 18.8 dB, more than 10 GHz, appropriately, achieved at the Ansys HFSS simulator so that the simulated results are similar to each other regarding the proposed first antenna.

7.7.2 THz Slotting Circular Patch Microstrip Array Antenna with SIW

A couple of THz slotting circular patch microstrip array antennas with SIW were designed with 20 × 32 slotting radiators at hybrid fed for gain and bandwidth enhancement, while the second was with bandstop FSSs for gain enhancement besides the 20 × 32 slotting circular radiators at hybrid fed. The designed THz slotting circular patch microstrip array antennas with SIW are based on rolled copper Rogers RO3003 laminate, which has the following parameters of a copper thickness (t), substrate height (h), loss tangent (tanδ), and dielectric constant ε_r of 8.75 μm, 127 μm, (0.0019, 3.02) @ 100 GHz, respectively, for the resonance frequency

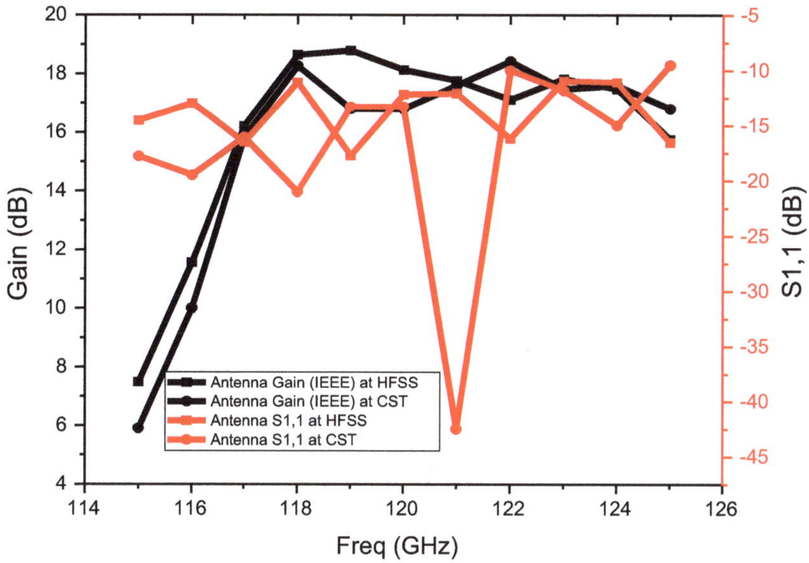

Fig. 7.18 Simulation results from a solvers comparison of the gains and the S_{11} for the offered first THz low SLLs microstrip array antenna

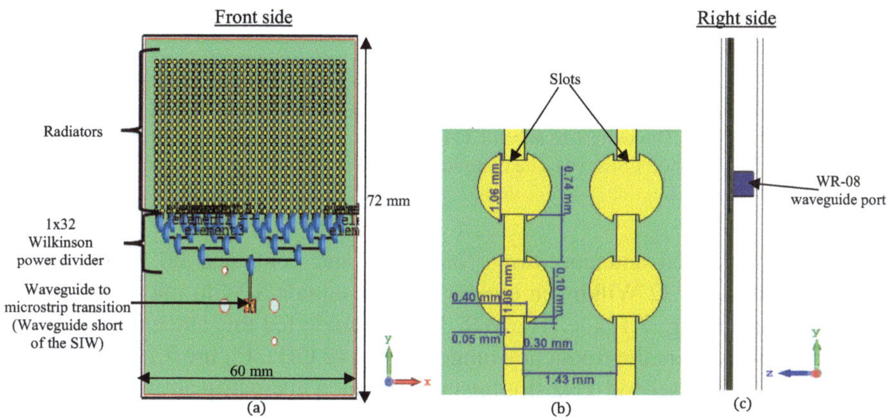

Fig. 7.19 The first THz slotting circular patch microstrip array antenna with SIW, (**a**) The complete configuration, (**b**) The Zoomed partial configuration of a hybrid fed circular radiators, (**c**) Right side

beyond 114.7 GHz. The magnitude of the substrate was 72 × 60 mm^2 that equals $27.53\lambda_0 \times 22.94\lambda_0$ @ 114.7GHz. Figure 7.19a presents the first THz slotting circular patch microstrip array antennas with SIW configuration. While Fig. 7.19b presents zoomed partial structure of the radiation elements, and Fig. 7.19c presents the right size of this antenna.

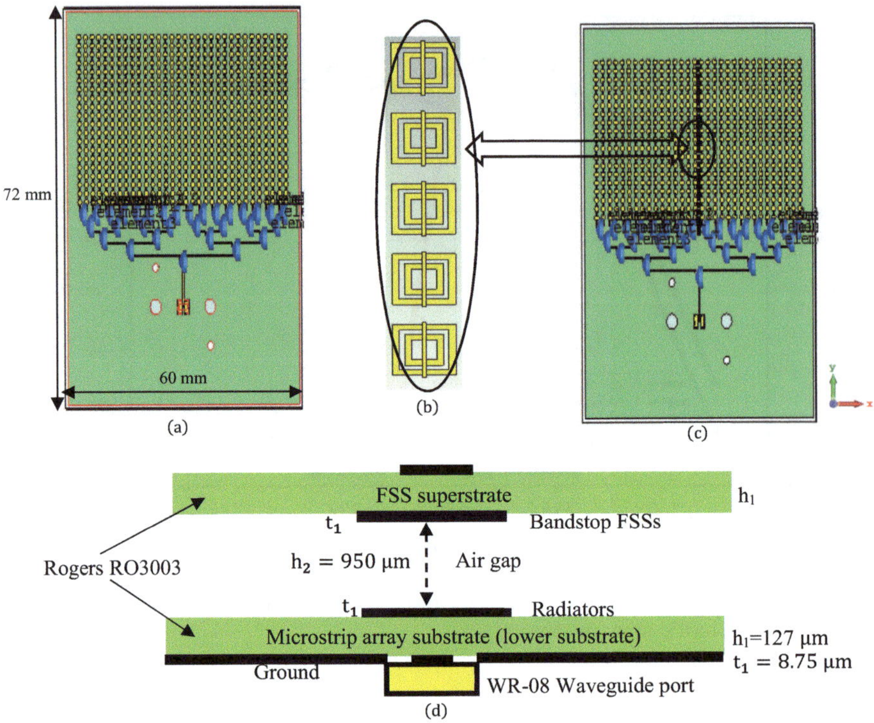

Fig. 7.20 The configuration of the second THz slotting circular patch microstrip array antenna with SIW and the designed FSSs, (**a**) Microstrip array antenna substrate, (**b**) Part of FSS superstrate with part of UCell FSSs, (**c**) The complete second THz slotting circular patch microstrip array antenna with SIW and the designed FSSs, (**d**) The complete second microstrip array antenna with SIW and FSSs in side view

The first designed antenna was at one PCB and contained 20 × 32 circular radiators with a 1 × 32 Wilkinson divider and waveguide/SIW to microstrip transition, as shown in Fig. 7.19a–c. The WR-08 waveguide fed this antenna. Moreover, this waveguide covers the frequency range of 90–140 GHz. Figure 7.20 exhibits the configuration of the second THz slotting circular patch microstrip array antenna with SIW and the designed FSSs. Where Fig. 7.20a shows the second THz slotting circular patch microstrip array antenna with SIW without the FSSs, Fig. 7.20b shows the part of the FSS superstrate with the designed FSSs, Fig. 7.20c shows the second THz slotting circular patch microstrip array antenna with SIW and the designed FSSs, and Fig. 7.20d shows the complete second THz slotting circular patch microstrip array antenna with SIW and FSSs in side view.

The second antenna was designed with two PCBs and contained 20 × 32 circular radiators with 1 × 32 Wilkinson dividers, SIW, waveguide/SIW to microstrip transition, and 26 UCell bandstop FSSs, as shown in Fig. 7.20a–d. This antenna is established on two rolled copper Rogers RO3003 substrates with the same substrate parameters as the first THz slotting circular radiators microstrip array antenna with

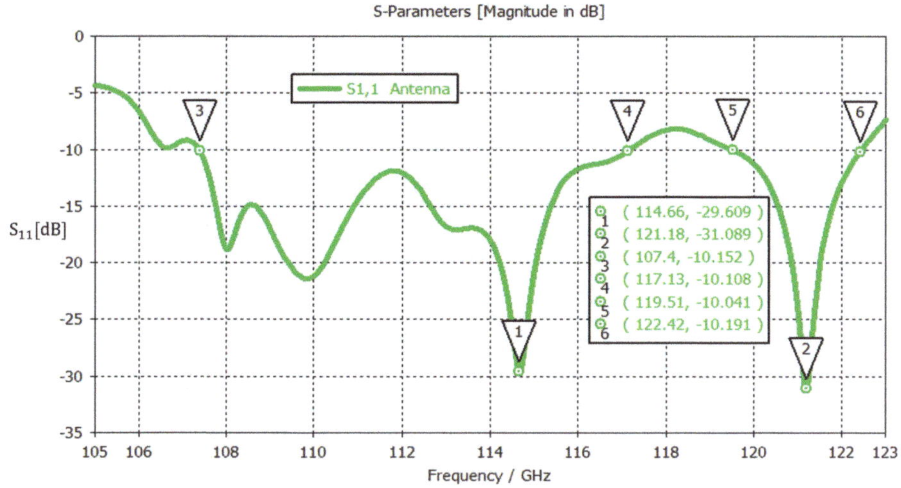

Fig. 7.21 Simulation result of the S_{11} for the first THz slotting circular patch microstrip array antenna with SIW

SIW and with an air gap of $h_2 = 950$ μm between both PCBs. The WR-08 waveguide fed this antenna. All magnitudes of this antenna, including the numbers of the Unit-cell bandstop FSSs, and the air gap distance (h_2) were optimized to get this antenna's best gain and bandwidth with reasonable time simulation. The simulated result of the S_{11} for the offered first antenna is shown in Fig. 7.21. From this figure it has been shown that the BW ($S_{11} \leq -10$ dB) = 12.86 GHz (11.3%). Moreover, the frequency of the resonance of this offered first antenna is 114.66 GHz and 121.18 GHz.

Figures 7.22a–c show the simulation results of the 3D far-field radiation pattern, gain, and directivity at the E-plane and H-plane of the designed first THz slotting circular patch microstrip array antenna with SIW at a frequency of 107 GHz at the polar plots, respectively. From Fig. 7.22a, it has been shown that the directivity is 23.3 dBi, and the total efficiency is −1.9 dB for 107 GHz. From Fig. 7.22b, it has been shown that the gain is 24.3 dB in the E-plane with the direction of the main lobe at 0°, while the –3 dB angular width is 2.4° and the SLL is −14.3 dB. The gain is 24.7 dB in the H-plane with the direction of the main lobe at −1°, while the –3 dB angular width is 4.8° and the SLL is −7 dB. From Fig. 7.22c, it has been shown that the directivity is 25.5 dBi in the E-plane with the direction of the main lobe at 0°, while the –3 dB angular width is 2.4° and the SLL is −14.3 dB. The directivity is 25.9 dBi in the H-plane with the direction of the main lobe at −1°, while the –3 dB angular width is 4.8° and the SLL is −7 dB.

The simulated result of the S_{11} for the offered second antenna is shown in Fig. 7.23. From this figure, it has been shown that the bandwidth($S_{11} \leq -10$ dB) = 12.47 GHz(11.13%). Moreover, the frequency of the resonance of this offered second antenna is 114.47 GHz and 121.3 GHz.

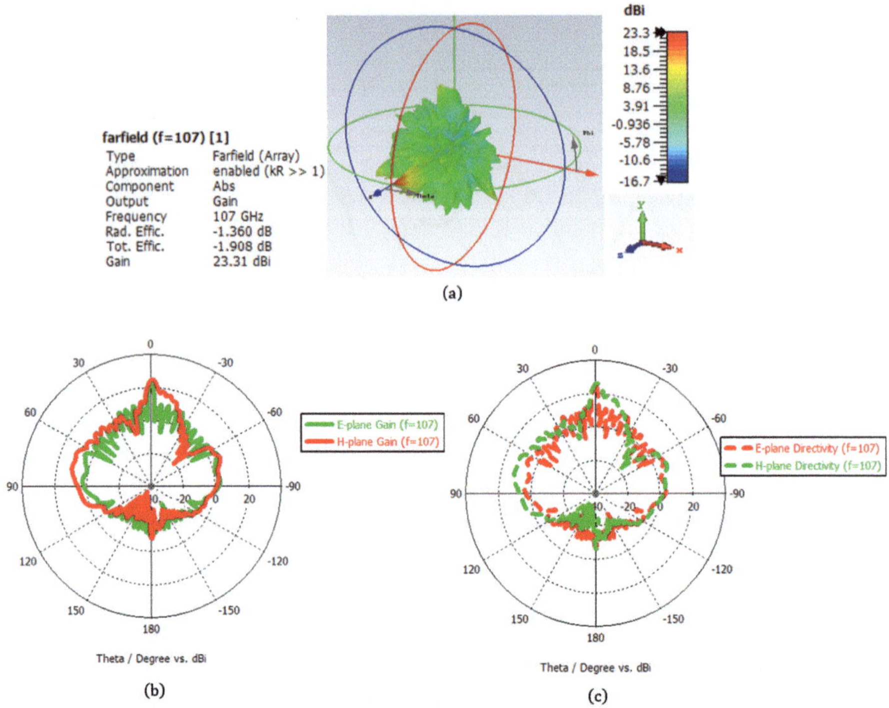

Fig. 7.22 Simulation results of the 3D far-field radiation pattern, E-plane, and H-plane of the first THz slotting circular patch microstrip array antenna with SIW (**a**) 3D far-field, (**b**) Gain, and (**c**) Directivity

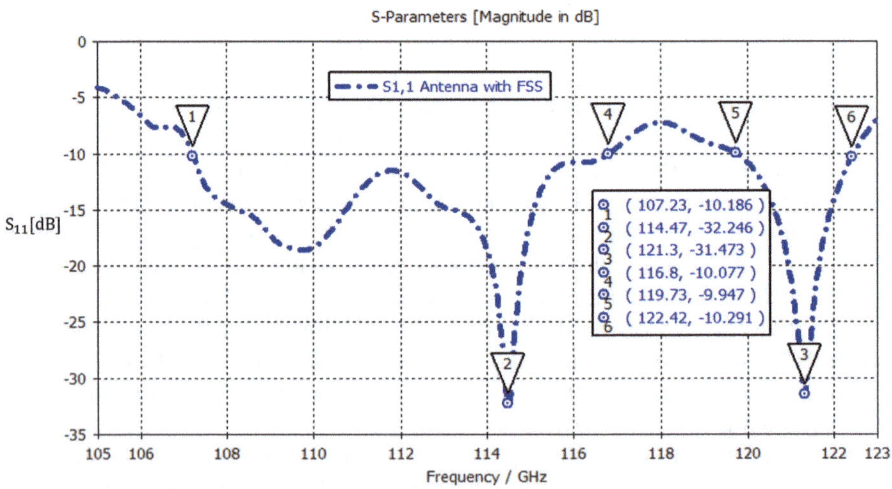

Fig. 7.23 Simulation result of the S_{11} for the second THz slotting circular patch microstrip array antenna with SIW and FSSs

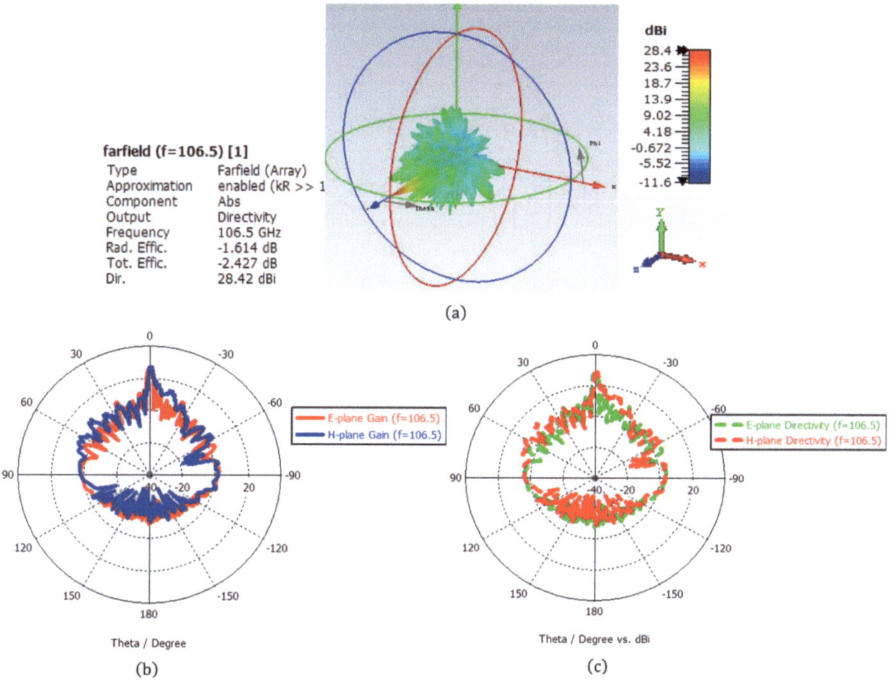

Fig. 7.24 Simulation results of the 3D far-field radiation pattern, E-plane, and H-plane of the second THz slotting circular patch microstrip array antenna with SIW (**a**) 3D far-field, (**b**) Gain, and (**c**) Directivity

Figures 7.24a–c show the simulation results of the 3D far-field radiation pattern, gain, and directivity at the E-plane and H-plane of the designed first THz slotting circular patch microstrip array antenna with SIW at a frequency of 106.5 GHz at the polar plots, respectively. From Fig. 7.24a, it has been shown that the directivity is 28.42 dBi, and the total efficiency is −2.43 dB for 106.5 GHz. From Fig. 7.24b, it has been shown that the gain is 26.8 dB in the E-plane with the direction of the main lobe at 0°, while the –3 dB angular width is 2.4° and the side lobe level (SLL) is – 14.6 dB. The gain is 27 dB in the H-plane with the direction of the main lobe at −1°, while the –3 dB angular width is 1.9° and the SLL is −8.7 dB. From Fig. 7.24c, it has been shown that the directivity is 28.4 dBi in the E-plane with the direction of the main lobe at 0°, while the –3 dB angular width is 2.4° and the SLL is −14.6 dB. The directivity is 28.7 dBi in the H-plane with the direction of the main lobe at −1°, while the –3 dB angular width is 1.9° and the SLL is −8.7 dB.

The simulation results from comparisons between offered first and second antennas are shown in Figs. 7.25a, b. From Fig. 7.25a, It has been shown that the gain with FSS was more significant by more than 1.7 dB between 106.2–107.5 GHz vs. without FSS, and maximal change was 3.06 dB at 106.5 GHz, i.e., adding the FSS is enhancing the gain of the antenna at a small part of the frequency band. Figure 7.25b shows that the bandwidth is reduced by 0.39 GHz (−0.17%) by adding the FSS to the antenna.

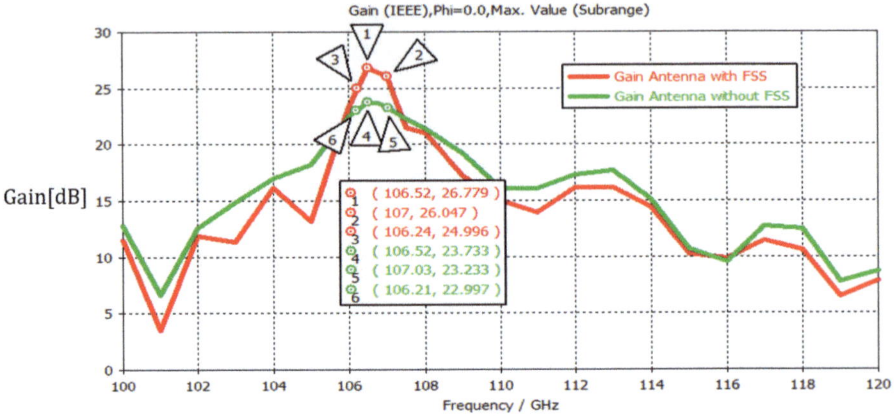

(a)

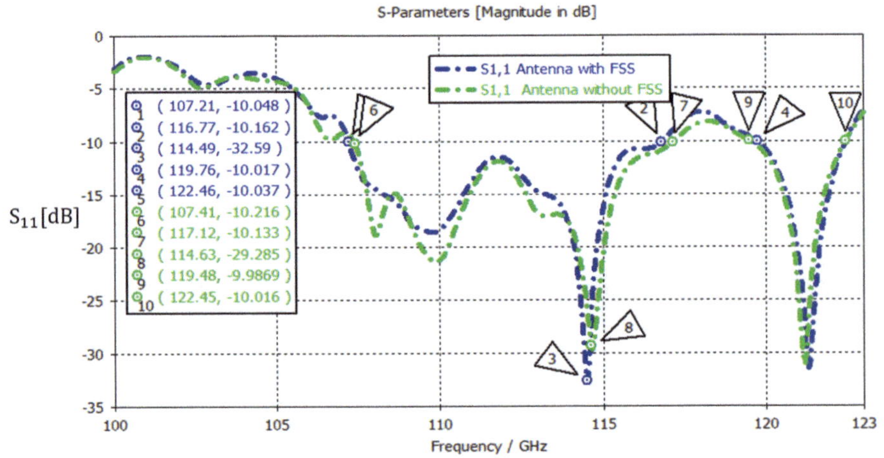

(b)

Fig. 7.25 The simulation results from the comparison between first and second antennas, (**a**) Gain, and (**b**) S_{11}

Figure 7.26 exhibits the validation/comparison of the gain and S_{11} of the offered first THz slotting circular microstrip array antenna with SIW apperceived from the CST MWS and Ansys HFSS. This figure exhibits that the maximal gain and bandwidth apperceived from the Ansys HFSS was about 22.8 dB, 13.1 GHz, while the maximal gain and bandwidth apperceived from the CST MWS was 23.3 dB, 12.86 GHz. So, it has been shown that an excellent agreement between simulation results is apperceived concerning the offered first THz slotting circular patch microstrip array antenna with SIW.

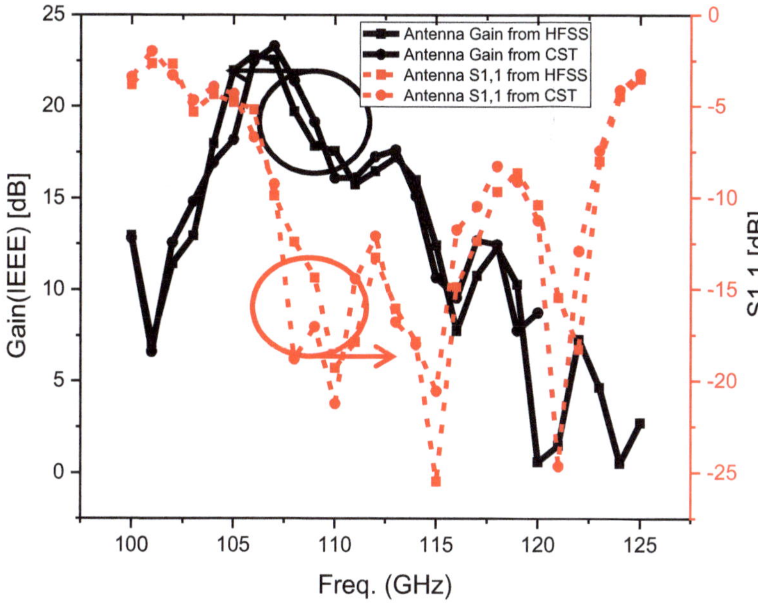

Fig. 7.26 Simulation results from a comparison of the gain and the S_{11} for the first THz slotting circular patch microstrip array antenna with SIW simulated with the Ansys HFSS and CST MWS simulators

7.7.3 THz Parasitic Patches Microstrip Array Antenna

Several THz parasitic patches microstrip array antennas were designed with 26 × 32 radiators and 52 × 32 parasitic patches at hybrid fed for gain and bandwidth enhancement. The second was with bandstop FSSs for gain enhancement besides the 26 × 32 radiators and 52 × 32 parasitic patches at hybrid fed. Moreover, the chosen substrate for these THz antennas of a Taconic TLY-5 microstrip substrate, which the $(\tan\delta, \varepsilon_r)$ is measured up to 500 GHz [42, 43], and the chosen Taconic TLY-5 parameters of tan δ, ε_r, copper thickness (t), laminate height (h): (0.00943, 2.253) at 325 GHz 17.5 μm, 127 μm, appropriately, which are $(h = 127\mu m \leq 0.125\lambda_{high})$, $(t = 17.5\ \mu m \leq 0.125\lambda_{high}, \varepsilon_r = 2.253 < 3.5$, $\tan\delta = 0.00943 < 0.0095$), will guarantee that the conduction losses, the substrate losses, and the surface radiation losses are reduced to a minimum, which is the main concern at the THz band [6, 44], and the simulation results should be as much as can be close with empirical validation. The magnitude of the substrate was 55 × 55 mm^2 that equals $20.89\lambda_0 \times 20.89\lambda_0$ @ 114 GHz. The first antenna is presented in Fig. 7.27, where Fig. 7.27a shows the complete THz parasitic patches microstrip array antennas, while Fig. 7.27b shows zoomed partial part radiating and parasitic elements.

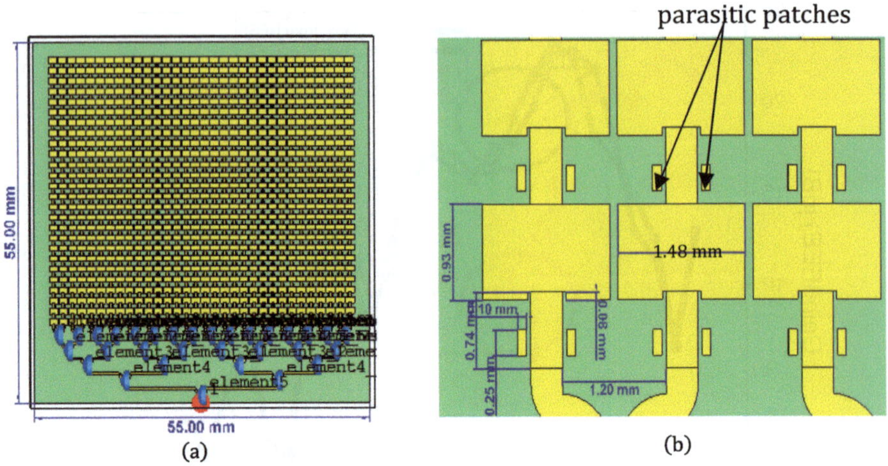

Fig. 7.27 Configuration of the first THz parasitic patches microstrip array antenna, (**a**) The complete first THz parasitic patches microstrip array antenna, (**b**) Zoomed partial part radiating and parasitic elements

The second antenna was designed at two PCBs layers with 25 Unit-cell bandstop FSSs, besides the 26 × 32 radiation elements with parasitic patches, and presented in Fig. 7.28. where Fig. 7.28a shows the second THz parasitic patches microstrip array antenna without the FSSs, where Fig. 7.28b shows the part of the FSS superstrate with the designed FSSs, Fig. 7.28c shows the second THz parasitic patches microstrip array antenna and the designed FSSs, and Fig. 7.28d shows the complete second THz parasitic patches microstrip array antenna and FSSs in side view. The two PCBs were established on Taconic TLY-5 microstrip substrates with the same substrate parameters as the first parasitic patches microstrip array antenna and with an air gap of $h_1 = 1390$ μm between two PCBs. All magnitudes of this antenna, including the numbers of the UCells bandstop FSSs, and the air gap distance (h_1) were optimized to get this antenna's best gain and bandwidth with reasonable time simulation.

The simulated result of the S_{11} for the offered first antenna is shown in Fig. 7.29. In this figure, the bandwidth of this antenna is shown as: bandwidth($S_{11} \leq -10$ dB) = 18.01 GHz(16.37%). Moreover, the frequency of the resonance of this offered first antenna is 114.07 GHz.

Figures 7.30a–c show the simulation results of the 3D far-field radiation pattern, gain, and directivity at the E-plane and H-plane of the designed first THz parasitic patches microstrip array antenna at a frequency of 106 GHz at the polar plots, respectively. From Fig. 7.30a, it has been shown that the directivity is 25.01 dBi, and the total efficiency is −1.51 dB for 106 GHz. From Fig. 7.30b, it has been shown that the gain is 20.4 dB in E-plane with the direction of the main lobe at 0°, while the –3 dB angular width is 2.7° and the SLL is −11.1 dB. The gain is 24.1 dB in the

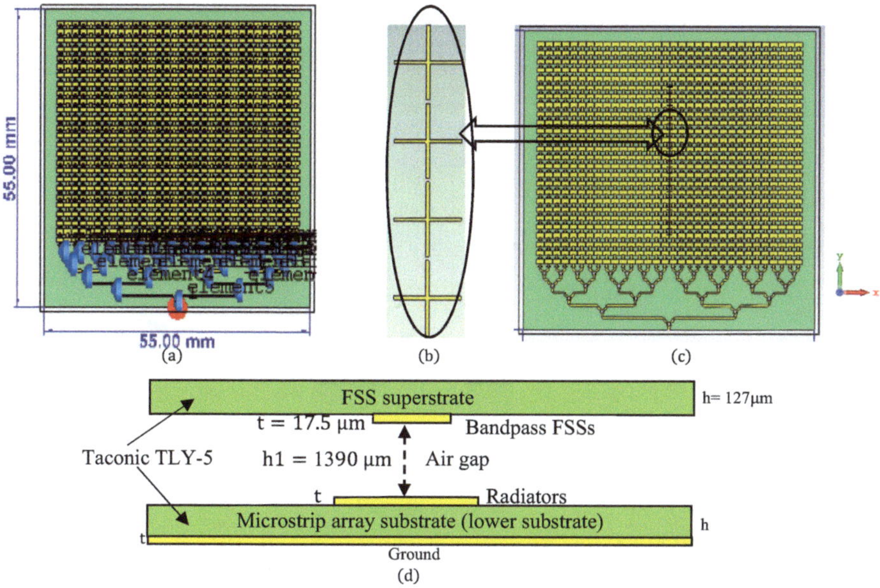

Fig. 7.28 Configuration of the complete second parasitic patches microstrip array antenna with FSSs (**a**) Microstrip array antenna substrate (lower substrate), (**b**) Part of Unit-cell bandstop FSSs, (**c**) The complete second THz parasitic patches microstrip array antenna with FSSs, and (**d**) Side view

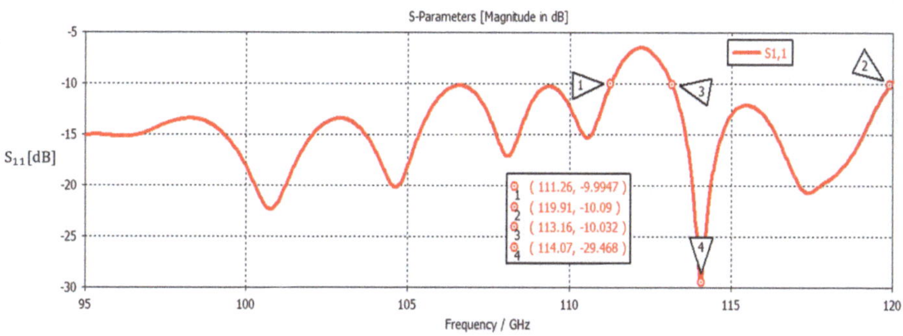

Fig. 7.29 Simulation result of the S_{11} for the first THz parasitic patches microstrip array antenna

H-plane with the direction of the main lobe at −6°, while the –3 dB angular width is 8.3° and the SLL is −11.6 dB. From Fig. 7.30c, it has been shown that the directivity is 21.6 dBi in the E-plane with the direction of the main lobe at 0°, while the –3 dB angular width is 2.7° and the SLL is −11.1 dB. The directivity is 25.2 dBi in the H-plane with the direction of the main lobe at −6°, while the –3 dB angular width is 8.3° and the SLL is −11.6 dB.

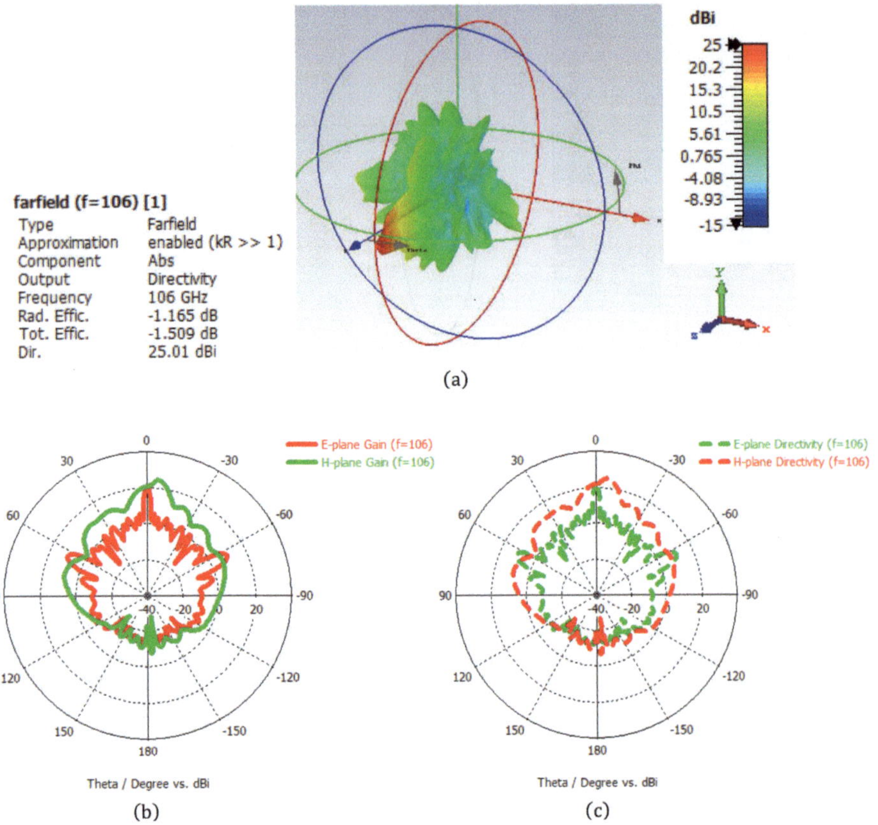

Fig. 7.30 Simulation results of the 3D far-field radiation pattern, E-plane, and H-plane of the first THz parasitic patches microstrip array antenna, (**a**) 3D far-field, (**b**) Gain, (**c**) Directivity

The simulated result of the S_{11} for the offered second antenna is shown in Fig. 7.31. In this figure, the bandwidth of this antenna is shown as: bandwidth ($S_{11} \leq -10$ dB) = 18.46 GHz(16.78%). Moreover, the frequency of the resonance of this offered second antenna is 118.09 GHz.

Figures 7.32a–c show the simulation results of the 3D far-field radiation pattern, gain, and directivity at the E-plane and H-plane of the designed second THz parasitic patches microstrip array antenna at a frequency of 111 GHz at the polar plots, respectively. From Fig. 7.32a, it has been shown that the directivity is 22.7 dBi, and the total efficiency is −1.68 dB for 111 GHz. From Fig. 7.32b, it has been shown that the gain is 21.4 dB in the E-plane with the direction of the main lobe at 0°, while the –3 dB angular width is 2.6° and the SLL is −12.2 dB. The gain is 22.5 dB in the H-plane with the direction of the main lobe at 2°, while the –3 dB angular width is 7.7°, and the SLL is −4 dB. From Fig. 7.32c, it has been shown that the directivity is

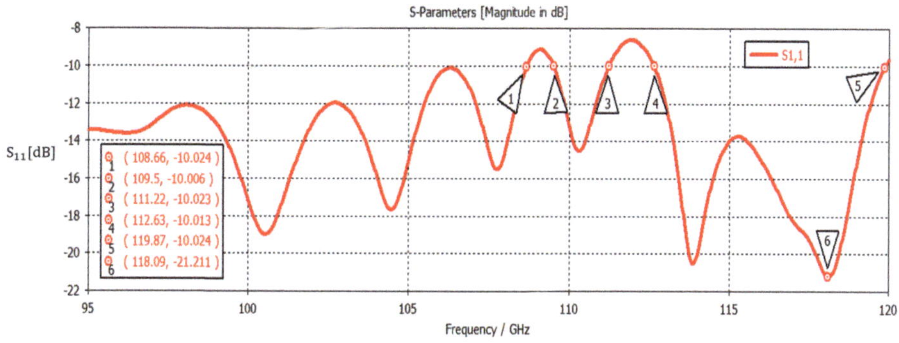

Fig. 7.31 Simulation result of the S_{11} for the second THz parasitic patches microstrip array antenna

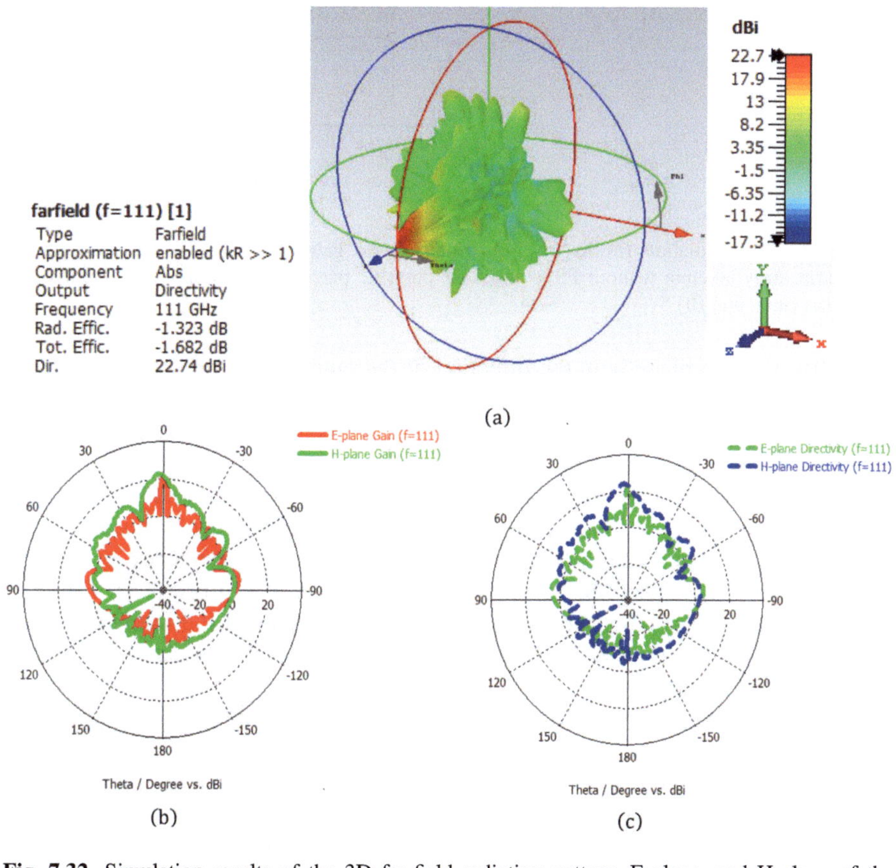

Fig. 7.32 Simulation results of the 3D far-field radiation pattern, E-plane, and H-plane of the second THz parasitic patches microstrip array antenna (**a**) 3D far-field, (**b**) Gain, and (**c**) Directivity

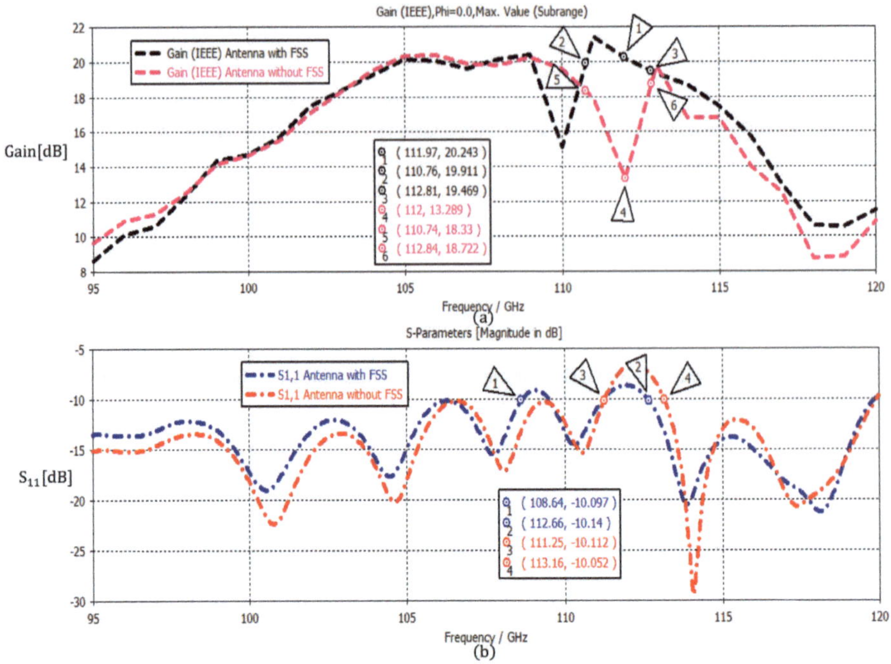

Fig. 7.33 The simulation results in comparison of the gain and S_{11} THz parasitic patches microstrip array antenna without FSSs and THz parasitic patches microstrip array antenna with FSSs, (**a**) Gain, and (**b**) S_{11}

21.6 dBi in the E-plane with the direction of the main lobe at 0°, while the –3 dB angular width is 2.7° and the SLL is −11.1 dB. The directivity is 22.7 dBi in the H-plane with the direction of the main lobe at 2.6°, while the –3 dB angular width is 2.6° and the SLL is −12.2 dB.

The simulation results from comparisons between offered first and second antennas are shown in Figs. 7.33a, b. From Fig. 7.33a, it has been shown that the gain with FSS was more significant by more than 0.7 dB between frequency of 110.7–-112.8 GHz vs. without FSS, and maximal change was 6.95 dB at 112 GHz, i.e., adding the FSS enhancing the gain of the antenna, at part of the frequency band. From Fig. 7.33b, it has been shown that 0.45 GHz increases bandwidth due to adding FSSs to the antenna.

Figure 7.34 exhibits the validation/comparison of the gain and S_{11} of the offered first THz parasitic patches microstrip antenna apperceived from the CST MWS and Ansys HFSS simulators at 95–120 GHz. This figure exhibits that the maximal gain and bandwidth apperceived from the Ansys HFSS was about 19.1 dB, 19 GHz, while the maximal gain and bandwidth apperceived from the CST MWS was 20.4 dB, 18 GHz. So, it has been shown that an acceptable agreement between simulation results is achieved regarding the first THz parasitic patches microstrip array antenna at 95–120 GHz.

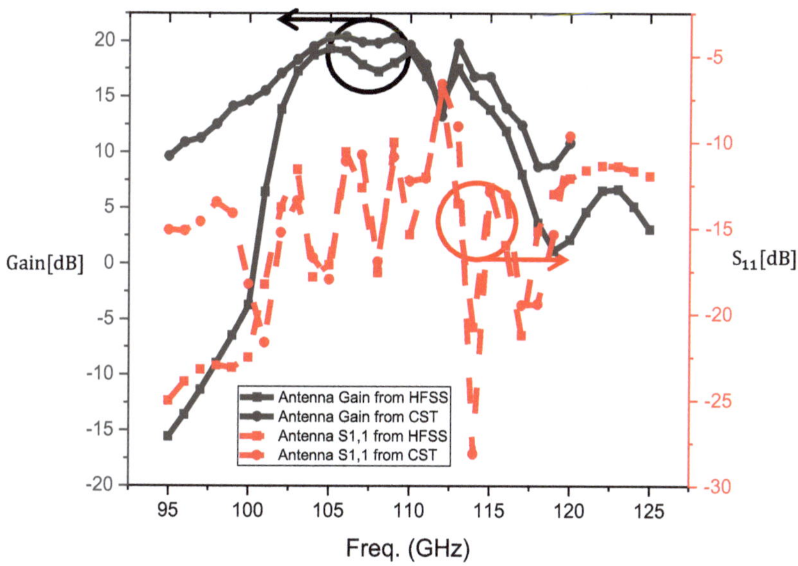

Fig. 7.34 The simulation results from a comparison of the gain and the S_{11} for the first THz parasitic patches microstrip array antenna simulated with the Ansys HFSS and CST MWS simulators

7.7.4 THz Log-Periodic Microstrip Array Antenna

Several THz log-periodic microstrip array antennas were designed at a resonance frequency of 137 GHz with hybrid fed for gain and bandwidth enhancement. The first was with a log-periodic microstrip antenna with 24 × 16 radiators. The second was with a log-periodic microstrip antenna with 24 × 32 radiators, while the third was with bandstop FSSs for gain enhancement besides the log-periodic microstrip antenna with 24 × 32 radiators. Moreover, the chosen substrate of Isola Astra MT77 microstrip substrate, the $(\tan\delta, \varepsilon_r)$, is measured up to 125 GHz, which is relatively close to the resonance of this antenna design, and the chosen Isola Astra MT77 parameters of tan δ, ε_r, copper thickness (t), laminate height (h): (0.001, 2.82) at 125 GHz 17.5 μm, 127 μm, appropriately, which are ($h = 127\ \mu m \leq 0.125\lambda_{high}$), ($t = 17.5\ \mu m \leq 0.125\lambda_{high}$, $\varepsilon_r = 2.82 < 3.5$, $\tan\delta = 0.001 < 0.0095$), will guarantee that the conduction losses, the substrate losses, and the surface radiation losses are reduced to a minimum, which is the main concern at the THz band [6, 44], and the simulation results should be as much as can be close with empirical validation. The first antenna magnitude of the substrate 39 × 34 mm^2 that equals $17.85\lambda_0 \times 15.56\lambda_0$ @ 137.3 GHz is presented in Fig. 7.35a, b. Figure 7.35a shows the first THz log-periodic microstrip array antenna, while Fig. 7.35b shows one column of this array.

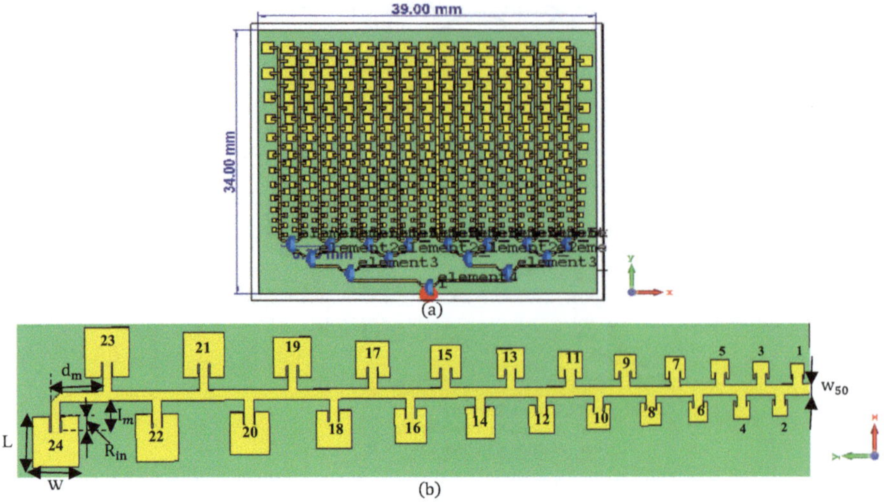

Fig. 7.35 Configuration of the designed first THz log-periodic microstrip array antenna, (**a**) The complete first antenna, and (**b**) One array column

Following Eqs. (7.21), (7.22), (7.23), (7.24), (7.25), (7.26), (7.27), (7.28), and (7.29) and CST MWS simulator macros, these values were apperceived and presented in Table 7.5.

The magnitudes of each log-periodic radiator patch and feed line at the first offered THz antenna were apperceived following optimization with the CST MWS simulator to get the best gain and bandwidth of this antenna. The second antenna is presented in Fig. 7.36. The magnitude of the substrate was 76 × 36 mm^2, which equals $34.8\lambda_0 \times 16.5\lambda_0$ @ 137.3 GHz. The magnitudes of each log-periodic radiator patch and feed line at the second offered THz antenna was apperceived following optimization with the CST MWS simulator to get the best gain and bandwidth of this antenna.

The third THz antenna was designed at two PCBs with 26 Unit-cell bandstop FSSs, besides the 26 × 32 log-periodic radiators and a 1 × 32 Wilkinson power divider, and presented in Fig. 7.37a–d, where Fig. 7.37a shows the third THz log-periodic microstrip array antenna without the FSSs, Fig. 7.37b shows the part of the FSS superstrate with the designed FSSs, Fig. 7.37c shows the third THz log-periodic microstrip array antenna and the designed FSSs, and Fig. 7.37d shows the complete second THz parasitic patches microstrip array antenna and FSSs in side view. The two PCBs were established on Isola Astra MT77 microstrip substrates with the same substrate parameters as the first THz log-periodic microstrip array antenna with an air gap of $h_1 = 1450$ μm between two PCBs. All magnitudes of this antenna, including the log-periodic radiators sizes, numbers of the Unit-cells bandstop FSSs, and the air gap distance (h_1) were optimized to get this antenna's best gain and bandwidth with reasonable time simulation.

Table 7.5 The 24 patches log-periodic microstrip array antenna magnitudes

Patch number	Freq. (GHz)	W = L (mm)	Im (Quarter wavelength) (mm)	d_m (Half wavelength) (mm)	R_{in} (for 50 Ω) (mm)
24	54.15	1.545	0.893	1.701	0.461
23	56.85	1.469	0.851	1.62	0.438
22	59.69	1.395	0.810	1.543	0.416
21	62.68	1.325	0.772	1.469	0.395
20	65.82	1.258	0.735	1.399	0.375
19	69.1	1.195	0.7	1.333	0.356
18	72.56	1.134	0.667	1.269	0.338
17	76.19	1.077	0.635	1.209	0.321
16	80	1.022	0.605	1.15	0.305
15	84	0.969	0.576	1.096	0.289
14	88.2	0.919	0.548	1.044	0.274
13	92.61	0.874	0.522	0.994	0.261
12	97.24	0.826	0.497	0.947	0.246
11	102.1	0.783	0.474	0.902	0.233
10	107.2	0.741	0.451	0.859	0.221
9	112.57	0.702	0.429	0.819	0.209
8	118.2	0.664	0.409	0.779	0.198
7	124.1	0.629	0.389	0.742	0.187
6	130.31	0.595	0.371	0.707	0.177
5	136.83	0.562	0.353	0.673	0.167
4	143.67	0.531	0.336	0.641	0.158
3	150.82	0.502	0.320	0.611	0.149
2	158.4	0.474	0.305	0.582	0.141
1	166.31	0.447	0.291	0.554	0.133

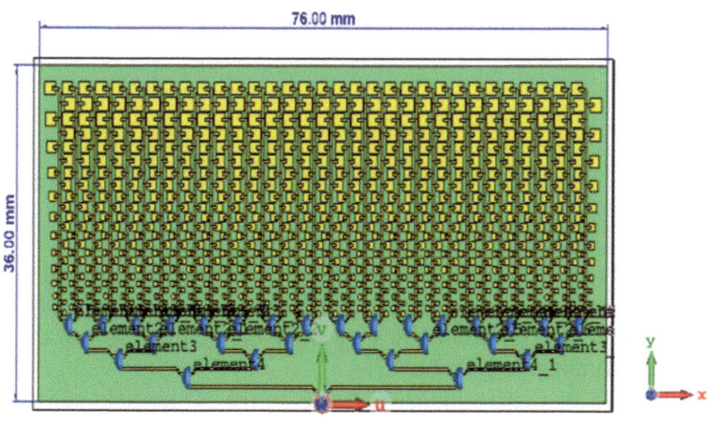

Fig. 7.36 Configuration of the designed second THz log-periodic microstrip array antenna

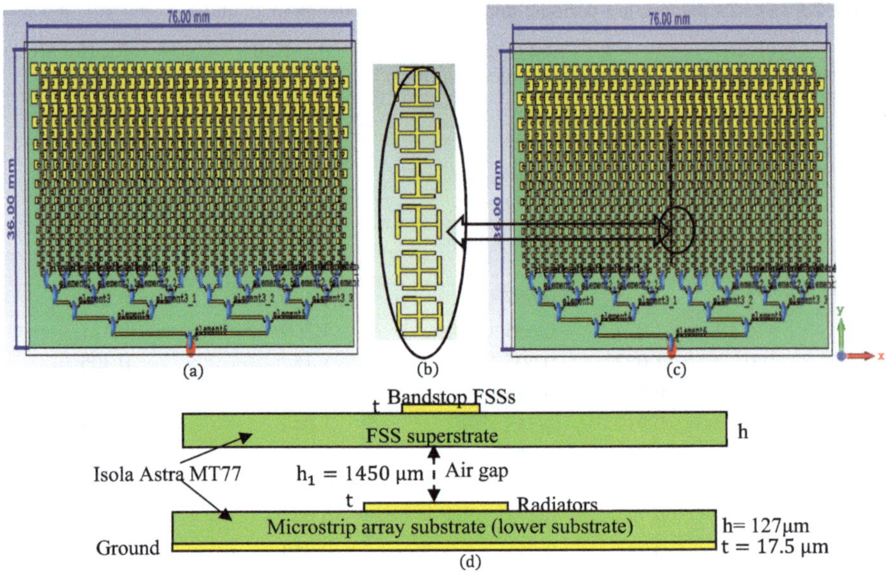

Fig. 7.37 Configuration of the designed third THz log-periodic microstrip array antenna with FSSs, (**a**) THz log-periodic microstrip array antenna at the lower substrate, (**b**) Part of Unit-cell FSSs at the superstrate, (**c**) The complete third THz log-periodic microstrip array antenna with FSSs, and (**d**) Side view

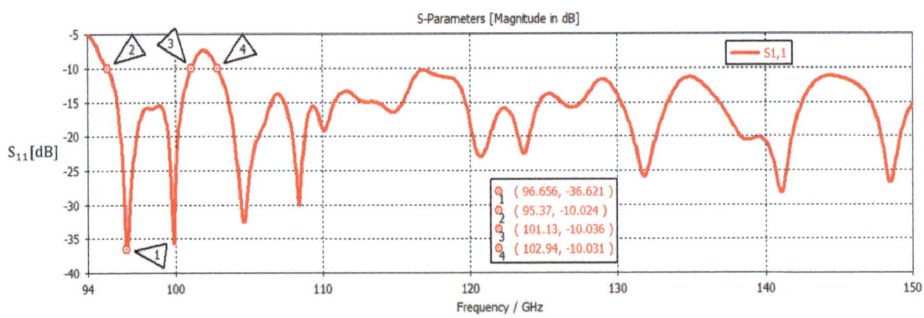

Fig. 7.38 Simulation result of the S_{11} for the first THz log-periodic microstrip array antenna

The simulated result of the S_{11} for the offered first antenna is shown in Fig. 7.38. From this figure, it has been shown that the bandwidth of this antenna: bandwidth ($S_{11} \leq -10$ dB) > 52.6 GHz(>42.86%). Moreover, the frequency of the resonance of this offered first antenna is 96.66 GHz.

Figures 7.39a–c show the simulation results of the 3D far-field radiation pattern, gain, and directivity at the E-plane and H-plane of the designed first THz log-periodic microstrip array antenna at a frequency of 100 GHz at the polar plots,

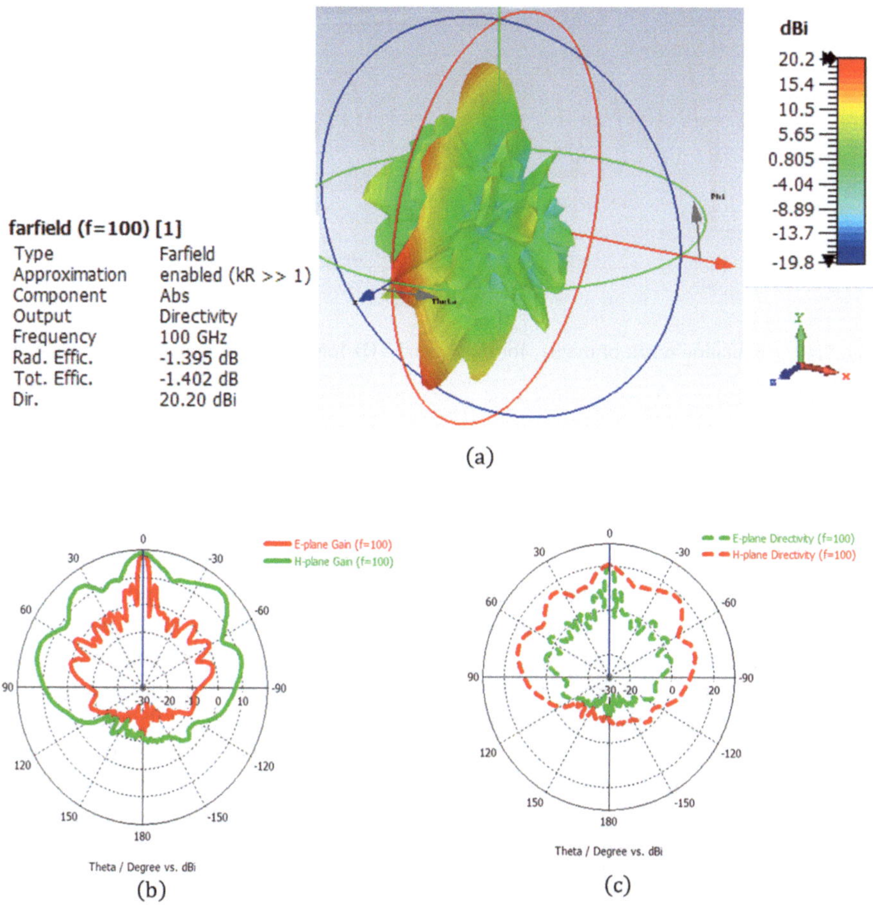

Fig. 7.39 Simulation results of the 3D far-field radiation pattern, E-plane, and H-plane of the first THz log-periodic microstrip array antenna, (**a**) 3D far-field, (**b**) Gain, (**c**) Directivity

respectively. From Fig. 7.39a, it has been shown that the directivity is 20.2 dBi, and the total efficiency is −1.4 dB for 100 GHz. From Fig. 7.39b, it has been shown that the gain is 18.8 dB in E-plane with the direction of the main lobe at 0°, while the –3 dB angular width is 4.2° and the SLL is −11.2 dB. The gain is 18.8 dB in the H-plane with the direction of the main lobe at 0°, while the –3 dB angular width is 14°, and the SLL is –2.2 dB. From Fig. 7.39c, it has been shown that the directivity is 20.2 dBi in the E-plane with the direction of the main lobe at 0°, while the –3 dB angular width is 4.2° and the SLL is −11.2 dB. The directivity is 20.2 dBi in the H-plane with the direction of the main lobe at 0°, while the –3 dB angular width is 14° and the SLL is –2.2 dB.

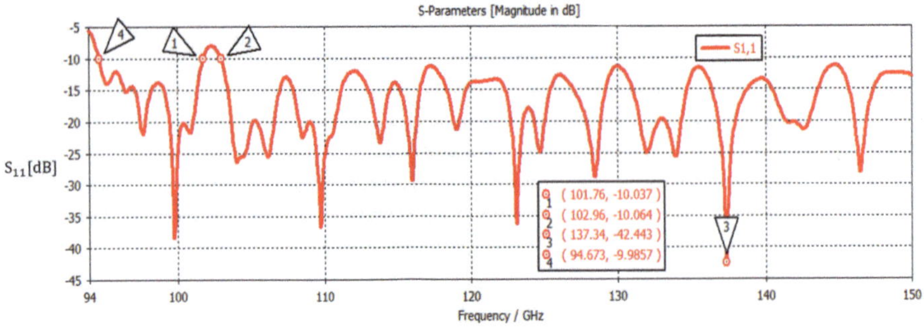

Fig. 7.40 Simulation result of the S_{11} for the second THz log-periodic microstrip array antenna

The simulated result of the S_{11} for the offered second antenna is shown in Fig. 7.40. This figure shows the bandwidth of this antenna: bandwidth($S_{11} \leq -10$ dB) > 54.3 GHz(>44.41%). Moreover, the frequency of the resonance of this offered second antenna is 137.34 GHz.

Figures 7.41a–c show the simulation results of the 3D far-field radiation pattern, gain, and directivity at the E-plane and H-plane of the designed first THz log-periodic microstrip array antenna at a frequency of 100 GHz at the polar plots, respectively. From Fig. 7.41a, it has been shown that the directivity is 21.3 dBi, and the total efficiency is −2.06 dB for 100 GHz. From Fig. 7.41b, it has been shown that the gain is 19.8 dB in E-plane with the direction of the main lobe at 0°, while the –3 dB angular width is 2° and the SLL is −11.5 dB. The gain is 19.8 dB in the H-plane with the direction of the main lobe at 0°, while the –3 dB angular width is 16.4°, and the SLL is −1.8 dB. From Fig. 7.41c, it has been shown that the directivity is 21.8 dBi in the E-plane with the direction of the main lobe at 0°, while the –3 dB angular width is 2° and the SLL is −11.5 dB. The directivity is 21.8 dBi in the H-plane with the direction of the main lobe at 0°, while the –3 dB angular width is 16.4°, and the SLL is −1.8 dB.

The simulated result of the S_{11} for the offered third antenna is shown in Fig. 7.42. In this figure, it has been shown that the bandwidth of this antenna: bandwidth ($S_{11} \leq -10$dB) > 53.9 GHz(>44.04%). Moreover, the frequency of the resonance of this offered third antenna is 128.46 GHz.

Figures 7.43a–c show the simulation results of the 3D far-field radiation pattern, gain, and directivity at the E-plane and H-plane of the designed first THz log-periodic microstrip array antenna at a frequency of 100 GHz at the polar plots, respectively. From Fig. 7.43a, it has been shown that the directivity is 23.5 dBi, and the total efficiency is −2.1 dB for 100 GHz. From Fig. 7.43b, it has been shown that the gain is 21.4 dB in the E-plane with the direction of the main lobe at 0°, while the –3 dB angular width is 2° and the SLL is −10.9 dB. The gain is 21.4 dB in the H-plane with the direction of the main lobe at 1°, while the –3 dB angular width is 19.3°, and the SLL is −2.9 dB. From Fig. 7.43c, it has been shown that the directivity is 23.5 dBi in the E-plane with the direction of the main lobe at 0°,

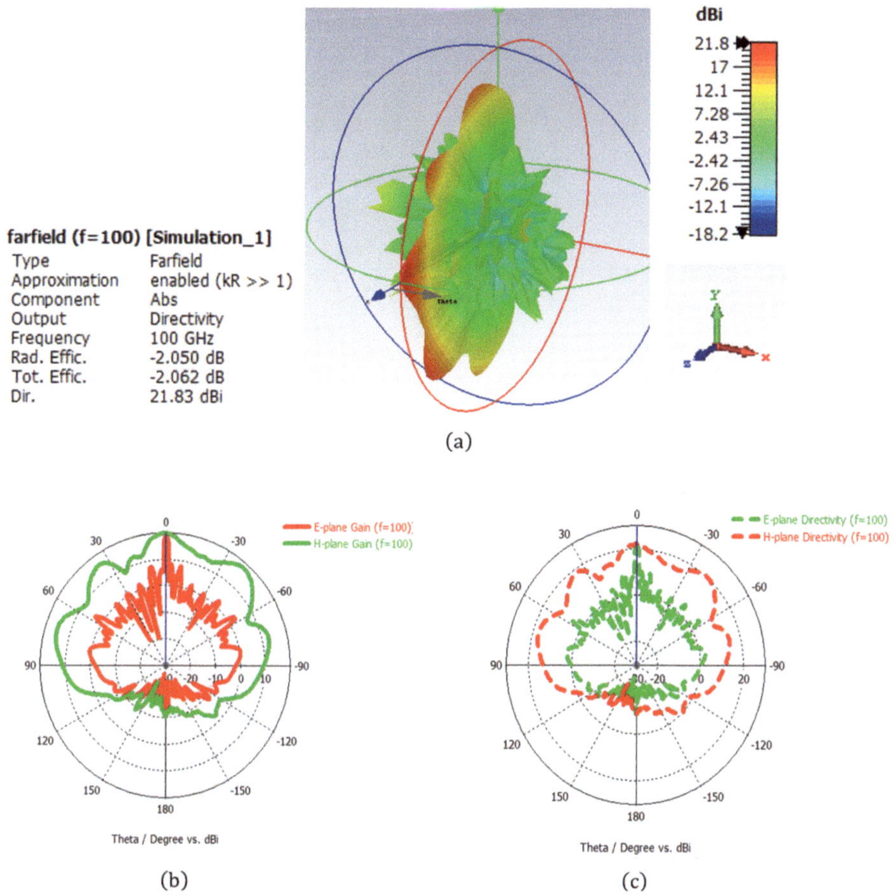

Fig. 7.41 Simulation results of the 3D far-field radiation pattern, E-plane, and H-plane of the second THz log-periodic microstrip array antenna (**a**) 3D far-field, (**b**) Gain, and (**c**) Directivity

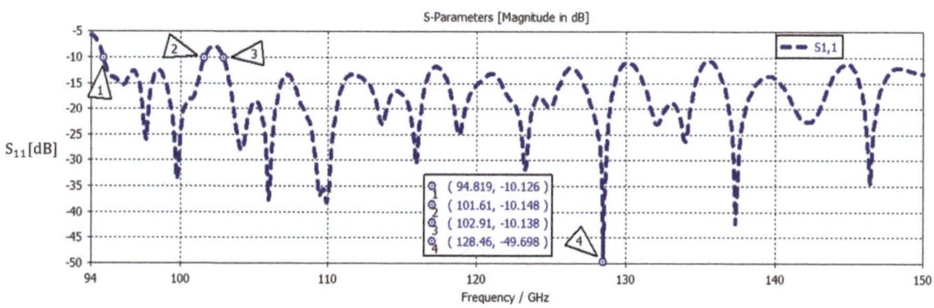

Fig. 7.42 Simulation result of the S_{11} for the third THz log-periodic microstrip array antenna

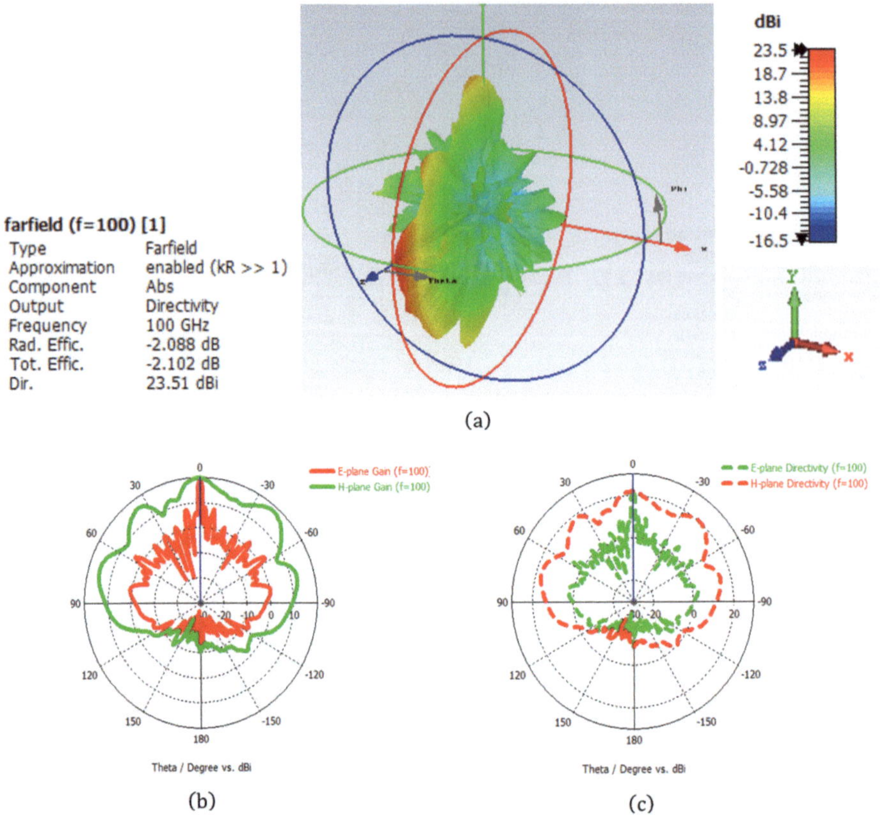

Fig. 7.43 Simulation results of the 3D far-field radiation pattern, E-plane, and H-plane of the third THz log-periodic microstrip array antenna (**a**) 3D far-field, (**b**) Gain, and (**c**) Directivity

while the –3 dB angular width is 2° and the SLL is −10.9 dB. The directivity is 23.5 dBi in the H-plane with the direction of the main lobe at 1°, while the –3 dB angular width is 19.3°, and the SLL is −2.9 dB.

The simulation results from comparisons between offered first up to third antennas are shown in Figs. 7.44a–c. From Fig. 7.44a, it has been shown that the gain is proportional to the number of microstrip array antenna columns. For each doubling of the number of columns, the antenna gain increase by 1 dB at 95–108 GHz. In contrast, by multiplying the number of the radiation column with the FSSs technique, a 3 dB improvement can be achieved at the antenna gains. It has been shown that the antenna gain with the designed FSS (third antenna) was more significant by more than about 1.5 dB, between at 95–108.4 GHz vs. without the designed FSS (second antenna), and maximal change was 3.12 dB @ 102.5 GHz, i.e., adding the FSS enhances the antenna's gain at 95–109 GHz. From Fig. 7.44b, it has been shown that the bandwidth is decreased by more than 0.4 GHz when adding the FSS feature

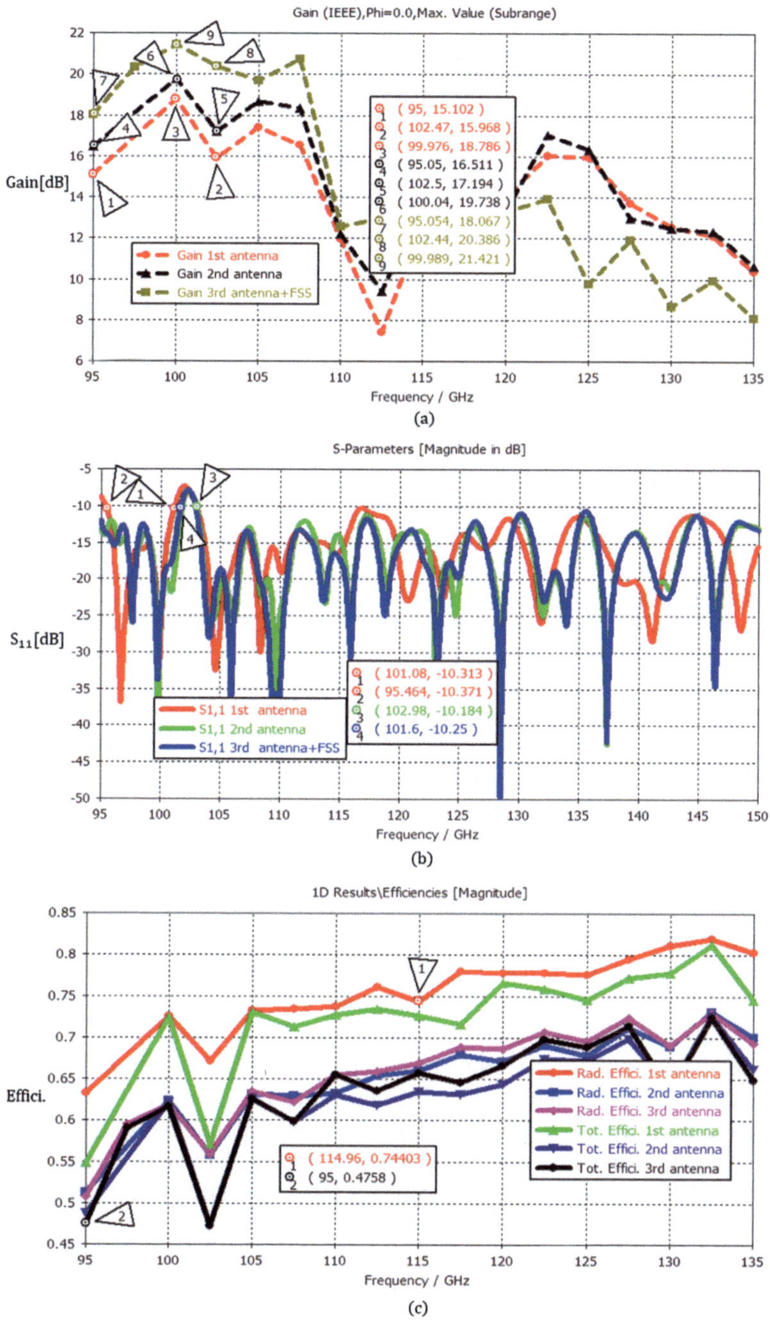

Fig. 7.44 The simulation results from a comparison of the gain, S_{11},radiation efficiency and total efficiency between first up to third THz log-periodic microstrip antennas, (**a**) Gain, (**b**) S_{11}, and (**c**) Radiation efficiency and total efficiency

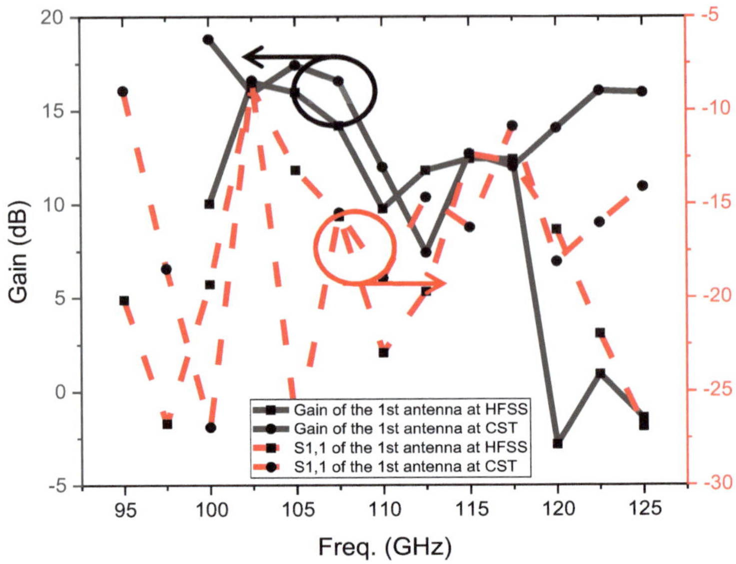

Fig. 7.45 Simulation results comparison of the gain and the S_{11} of the first THz log-periodic microstrip array antenna simulated with the Ansys HFSS and CST MWS simulators

to the antenna. Figure 7.44c exhibits those increasing radiating elements decreases the radiating and total efficiencies. Therefore, adding the FSS elements to the antenna may have been shown to decrease its radiating and total efficiencies.

Figure 7.45 exhibits the validation/comparison of the gain and S_{11} of the offered first THz log-periodic microstrip array antenna apperceived from the CST MWS and Ansys HFSS simulators at 95–125 GHz. This figure exhibits that the maximal gain and bandwidth apperceived from the Ansys HFSS was about 16.5 dB, 27.5 GHz, while the maximal gain and bandwidth apperceived from the CST MWS was 18.8 dB, 27.5 GHz. So, it has been shown that an acceptable agreement between simulation results is achieved regarding the first log-periodic microstrip array antenna at 102.5–117.5 GHz.

Figures 7.46a, b show the simulation results of the S_{11} and the gain for the first antenna with or without etching tolerance accuracy of ± 10 μm [45]. This deviation is supposed to be when this antenna is fabricated for experimental verification. These figures show that the first antenna's bandwidth and gain may change by about 0.6 GHz, 1–1.81 dB when this first antenna is fabricated due to the etching tolerance as described.

Figures 7.47a, b show the simulation results of the S_{11} and the gain for the first antenna when the ε_r and tanδ of Isola Astra MT 77 microstrip substrate are set at the CST MWS simulator for a frequency of 20 GHz or frequency of 125 GHz, where 20 GHz $\varepsilon_r = 3$ and $\tan\delta = 0.0017$ and for 125 GHz $\varepsilon_r = 2.82$ and

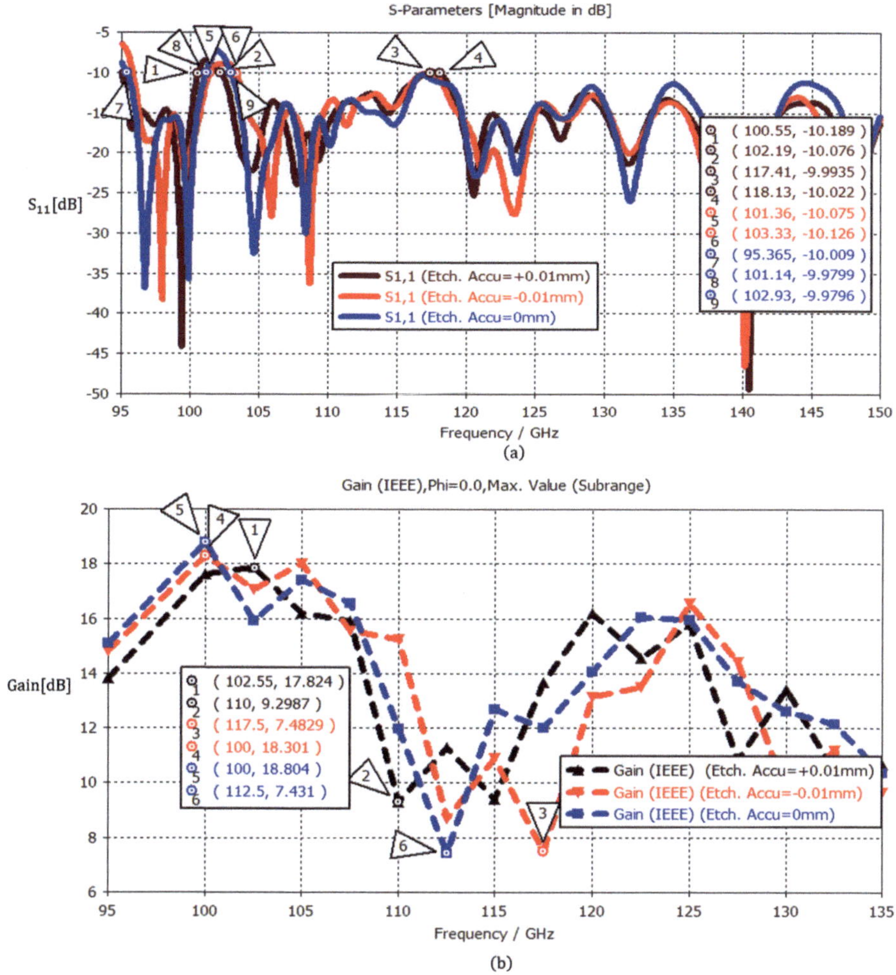

Fig. 7.46 Simulation results from the comparison of the gain and bandwidth of the first THz log-periodic microstrip array antenna simulated due to fabrication tolerance error, (**a**) Comparison of simulation results of S_{11}, (**b**) Comparison of simulation results of gain

$\tan\delta = 0.0025$ [46]. This deviation is supposed to be when this antenna is fabricated for experimental verification. These figures show that the first antenna's bandwidth and gain may change by about 3.2 GHz, 0.05–0.2 dB when this first antenna is fabricated due to the deviation of the ε_r and $\tan\delta$ of Isola Astra MT 77 microstrip substrate is frequencies-dependent.

The performance summary of these designed and simulate THz antennas presented above in this chapter is presented in Table 7.6.

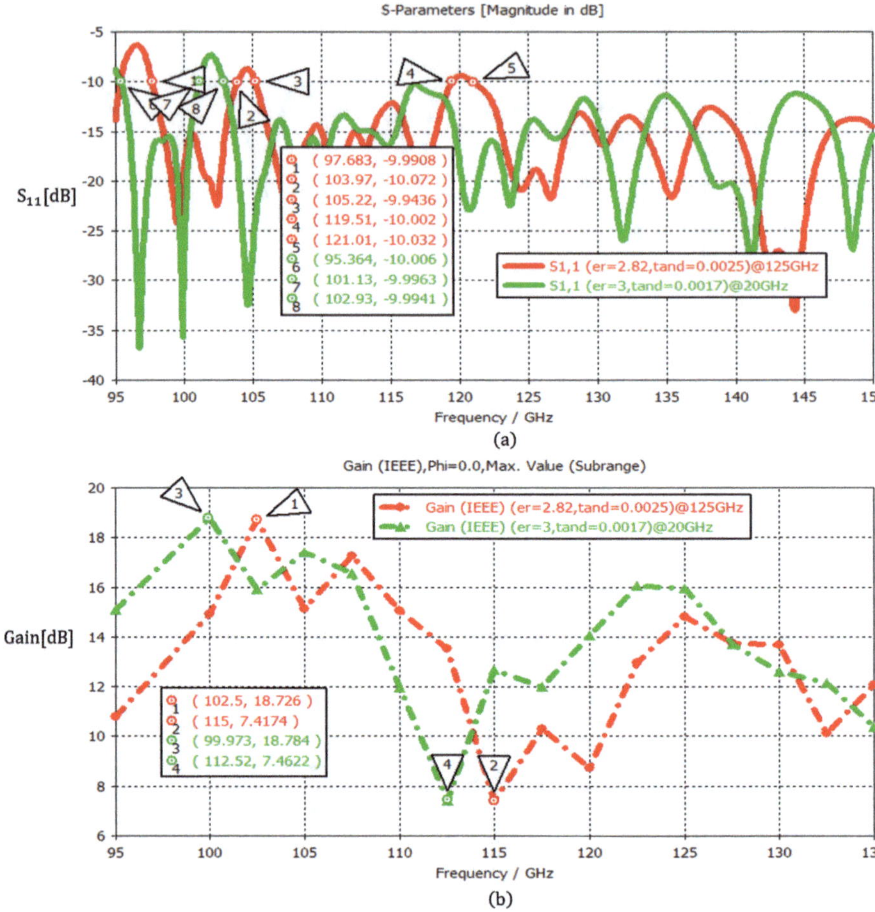

Fig. 7.47 Simulation results from the comparison of the gain and bandwidth of the first antenna due to the deviation of ε_r and tanδ of Isola Astra MT 77 microstrip substrate, (**a**) Comparison of simulation results of S_{11}, (**b**) Comparison of simulation results of gain

7.7.5 *Discussion*

The antenna is one of the main components for successfully deploying cellular communication in the 6G at THz frequencies. Therefore, it is necessary to use techniques such as microstrip array, slotting patches, parasitic patches, log-periodic, and FSSs to increase the gain and the impedance bandwidth of the Unit-cell microstrip antenna [3–7, 18–23], as described in detail in this chapter. This chapter shows the design and simulation of the THz low SLLs microstrip array antenna, THz slotting circular patch microstrip array and SIW, THz Parasitic patches microstrip array antenna, and THz log-periodic microstrip array antenna. Each

Table 7.6 The performance summary of the designed and simulate THz antenna in this chapter

THz antennas technology	Gain and bandwidth enhancement techniques	Freq. nom. (GHz)	bandwidth (GHz)	Maximum gain (dB)
Low SLLs microstrip array antenna	Hybrid-fed	118.2[a]	>7.86[a]	17.54[a]
Low SLLs microstrip array antenna with FSSs	Hybrid-fed and FSSs	123.9	>6.15	21.16
Slotting circular patch microstrip array antenna with SIW	Hybrid-fed and slotting circular patch	121.18[a]	12.86[a]	23.73[a]
Slotting circular patch microstrip array antenna with SIW and FSSs	Hybrid-fed, slotting circular patch, and FSSs	121.3	12.47	26.78
Parasitic patches microstrip array antenna	Hybrid-fed and parasitic patches	114.07[a]	18.01[a]	20.4[a]
Parasitic patches microstrip array antenna and FSSs	Hybrid-fed, parasitic patches, and FSSs	118.09	18.46	21.8
Log-periodic 24 × 16 microstrip array antenna	Hybrid-fed and log-periodic	96.66[a]	52.6[a]	18.78[a]
Log-periodic 24 × 32 microstrip array antenna	Hybrid-fed and log-periodic	137.34	54.3	19.74
Log-periodic 24 × 32 microstrip array antenna and FSSs	Hybrid-fed	128.46	53.9	21.42

[a]Simulated and compared to a different type of antenna solver/simulator

model contains at least two models, one without FSSs and the second with FSSs for gain enhancement. The performance summary is shown in Table 7.6. This table shows that the gain enhancement due to adding FSSs is up to about 2 dB.

In contrast, the primary enhancement is by the hybrid-fed array, which increases the gain of the Unit-cell microstrip radiator up to 16 dB compared to the Unit-cell microstrip antenna, while the bandwidth enhancement is about 690% (47.45 GHz) compared to the Unit-cell microstrip radiator with impedance bandwidth of 5% (6.85 GHz). In addition, the simulation results show that as the antenna's gain increases, the antenna's bandwidth decreases. Therefore, it is necessary to find a suitable compromise regarding each application's gain and impedance bandwidth. From error analysis, the proposed THz log-periodic microstrip array antenna gain and the bandwidth because of the predictable fabrication etching tolerance accuracy ±10 μm [45], the microstrip laminates frequency-pendent dielectric properties (ε_r) and the frequency-pendent losses (tanδ) may change by about 1.81 dB, 3.2 GHz, because of the antennas' high frequency working beyond 100 GHz. Furthermore, from the simulators' comparison between CST MWS and Ansys HFSS [46–50], the designed antenna gains can change up to 2.3 dB. At the same time, the impedance bandwidths can alter up to 1 GHz, so to conclude the discussion section, a good agreement was achieved despite possible changes due to future fabrication tolerance error and deviation of ε_r and tan δ of the designed THz antennas microstrip laminates.

7.8 Open Research Challenges

Based on this chapter, THz gain and bandwidth enhancement techniques for planar microstrip antennas for 6G have several open research challenges:

1. Further investigation and improvements on high gain and wideband THz microstrip array antennas.
2. Additional analysis and evolutions on THz microstrip array antennas with a gain of more than 26.8 dB and impedance bandwidth beyond 54.3 GHz.

7.9 Summary

This chapter discusses THz gain and bandwidth enhancement techniques for planar microstrip antennas. Four models were designed, and the simulation results are presented in this chapter. Each model contained at least one model without FSSs and one within FSSs for gain enhancement. The first model was a THz low SLLs microstrip array antenna for a resonance frequency of 118.2 GHz, and the achieved gain and impedance bandwidth was 17.54 dB and > 7.86 GHz. The sub-model of this design was a THz low SLLs microstrip array antenna with bandstop FSSs for a resonance frequency of 118.2 GHz, and the achieved gain and impedance bandwidth was 21.16 dB and > 6.15 GHz. The second model was a THz slotting circular patch microstrip array antenna with SIW for a resonance frequency of 121.18 GHz, and the achieved gain and impedance bandwidth was 23.73 dB and 12.86 GHz. The sub-model of this design was a THz slotting circular patch microstrip array antenna with SIW and bandstop FSSs for a resonance frequency of 121.3 GHz, and the achieved gain and impedance bandwidth was 26.78 dB and 12.47 GHz. The third model was a THz parasitic patch microstrip array antenna for a resonance frequency of 114.07 GHz, and the achieved gain and impedance bandwidth was 20.4 dB and 18.01 GHz. The sub-model of this design was a THz parasitic patch microstrip array antenna with bandstop FSSs for a resonance frequency of 118.09 GHz, and the achieved gain and impedance bandwidth was 21.8 dB and 18.46 GHz. The fourth model was a THz log-periodic 24 × 16 microstrip array antenna for a resonance frequency of 96.66 GHz, and the achieved gain and impedance bandwidth was 18.78 dB and 52.6 GHz. The first sub-model of this antenna was a THz log-periodic 24 × 32 microstrip array antenna for a resonance frequency of 137.34 GHz, and the achieved gain and impedance bandwidth was 19.74 dB and 54.3 GHz. While the second sub-model of this antenna was a THz log-periodic 24 × 32 microstrip array antenna with bandstop FSSs for a resonance frequency of 128.46 GHz, and the achieved gain and impedance bandwidth was 21.42 dB and 53.9 GHz. From error analysis, the proposed THz log-periodic microstrip array antenna gain and the bandwidth because of the predictable fabrication etching tolerance accuracy ±10 μm, the microstrip laminates frequency-pendent dielectric properties (ε_r) and the frequency-pendent losses (tanδ) may change by about 1.81 dB, 3.2 GHz, because of the antennas' high frequency working beyond 100 GHz. The simulation results of the primary models were verified with another kind of antenna simulator, and a good agreement was achieved.

References

1. E. C. Strinati, S. Barbarossa, J. L. G. Jimenez, D. Kténas, N. Cassiau, and C. Dehos, "6G: The Next Frontier. From Holographic Messaging to Artificial Intelligence Using Subterahertz and Visible Light Communication," *IEEE Vehicular Technology Magazine*, vol. 14, no. 3, pp. 42–50, Sept. 2019.
2. T. S. Rappaport, Y. Xing, O. Kanhere, S. Ju, A. Madanayake, S. Mandal, A. Alkhateeb, and G. C. Trichopulos, "Wireless communications and applications above 100 GHz: Opportunities and challenges for 6G and beyond," *IEEE Access*, vol. 7, pp. 78729–78757, May 2019.
3. D. Liu, B. Gaucher, U. Pfeiffer, and J. Grzyb, *Advanced Millimeter-Wave Technologies: Antennas, Packaging, and Circuits*. John Wiley and Sons, UK, pp. 537–648, 2009.
4. D. G. Fang, *Antenna Theory and Microstrip Antennas*. Taylor & Francis Group, USA, pp. 33–84, 2010.
5. Z. N. Chen and M. Y. W. Chia, *Broadband Planar Antennas Design and Applications*. John Wiley & Sons: England, 2006.
6. G. Kouemou, *Radar Technology*. IntechOpen, India, pp. 1–430, 2009.
7. J. Maharjan, and D. Y. Choi, "Four-element microstrip patch array antenna with corporate-series feed network for 5G communication," *International Journal of Antennas and Propagation*, vol. 2020, pp. 8760297/1–12, 2020.
8. J. D. Preez and S. Sinha, *Millimetre-Wave Antennas Configurations and Application*. Springer Signals and Communication Technology, Switzerland, pp. 1–81, 2016.
9. C. A. Balanis, *Antenna Theory Analysis, and Design*. John Wiley & Sons, 4th Edition, USA, 2016.
10. Y. Q. Guo, Y. M. Pan, and S. Y. Zheng, "Design of series-fed, single-layer, and wideband millimeter-wave microstrip arrays," *IEEE Transactions on Antennas and Propagation*, vol. 68, no. 10, pp. 7017–7026, Oct. 2020.
11. R. Chopra and G. Kumar, "Series-and corner-fed planar microstrip antenna arrays," *IEEE Transactions on Antennas and Propagation*, vol. 67, no. 9, pp. 5982–5990, Sep. 2019.
12. J. Xu, W. Hong, Z. H. Jiang, and H. Zhang, "Millimeter-wave broadband substrate integrated magneto-electric dipole arrays with corporate low-profile microstrip feeding structures," *IEEE Transactions on Antennas and Propagation*, vol. 68, no. 10, pp. 7056–7067, Oct. 2020.
13. Q. Zhu, K. B. Ng, C. H. Chan, and K. M. Luk, "Substrate-integrated-waveguide-fed array antenna covering 57–71 GHz band for 5G applications," *IEEE Transactions on Antennas and Propagation*, vol. 65, no. 12, pp. 6298–6306, Dec. 2017.
14. Y. Qiu and B. Li, "E-band microstrip patch array antenna based on hybrid feeding network," *Proceedings of IEEE International Conference on Microwave and Millimeter Wave Technology (ICMMT)*, Shanghai, China, 20–23 Sep. 2020, pp. 1–3.
15. J. Wu, Y. J. Cheng, and Y. Fan, "A wideband high-gain high-efficiency hybrid integrated plate array antenna for V-band inter-satellite links," *IEEE Transactions on Antennas and Propagation*, vol. 63, no. 4, pp. 1225–1233, Apr. 2015.
16. R. J. Mailloux, *Phased Array Antenna Handbook*. Artech House, 2nd Edition, USA, pp. 93–152, 2005.
17. D. M. Pozar, *Microwave Engineering*. John Wiley & Sons, 4th Edition, USA, pp. 328–332, 2012.

18. M. H. Prio, M. M. U. Rashid, L. C. Paul, and A. K. Sarkar, "Total efficiency comparison of different shaped microstrip patch antennas having defected ground structure," *Proceedings of IEEE International Conference on Electrical & Electronic Engineering (ICEEE)*, Rajshahi, Bangladesh, 04–06 Nov. 2015, pp. 265–268.
19. K. R. Jha and G. Singh, "Effect of low dielectric permittivity on microstrip antenna at terahertz frequency," *Optik-International Journal for Light and Electron Optics*, vol. 124, no. 22, pp. 5777–5780, Apr. 2013.
20. R. Shafique, K. Kanwar, F. Hussain, R. Morales-Menendez, M. K. Siddiqui, H. J. Arain, and M. Mehrotra, "Comparison of different feeding techniques for a patch antenna at an X frequency band to evaluate its quantitative impact on the antenna's parameters," *Journal of Applied Research and Technology*, vol. 18, no. 6, pp. 342–361, July 2021.
21. S. S. M. Khaleghi, G. Moradi, R. S. Shirazi, and A. Jafargholi, "Microstrip line impedance matching using ENZ metamaterials, design, and application," *IEEE Transactions on Antennas and Propagation*, vol. 67, no. 4, pp. 2243–2251, April 2019.
22. K. D. Xu, H. Xu, Y. Liu, J. Li, and Q. H. Liu, "Microstrip patch antennas with multiple parasitic patches and shorting vias for bandwidth enhancement," *IEEE Access*, vol. 6, pp. 11624–11633, Jan. 2018.
23. R. A. Panda and D. Mishra, "Log-periodic enhancement of a novel patch with square DGS for Ku-band applications," *Journal of King Saud University - Engineering Sciences*, pp. 1–8, March 2021.
24. G. Ahmad, M. I. Babar, M. Irfan, M. Ashraf, T. Jan, and S. W. Shah, "Bandwidth enhancement of patch antenna through various techniques for Ku-band application," *Proceedings of the Pakistan Academy of Sciences: Pakistan Academy of Sciences A. Physical and Computational Sciences*, vol. 55, no. 1, pp. 109–116, March 2018.
25. P. A. Ambresh, P. M. Hadalgi, and P. V. Hunagund, "Effect of slots on microstrip patch antenna characteristics," *Proceedings of IEEE International Conference on Computer, Communication and Electrical Technology (ICCCET)*, Tirunelveli, India, 18–19 March 2011, pp. 239–241.
26. D. M. Pozar and B. Kaufman, "Increasing the bandwidth of a microstrip antenna by proximity coupling," *Institution of Engineering and Technology Electronics Letters*, vol. 23, no. 8, pp. 368 –369, April 1987.
27. M. K. A. Rahim and P. Gardner, "Microstrip bandwidth enhancement using a log-periodic technique with inset feed," *Journal Technology of University Technology Malaysia (UTM)*, pp. 53–66, 2004.
28. M. K. A. Rahim and P. Gardner, "The design of nine elements quasi microstrip log periodic antenna," *Proceedings of IEEE RF and Microwave Conference*, Malaysia, 5–6 Oct. 2004, pp. 132–135.
29. D. Sharma and R. Kumar, "Design and analysis of five-element microstrip log-periodic antenna," *Proceedings of IEEE International Conference on Applications of Electromagnetism and Student Innovation Competition Awards (AEM2C)*, Taipei, 11–13 Aug. 2010, pp. 210–214.
30. S. Mumtaz, J. M. Jornet, J. Aulin, W. H. Gerstacker, X. Dong, and B. Ai, "Terahertz communication for vehicular networks," *IEEE Transactions Vehicular Technology*, vol. 66, no. 7, pp. 5617–5625, July 2017.
31. H. Song, and T. Nagatsuma, "Present and future of terahertz communications," *IEEE Transactions Terahertz Science Technology*, vol. 1, no. 1, pp. 256–263, Sep. 2011.
32. L. K. S. Granados, "Antennas for Millimeter-Wave Applications," Ph.D. Thesis, Electrical and Electronic Engineering, Santa María Tonantzintla University, Puebla, Mexico, pp. 1–89, 2016.
33. U. Nissanov, G. Singh, and A. Akinola, "Sixth-generation (6G) microstrip antenna with high-gain," *International Journal on Communications Antenna and Propagation*, vol. 11, no. 4, pp. 279–287, Aug. 2021.
34. U. Nissanov, "6G Rotman lens D-band beam-steering microstrip antenna," *Springer Nature Journal of Computational Electronics*, pp. 1–14, Jan. 2022.

35. U. Nissanov, G. Singh, and N. Kumar, "High gain microstrip array antenna with SIW and FSS for beyond 5G at THz band," *OPTIK: International Journal for Light and Electron Optics*, vol. 236c, pp. 166568/1–13, March 2021.
36. X. Ding, J. An, X. Bu, H. Han, J. Liu, and Z. S. He, "A 16 × 16-element slot array fed by double-layered gap waveguide distribution network at 160 GHz," *IEEE Access*, vol. 8, pp. 55372–55382, March 2020.
37. B. Aqlan, M. Himdi, H. Vettikalladi, and L. Le-Coq, "A 300-GHz low-cost high-gain fully metallic Fabry–Perot cavity antenna for 6G terahertz wireless communications," *Scientific Reports*, vol. 11, pp. 7703/1–9, April 2021.
38. D. Warmowska, K. A. Abdalmalak, L. E. G. Munoz, and Z. Raida, "High-gain, circularly-polarized THz antenna with proper modeling of structures with thin metallic walls," *IEEE Access*, vol. 8, pp. 125223–125233, July 2020.
39. H. H. Bae, T. H. Jang, H. Y. Kim, and C. S. Park, "Broadband 120-GHz L-probe differential feed dual-polarized patch antenna with soft-surface," *IEEE Transactions on Antennas and Propagation*, vol. 69, no. 10, pp. 6185–6195, Oct. 2021.
40. P. Lu, T. Haddad, B. Sievert, B. Khani, S. Makhlouf, S. Dülme, J. F. Estévez, A. Rennings, D. Erni, U. Pfeiffer, and A. Stöhr, "InP-based THz beam steering leaky-wave antenna," *IEEE Transactions on Terahertz Science and Technology*, vol. 11. no. 2, pp. 218–230, Mar. 2021.
41. J. Xiao, X. Li, Z. Qi, and Hua Zhu, "140-GHz TE_{340}-mode substrate integrated cavities-fed slot antenna array in LTCC," *IEEE Access*, vol. 7, pp. 26307–26313, Feb. 2019.
42. Z. W. Miao, Z. C. Hao, Y. Wang, B. B. Jin, J. B. Wu, and W. Hong, "A 400-GHz wideband high-gain quartz-established single-layered folded reflectarray antenna for terahertz," *IEEE Transactions on Terahertz Science and Technology*, vol. 9, no. 1, pp. 78–88, Nov. 2018.
43. Yi-Wen Wu, Zhang-Cheng Hao, Ming-Cui Tao, Xiang Wang, and Jia-Sheng Hong, "A simple and accurate method for extracting super wideband electrical properties of the printed circuit board," *IEEE Access*, vol. 7, pp. 57321–57331, April 2019.
44. D. Liu, B. Gaucher, U. Pfeiffer, and J. Grzyb, *Advanced Millimeter-Wave Technologies Antennas, Packaging, and Circuits.* John Wiley and Sons, UK, pp. 49–80, 2009.
45. J. Francey and T. Bateman, "PCB technology requirements for millimeter-wave interconnect and antenna," *The PCB Magazine*, pp. 36–42, Apr. 2017.
46. Isola Astra MT77- A Teflon Replacement, pp. 1–70, July 2017, http://www.microwavejournal.com/ext/resources/Webinars/2015a/SLIDES_Isola_21jul15.pdf.
47. U. Nissanov, G. Singh, P. Kumar, and N. Kumar, "High-gain terahertz microstrip array antenna for future generation cellular communication," *Proceedings of IEEE International Conference on Artificial Intelligence, Big Data, Computing and Data Communication (icABCD 2020)*, Durban, South Africa, 6–7 Aug. 2020, pp. 1–6.
48. U. Nissanov, G. Singh, and A. Akinola, "Low sidelobe levels (SLLs) microstrip antennas for THz bio-sensors and THz communications," *Sensors International*, vol. 2, pp. 100097/1–10, April 2021.
49. U. Nissanov and G. Singh, "MmWave/THz reconfigurable ultra-wideband (UWB) microstrip antenna," *Progress in Electromagnetic Research C*, vol. 111, pp. 207–224, April 2021.
50. U. Nissanov, G. Singh, E. Gelbart, and N. Kumar, "Highly directive microstrip array antenna with FSS for future generation cellular communication at THz band," *Wireless Personal Communications*, vol. 118, no. 1, pp. 599–617, Jan. 2021.

Chapter 8
Multi-beam and Beamforming Terahertz Array Antenna for 6G Communication

8.1 Introduction

The THz regime of the spectrum maintains a substantial pioneer development for applications, counting THz imaging and sensing, unintrusive THz medical imaging, high-data-rate THz wireless communications, and THz security screening. Nonetheless, due to THz waves' extraordinary natures and constraints, most of these applications *build* entirely on the accessibility of THz multi-beam, i.e., THz ultra-massive multiple-input-multiple-output (UM-MIMO) antennas and THz beamforming antennas, i.e., THz beam-steering array antennas. Thus, the advanced THz multi-beam UM-MIMO antenna and THz beamforming array antenna techniques produce a range of practical and unique applications [1–3].

8.2 Terahertz Beamforming Antennas

A THz beamforming/beam-steering antenna is an electronically scanned antenna array controlled by a computer through phase shifters. It generates a beam of THz waves, which may electronically steer to a spot in diverse directions with no antenna displacement, including several radiating antenna patches arranged at corporate/parallel and serial patterns. Furthermore, they are close to each other, enabling a more directional beam. The radiation pattern is attained by summarizing/superposing the radiation of each antenna element. The power from the transmitter via the source feeds the separate antennas utilizing the feed network and power splitter with the proper phase through phase shifters for the desired beam shape. Thus, the electromagnetic waves of the different antennas connect jointly to enhance the radiation in the required spot [4, 5].

U. Nissanov, G. Singh, *Antenna Technology for Terahertz Wireless Communication*,
https://doi.org/10.1007/978-3-031-35900-2_8

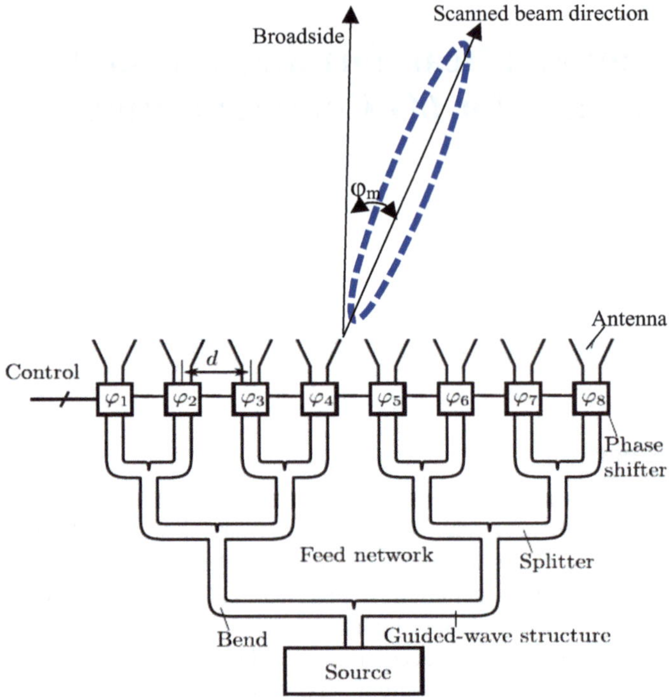

Fig. 8.1 Basic diagram of the beamforming/beam-steering phased array antenna, based on [2]

Figure 8.1 exhibits the basic diagram of the beamforming phased array antenna. The main problems of beam-steering phased array antennas are the mutual coupling between radiators, which behaves differently in neighboring elements, the substrate losses, and the feeding lines losses, which are also radiating energy. The following equation must be followed to remove the grating lobes' appearance [5]:

$$\frac{\mathrm{d}}{\lambda_0} \leq \frac{1}{1 + \sin \varphi_{\mathrm{m}}} \tag{8.1}$$

where φ_m is the maximal scan angle at (degrees), determined by the radiation pattern of the individual radiator, while d is the distance between each column of radiating antennas in (m), the scan angle φ_m more than $\pm 60°$ creates more considerable mutual coupling between radiating antenna elements, which causes "holes" at the radiating pattern [5].

8.3 Terahertz Beamforming

As discussed in the previous chapters, THz sources endure the low output power level that would inhibit them from numerous THz applications [6–8]. Therefore, one way to enhance the THz power is by beamforming. This way, the beam boost is directional and inquires to a distinct radiation direction within a consistent total radiated power. Therefore, beamforming can be inferred as a particular proceeding of dimensional power joining that multiplies the power transmitted to a specific direction by helping the directivity in a controlled approach. Furthermore, beamforming assists the base station in revealing an appropriate path to transmit data to the user and lowers interference within accessible users over the path [9].

A dimensional power combining an array of N radiators within a leveled phase will cause an N-factor boost in absolute power and an N^2-factor boost in power transmitted to a specific direction equalized to an individual radiator element event. Therefore, the development of a beam can be obtained within an ordinary dimensional combination within an N-element radiator array. Nevertheless, beamforming can control the direction of the beam by exploiting the phase delay for each radiator element of the array antenna. Beamforming permits rapid electrical beam steering by phased array rather than using a stationary component, aside from the necessity for cumbersome spinning mechanical scanners [10, 11].

There are two accepted ways of beamforming technology [12]. The first technology is a mechanical scanning method and is classical beam-steering technology in the microwave band, and it has lower performance than electronic beam-steering in precision and scanning speed. However, with the evolution of electromechanical technology, particularly the microelectromechanical system (MEMS) technology, electromechanical scanning technology now exhibits much potential for beam-steering. The second technology is an electronic beam-steering phased array with a much higher scanning speed, for electronic beam-steering technology. The main issue is flexibly modulating the phase shift of array elements and designing and fabricating phase shifters at the THz band. The leading solution for designing and fabricating phase shifters is to modulate the phase at a relatively low frequency, 90–100 GHz [13], and then convert it to the THz band. Electronic beam steering has several benefits, including high sensitivity, high directivity, multiple beamforming, fast beam scanning, and spatial filtering.

8.3.1 Phased-Array Beamforming

Equalized within the mechanical scanning technique, nonmechanical scanning techniques such as the phased-array technique hold a higher scanning speed. Therefore, the critical enigma is flexibly modulating the array elements' phase shifts for beam steering phased-array-based technology. Even though engineers have experienced significant challenges in designing THz phase shifters, there are many more ways to achieve them, for instance, shifting the phase before altering the EM wave to the THz, by taking the superiority of optical technology, and using programmable metamaterials (MtMs) [12].

8.3.2 Microelectromechanical System Beamforming

The microelectromechanical systems (MEMS) beamforming based on MEMS switch technology supplies a significantly smaller size, improved isolation, and reduced insertion loss than electromechanical switches. MEMS switches are usually a silicon-based integrated circuit technology within movable mechanical components that utilize etching silicon dioxide stratum. MEMS switches contain a signal line with output and input ports between the first and second ground layers and a beam to control the implementation. These switches can be used at frequencies beyond 100 GHz for THz 6G wireless communication beamforming networks [14].

8.4 Terahertz Beamforming Phase Shifting Technologies

Three sections can split up the THz beamformer phase shifting. Analog beamformers supply a tuneable continuous phase through a maximum and minimum phase shift angle using a delay line such as the Rotman Lens [15] or the Butler matrix [16]. However, digital beamformers require different control voltages to straighten the control circuit, which causes less precise beam-steering, pendent on the number of phase bits [17]. Therefore, the third THz beamformers technology combines analog and digital beamformers. Under the components that exist in contemporary technology, the central problem of the beamformers at the THz band is the limited phase angle range up to about ±(0–40°) [18].

8.4.1 Analog Beamforming Phase Shifting

Analog beamformer phase shifting technologies contain a matrix of constant phase shifters. Theoretically, we can shape a beam in a particular direction by controlling the phase of every transmitted signal from each radiator antenna. Analog beamformers typically consist of radio-frequency (RF) switching matrices and, subsequently, the stock of phase shifters to enable the beam steering and link the needed beams to a particular signal path. Part of beamformers can constantly scan a transmitted beam in a given zone. Furthermore, analog beamformers can set zeros over the direction of interferences to nullify inadvisable user interference in real time [19]. This approach is presented in Fig. 8.2.

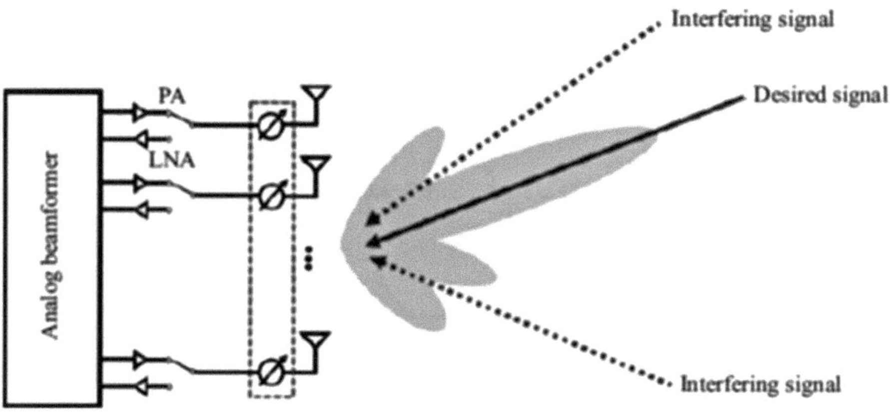

Fig. 8.2 Basic diagram of THz analog beamforming [19]

(a) *Terahertz Rotman Lens Beamforming*

Highly directional transmission is necessary to mitigate the THz band's high atmospheric absorption and path loss. Thus, tunable antenna arrays with efficacious beamformer network (BFN) and fair beam-steering abilities are used to merge signals of radiators into a pattern, which is further directive than the individual radiator pattern. A BFN generally comprises output and input ports organized along an arc. All input ports at the focal layer supply the desired phase and amplitude distribution so that the antenna beam is directed at a particular angle [20, 21]. Rotman lenses are BFN and sustain low-phase error, broadband, and broad-angle beam-steering. In addition, they are the true-time-delay (TTD) apparatus supplying non-dependent frequency beam-steering. The Rotam lens designer (RLD) software is founded on Geometrical Optics with the Rotman lens equations [22, 23] and analyzes the Rotman Lens with a rapid design tool. The analysis with the RLD software is based on the following assumptions: transmission line, material dispersion, and parasitic coupling are of a lesser degree. Radiative leaks are not accounted for, and the dummy load is ideal without losses. The following Fig. 8.3 shows the Rotman lens configuration.

(b) *Terahertz Rotman Lens Beamforming Design Equations*

The transmission line lengths and lens inner contour points are solved for using the path length comparison method as in [22, 23]:

$$\overrightarrow{F_1P} + W + N \sin \alpha = F + W_0 \tag{8.2}$$

$$\overrightarrow{F_2P} + W - N \sin \alpha = F + W_0 \tag{8.3}$$

$$\overrightarrow{F_0P} + W = G + W_0 \tag{8.4}$$

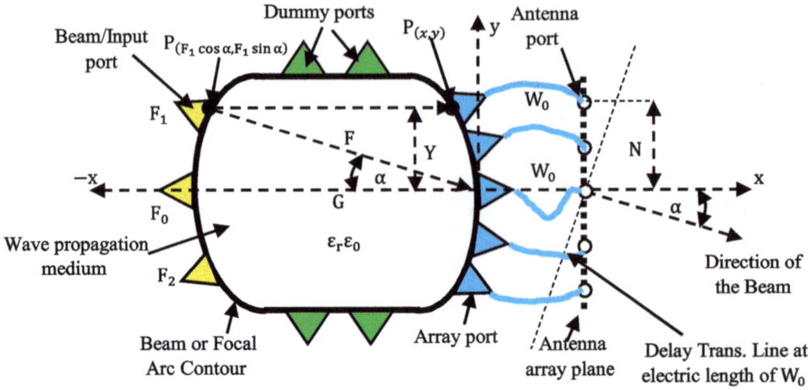

Fig. 8.3 THz Rotman lens beamforming configuration

$$\overrightarrow{(F_1P)}^2 = (F\ \cos\ \alpha + X)^2 + (F\ \sin\ \alpha - Y)^2 \tag{8.5}$$

$$\overrightarrow{(F_2P)}^2 = (F\ \cos\ \alpha + X)^2 + (F\ \sin\ \alpha + Y)^2 \tag{8.6}$$

$$\overrightarrow{(F_0P)}^2 = (G + X)^2 + (Y)^2 \tag{8.7}$$

Lens dimensions are then normalized by the off-axis focal length [22]:

$$\eta = N/F \tag{8.8}$$

$$g = G/F \tag{8.9}$$

$$x = X/F \tag{8.10}$$

$$y = Y/F \tag{8.11}$$

$$w = \frac{W - W_0}{F} \tag{8.12}$$

For x, y, w, we can get:

$$y = \eta(1 - w) \tag{8.13}$$

$$x^2 + y^2 + 2a_0x = w^2 + {b_0}^2\eta^2 - 2w \tag{8.14}$$

$$aw^2 + bw + c = 0 \tag{8.15}$$

where:

$$a = 1 - \eta^2 - \left(\frac{g - 1}{g - a_0}\right) \tag{8.16}$$

$$b = \left[2g\left(\frac{g-1}{g-a_0}\right) - \left(\frac{g-1}{(g-a_0)^2}\right){b_0}^2\eta^2 + 2\eta^2\right] - 2g \tag{8.17}$$

$$c = \frac{g{b_0}^2\eta^2}{g-a_0} - \frac{{b_0}^4\eta^4}{4(g-a_0)^2} - \eta^2 \tag{8.18}$$

where η is the element gapping – the gapping of the radiators along with the outer profile, g is the Focal Ratio-the ratio of on-axis focal distance to off-axis focal distance, lens width (G or F_0) are the length between the middle of the focal profile and the middle of the array profile, scan angle (α) is the direction of the main beam, where phase error is null. The RLD software solves these points each time F, η, α, g are modified.

When the lens profile is well thought out, the transmission line electrical length for standard microstrip design equations is used to find the line width of the transmission lines [24]:

$$\frac{W}{d} = \left\{\begin{array}{ll} \dfrac{8e^A}{e^{2A}-2}; & \text{for } \dfrac{W}{d} < 2 \\ \dfrac{2}{\pi}\left[B - 1 - \ln(2B-1) + \dfrac{\varepsilon_r - 1}{2\varepsilon_r}\left\{\ln(B-1) + 0.39 - \dfrac{0.61}{\varepsilon_r}\right\}\right]; & \text{for } \dfrac{W}{d} > 2 \end{array}\right\} \tag{8.19}$$

where:

$$A = \frac{Z_0}{60}\sqrt{\frac{\varepsilon_r - 1}{\varepsilon_r + 1}} + \frac{\varepsilon_r - 1}{\varepsilon_r + 1}\left(0.23 + \frac{0.11}{\varepsilon_r}\right) \tag{8.20}$$

$$B = \frac{377\pi}{Z_0\sqrt{\varepsilon_r}} \tag{8.21}$$

The ratio, which is used to calculate the effective dielectric constant (ε_{eff}) [24]:

$$\varepsilon_{eff} = \left(\frac{\varepsilon_r + 1}{2}\right) + \left(\frac{\varepsilon_r - 1}{2}\right)\frac{1}{\sqrt{1 + 12d/W}} \tag{8.22}$$

The transmission lines port impedance can be calculated by using [24]:

$$Z = \left\{\begin{array}{ll} \dfrac{60}{\sqrt{\varepsilon_{eff}}}\ln\left(\dfrac{8d}{W} + \dfrac{W}{4d}\right); & \text{for } \dfrac{W}{d} \le 1 \\ \dfrac{120\pi}{\sqrt{\varepsilon_{eff}}\left[\dfrac{W}{d} + 1.393 + 0.667\ln\left(\dfrac{W}{d} + 1.444\right)\right]}; & \text{for } \dfrac{W}{d} \ge 1 \end{array}\right\} \tag{8.23}$$

8.4.2 Digital Beamforming Phase Shifting

In a digital beamformer phase shifting, an array of transmitted beam radiation patterns is produced by processing the transmitted beam in the wanted direction and canceling the transmitted beam's radiation of interfering directions, all through the digital realm. This process is usually followed by finite impulse response filters [25], which can adaptively manage the weights of each antenna radiator. Theoretically, ideal beamforming can be achieved by optimizing the weights within ideal zeros arranged in the direction of the interfering signal. There are certain superiorities of digital beamforming phase shifting in large antenna array systems. These contain high-resolution direction of arrival (DOA) estimation, high system security, high spectral efficiency, and enhanced energy effectiveness [26]. In addition, beamforming phase shifting improves signal quality at the RX end. Meanwhile, this quality is coupled with high energy effectiveness, and the digital beamforming phase shifting technique can use the regularities of the UM-MIMO scheme to increase spectral effectiveness [27]. This approach is presented in Fig. 8.4.

8.4.3 Analog and Digital Beamforming Phase Shifting

Analog beamforming is simple to fulfill by applying improvements in a phased array scheme; nonetheless, interference elimination is not as well as within a digital domain beamformer. However, connecting each antenna radiator to a digital realm

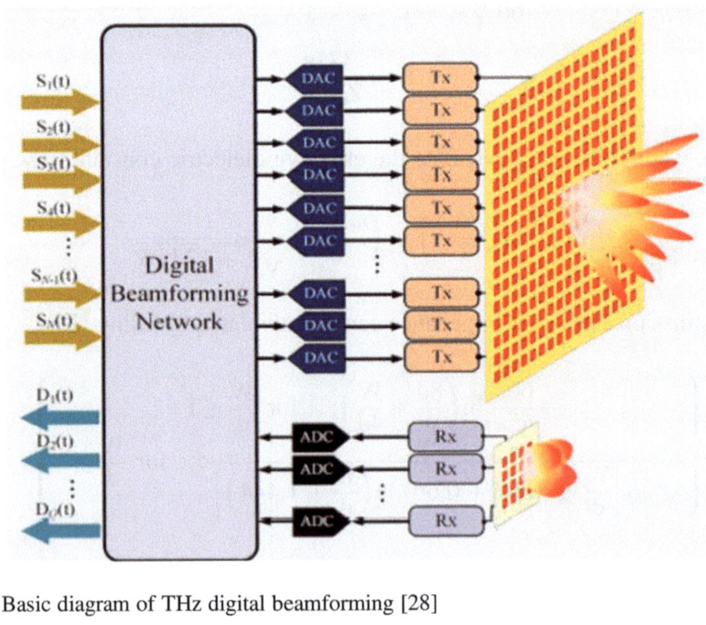

Fig. 8.4 Basic diagram of THz digital beamforming [28]

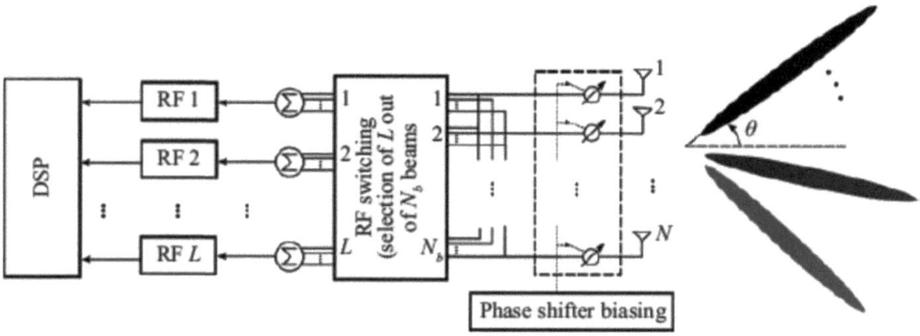

Fig. 8.5 Basic diagram of THz digital and analog (hybrid) beamforming [19]

needs complicated hardware setups. The analog and digital beamformer phase shifting profit the advantages of both analog and digital beamforming. Such a beamformer is sorted as a hybrid or analog/digital beamformer. The gain of this scheme is that digital beamforming techniques set up adequate digital signals to permit good-quality DOA estimation.

In contrast, the analog beamforming scheme facilitates fewer RF networks needed to beamform. This scheme ultimately forms it possible, in fundamental, to design commercial hardware for the comprehensive system. The analog components of the beamformer consist of DAC conversion and PA power boosting on transmit and LNA unit on receive [29, 30]. This approach is presented in Fig. 8.5.

8.5 Terahertz Multi-beam Ultra-Massive Multiple-Input, Multiple-Output

The THz multi-beam UM-MIMO is studied as an efficient remedy to conquer the capacity constraint in a 6G wireless communication network. Moreover, the unification of THz with the UM-MIMO technology can straighten things by supplying feasible paths to mitigate the ever-increasing capacity requirements. UM-MIMO is the state of the art of present MIMO system used in current 5G wireless networks that simultaneously bands hundreds and up to thousands of antennas at the base station and handles tens of users simultaneously. At the same time, the fundamental concept astern them is to merge the signals from each antenna to boost the system's total operation gain [31]. Therefore, to boost capacity and lighten the distance dilemma at THz-band 6G wireless communication link accomplishments the multipath signal spread and boosts the spectral effectual of a specified accumulated transmit power for the 6G wireless communication networks using multi-beam UM-MIMO antennas.

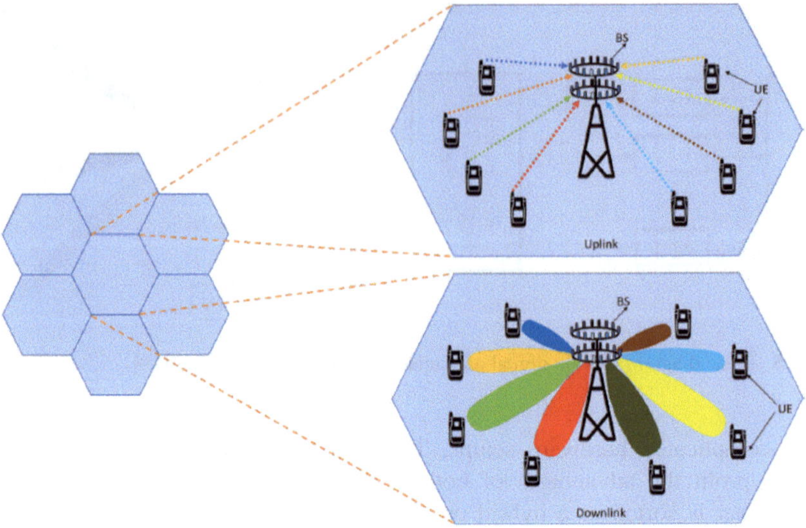

Fig. 8.6 Basic approach of THz multi-beam UM-MIMO uplink and downlink [34]

Furthermore, the benefits of the THz multi-beam UM-MIMO antennas can be expanded by increasing the antenna's directivity by using an efficacious diversity antenna or transmitting big approximate data flows to boost data rates and spectral efficacious [32, 33]. Therefore, multi-beam UM-MIMO is a crucial technology for future 6G systems, which would use vast antenna arrays to transmit data to multiple devices within a small cell in the same frequency band. This approach is presented in Fig. 8.6.

8.6 THz Multiple-Input, Multiple-Output Antenna Parameters

In order to accomplish a MIMO system, it is necessary to design an effective MIMO antenna. The most significant parameter for designing an effective MIMO antenna is the envelope correlation coefficient (ECC). It delivers data regarding the correlation within several antenna ports. For the case of the MIMO antenna, the value of ECC (ρ) must be pretty low ($\rho < 0.2$) when a higher value diminishes the spectral efficiency of any MIMO antenna. Two ways can evaluate ECC: (i) by applying S-parameters and (ii) by applying far-field parameters. The equation for calculating the ECC by applying S-parameters is as follows [35]:

$$\mathrm{ECC} = \rho_{ij} = \frac{\left|S_{ii}{}^{*}S_{ij} + S_{ji}{}^{*}S_{jj}\right|^2}{\left(1 - |S_{ii}|^2 - |S_{ji}|^2\right)\left(1 - |S_{jj}|^2 - |S_{ij}|^2\right)} \tag{8.24}$$

In the above equation, "i" and "j" denote the MIMO antennas ports, where i = 1 and j = 2, 3, and 4. S_{ii}^* and S_{ji}^* represent the conjugate of S_{ii} and S_{ji}. Another way to characterize and calculate the performance of the MIMO antenna system is the diversity gain (DG), which characterizes the losses that develop in the transmitted power. The DG of the antenna measures the effect of diversity in the communication channel. It is specified as the ratio of the signal-to-noise (SNR) of a multiport MIMO antenna to the SNR of a single-port antenna. The following formula is used to attain the DG of any MIMO antenna [35]:

$$DG(dB) = 10\sqrt{1 - (ECC)^2} \quad (8.25)$$

So, the DG and ECC are correspondents to each other. Therefore, a higher DG can be attained by decreasing the ECC values. In feasible applications, the value of DG =10 dB is sufficient for improved MIMO performance.

8.7 Problem Formulation

The shortened wavelength of THz signals facilitates the integration of an ultra-massive number of antenna radiators within transceivers in a tiny trace to shape multi-beam UM-MIMO antennas [36]. Furthermore, such antenna arrays are designed carefully to deliver highly directional gains, e.g., 30–50 dB [37]; thus, they can compensate for the severe absorption losses and spreading losses experienced at the THz band. Furthermore, multi-beam architectures can be an advantage for massive user multiplexing in a short distance. To mitigate the high spreading losses and severe absorption losses, a THz high gains antenna arrays and efficient beamforming engineering that attain high energy efficiency (EE) and spectrum efficiency (SE) are needed. Among variant beamforming schemes, hybrid beamforming schemes (HBSs) are predicted for THz communication networks because of their reduced hardware intricacy and ability to simultaneously achieve high EE and SE [38, 39].

8.8 Related Work

Sengupta et al. [40] demonstrated a THz beam steering 4 × 4 radiating array within distributed active radiator (DAR) microstrip antenna at 280 GHz. This antenna was fabricated with a 45 nm CMOS process. The BW, directivity, steering angles, and ERIP measurement results were 20 GHz (270–290 GHz), 16 dBi, ±40°, and 9.4 dBm at 281 GHz, respectively. Buadana et al. [41] demonstrated a THz digital control phase shifting beam steering within a 2 × 2 dielectric resonator antenna (DRA) array at 280 GHz. This antenna was fabricated with a 65 nm CMOS process.

The BW, steering angles, and ERIP measurement results were 16 GHz (272–288 GHz), ±15°, and 21 dBm at 280 GHz, respectively. Finally, Tang et al. [42] demonstrated a THz phased array membrane dipole antenna at 140 GHz. This antenna was fabricated with a 65 nm CMOS process. The steering angles and ERIP measurement results were +15°, −25 °, and 20.8 dBm at 140 GHz, respectively.

In comparison, Liu et al. [43] demonstrated a THz beam steering-based MEMS technology-based reflection grating, including numerous subwavelength reflectors within the COMSOL simulator. The BW and steering angles simulation results were 700 GHz (300–1000 GHz) and ±56.4°, while Ali et al. [44] demonstrated a THz reconfigurable graphene-based dual port two-band MIMO patch antenna within the CST MWS simulator for a tunable frequency range of 1450 GHz up to 2170 GHz. The resonance frequencies, isolation, and maximum directivity simulation results were 1450–2170 GHz, more than 30 dB, and 4.56 dBi, respectively. Das et al. [45] demonstrated a THz planar super wideband (SWB) includes two coplanar waveguides (CPW) fed half elliptical fourth-port MIMO fractal patch antenna within the CST MWS simulator for a frequency range of 800 GHz up to 10,000 GHz. The BW, isolation, and maximum directivity simulation results were 10.2 THz, more than 20 dB, and 8.6 dBi, respectively. Mohanty et al. [46] designed a fourth-port THz MIMO planar microstrip antenna for an operating frequency of 2380–11,180 GHz. The simulated peak gain, impedance BW, peak isolation, ECC, DG, and channel capacity loss (CCL) were >8 dB, 8800 GHz, 22 dB, <0.18, >9.4 dB, and <0.3 bits/s/Hz, respectively. This THz MIMO antenna also can be used in future 6G cellular communication. Finally, Vamsi et al. [47] designed a graphene-based THz 4-port MIMO circular polarization antenna for the 4670–4870 GHz frequency range. The isolation and impedance BW simulation results were >35 dB and 200 GHz, respectively. The performance summary of these THz antennas discussed above is presented in Table 8.1 according to the related work.

8.9 Simulation Results and Discussion

8.9.1 THz Beamforming Proximity Coupled Feed Microstrip Array Antenna

Several THz beamforming proximity-coupled feed microstrip array antennas were designed at a resonance frequency of 146.5 GHz with hybrid fed, proximity-coupled feed, and an E-shaped radiator for gain and BW enhancement. The first was with a beamforming microstrip antenna with 30 × 16 E-shaped radiators. The second was with a beamforming microstrip antenna with 30 × 16 E-shaped radiators, 20 bandstop FSSs, and 20 bandpasses FSSs for gain enhancement. Moreover, the first beamforming microstrip antenna was made with two substrates of Isola Astra® MT77 microstrip substrates, which have substrate height (h), copper thickness (t), dielectric constant (ε_r), and loss tangent (tanδ): 0.127 mm, 17.5 μm, (2.82, 0.001) at 125 GHz, respectively. So that the simulation results would be

Table 8.1 THz antennas classification and performance according to related work in this chapter

References	THz antenna	Freq. nom. (GHz)	BW (GHz)	Steering angles (°)	Maximum directivity (dBi)	Isolation (dB)
[40]	4 × 4 beam steering with DAR	280[a]	20 (270–290)[a]	±40[a]	16[a]	n/a
[41]	2 × 2 digital control beam steering DRA	280[a]	16 (272–288)[a]	±15[a]	n/a	n/a
[42]	Phased array membrane dipole	140[a]	n/a	15[a], –25[a]	n/a	n/a
[43]	Beam steering-based MEMS reflectors	650	700 (300–1000)	±56.4	n/a	n/a
[44]	2-port Reconfigurable graphene-based patch MIMO	(1450–2170)	n/a	n/a	n/a	>30
[45]	4-port SWB CPW fractal patch MIMO	5100	10,200 (800–10,000)	n/a	8.6[a]	>20
[46]	4-port THz MIMO planar microstrip	2380–11,180	8800	n/a	n/a	≤22
[47]	4-port THz graphene MIMO	4670–4870	200	n/a	n/a	>35

[a]Measured results

linked to the simulation for the frequency range of 132–160 GHz, the ε_r and tanδ was measured and applied at the simulators upon the measurements of this microstrip substrate that the manufacturer did at Fraunhofer Institute of Reliability and micro-integration IZM Berlin for the highest frequency of 125 GHz. This microstrip laminate was chosen because this substrate is used up to 170 GHz. So, with these laminate parameters that were (h $=$ 127μm $\leq$ 220 μm), ($t = 17.5\ \mu m \leq 220\ \mu m, \varepsilon_r = 2.82 < 4, \tan\delta = 0.001 < 0.01$), the conduction, substrate, and surface wave losses were made smaller [48–50]. The magnitudes of the substrate of the first antenna were 37 × 28 mm^2, which equals $17.57\lambda_0 \times 13.3\lambda_0$ @ 146.5 GHz. This antenna also included a truncated ground plane at magnitudes of 11.5 × 28 mm^2. There were 16 independent input ports with the same uniform amplitude when each port was set up with a different angle from its predecessor to steer the radiating beam and the CST MWS simulator setup to add whole the sources jointly (simultaneously). Figures 8.7a–d illustrate the first proposed THz beamforming proximity coupled feed microstrip array antenna.

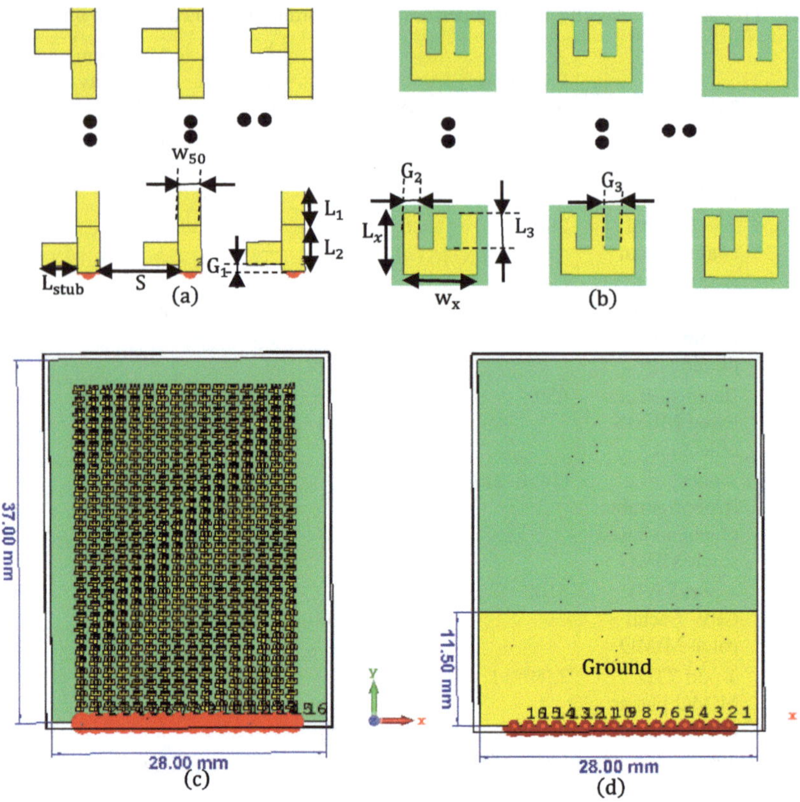

Fig. 8.7 Structure of the designed first proposed THz beamforming proximity coupled feed antenna, (**a**) Feed lines at feed layer, (**b**) Radiation elements at radiation layer, (**c**) Frond section, and (**d**) Back section

The second antenna was designed and simulated at four PCBs and included 30 × 16 E-shaped radiators at proximity coupled feed, FSS superstrate with 20 UCell bandpass FSSs, and 20 UCell bandstop FSSs at lower FSS substrate, presented in Figs. 8.8a–d. All PCBs are established on Isola Astra MT77 substrates with the same substrate parameters as the first THz beamforming microstrip antenna with an air gap of 1180 μm between substrates. This antenna also included a truncated ground plane at magnitudes of 11.5 × 28 mm^2. All magnitudes of this antenna, including the numbers of the UCell bandstop and bandpass FSSs, and the air gap distance (h_1) were optimized to get this antenna's best gain and BW with reasonable time simulation. There were 16 independent input ports with the same uniform amplitude when each port was set up with a different angle from its predecessor to steer the radiating beam and the CST MWS simulator setup to add whole the sources jointly (simultaneously).

The simulated result of the S_{11} for the offered first THz beamforming microstrip antenna is shown in Fig. 8.9. From this figure, it has been shown that the BW

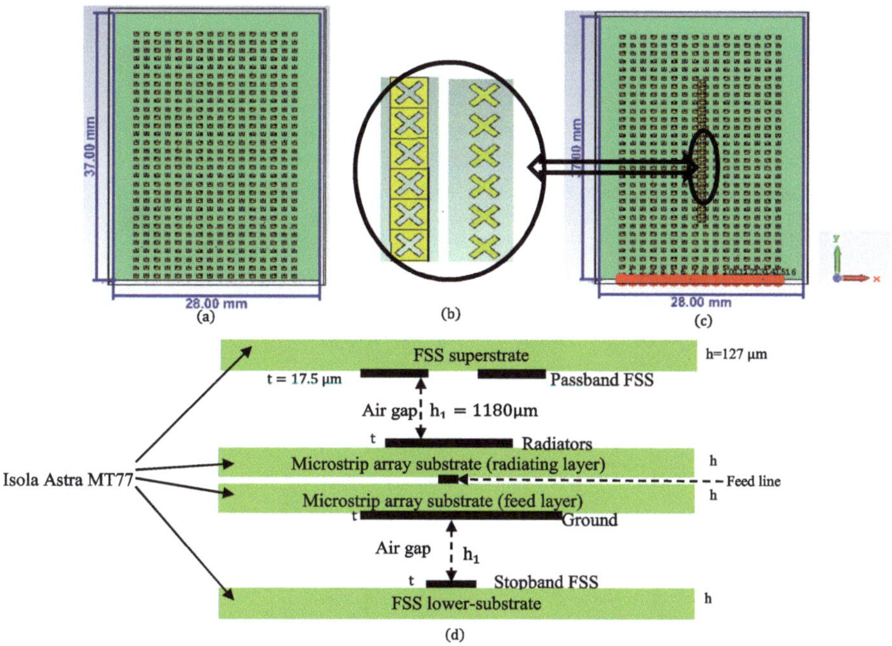

Fig. 8.8 Configuration of the designed second beamforming microstrip antenna with FSSs, (**a**) Front section, (**b**) Part of Unit-cell bandpass and bandstop FSSs at FSS superstrate and lower FSS substrate, (**c**) Complete second beamforming microstrip antenna, and (**d**) Side view of the complete second beamforming microstrip antenna with FSS

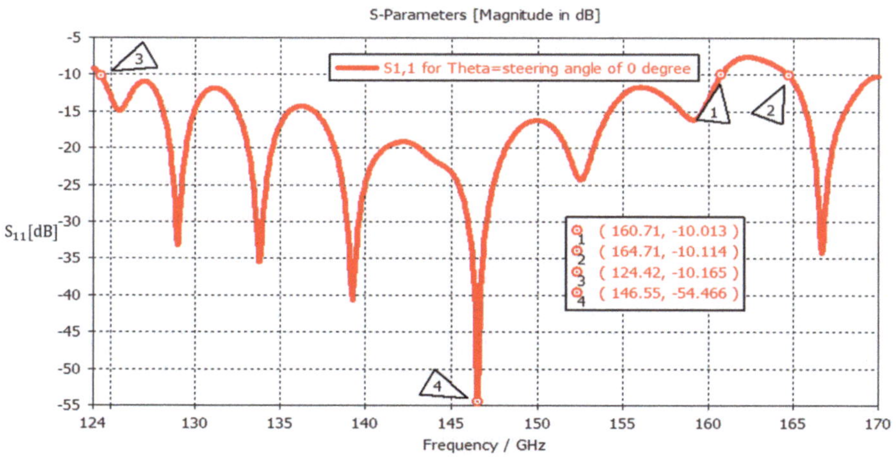

Fig. 8.9 Simulation results of the S_{11} for the first THz beamforming microstrip array antenna

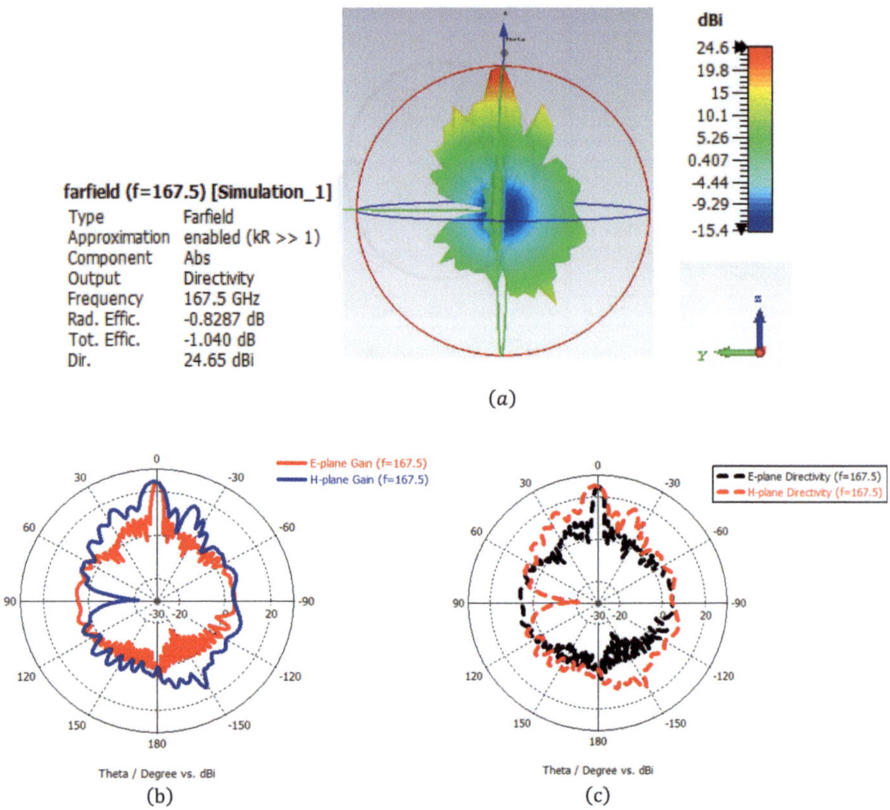

Fig. 8.10 Simulation results of directivity, E-plane, and H-plane of the first THz beamforming microstrip array antenna, (**a**) 3D far-field radiation pattern, (**b**) Gain, and (**c**) Directivity

($S_{11} \leq -10$ dB) > 41.8 GHz (>29.3%), moreover, the frequency of the resonance of this offered first antenna is 146.55 GHz.

Figures 8.10a–c show the simulation results of the 3D far-field radiation pattern, gain, and directivity at the E-plane and H-plane of the designed first THz beamforming microstrip array antenna at a frequency of 167.5 GHz at the polar plots, respectively. From Fig. 8.10a, it has been shown that the directivity is 24.6 dBi, and the total efficiency is −1.04 dB for 167.5 GHz. From Fig. 8.10b, it has been shown that the gain is 23.8 dB in the E-plane with the direction of the main lobe at 0°, while the –3 dB angular width is 4.4° and the side lobe level (SLL) is −15.1 dB. The gain is 24.2 dB in the H-plane with the direction of the main lobe at 2°, while the –3 dB angular width is 9.4°, and the SLL is −8 dB. From Fig. 8.10c, it has been shown that the directivity is 24.6 dBi in the E-plane with the direction of the main lobe at 0°, while the –3 dB angular width is 4.4° and the SLL is −15.1 dB. The directivity is 25 dBi in H-plane with the central lobe direction at 2°, while the –3 dB angular width is 9.4°, and the SLL is −8 dB.

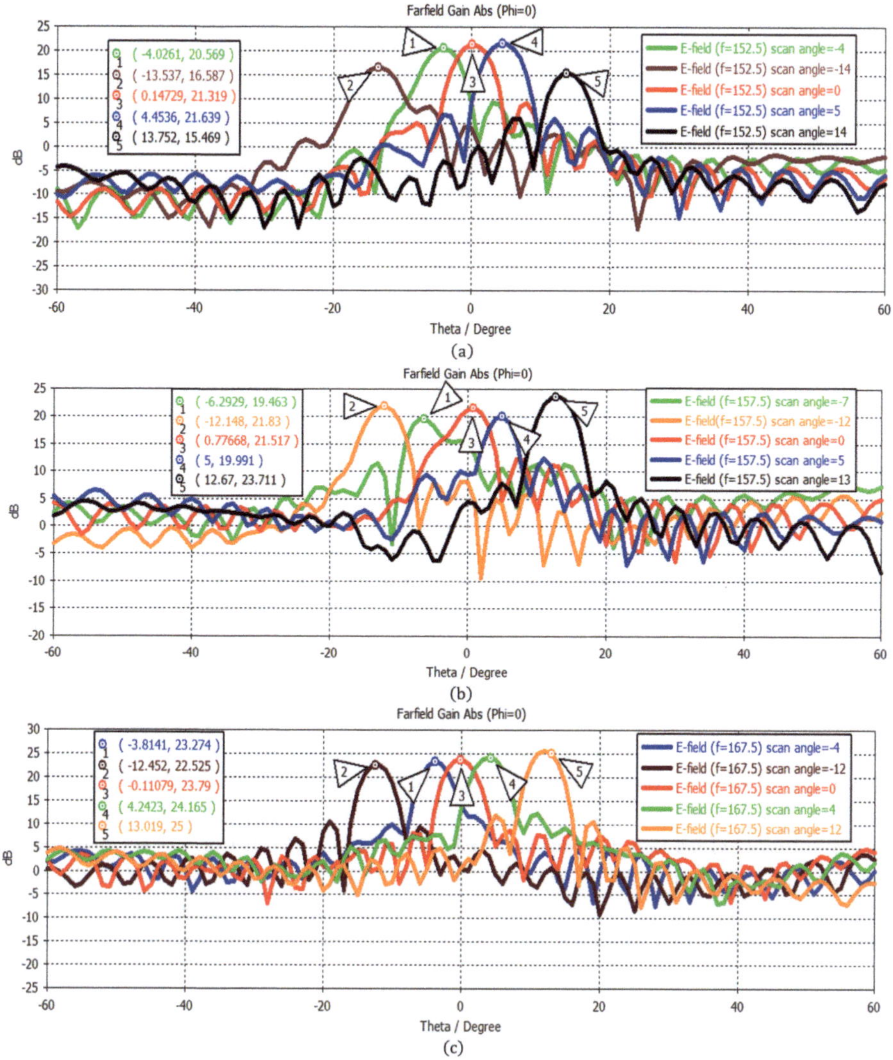

Fig. 8.11 Simulation results of the E-field from the first THz beamforming microstrip array antenna, (**a**) For 152.5 GHz, (**b**) For 157.5 GHz, and (**c**) For 167.5 GHz

The simulation results of the E-field depending on different steering angles at frequencies of 152.5, 157.5, and 167.5 GHz of the offered first THz beamforming microstrip antenna at 2D Cartesian plot results are shown in Fig. 8.11a–c. This figure exhibits that the E-field is between 15.47 and 25 dB, and the steering angles are between $-13.5°$ and $13.7°$.

The simulated result of the S_{11} for the offered second THz beamforming microstrip array antenna is shown in Fig. 8.12. From this figure it has been shown that the BW ($S_{11} \leq -10$dB) > 33.3 GHz(>22.79%), moreover, the frequency of the resonance of this offered second THz beamforming microstrip array antenna is 136.45 GHz.

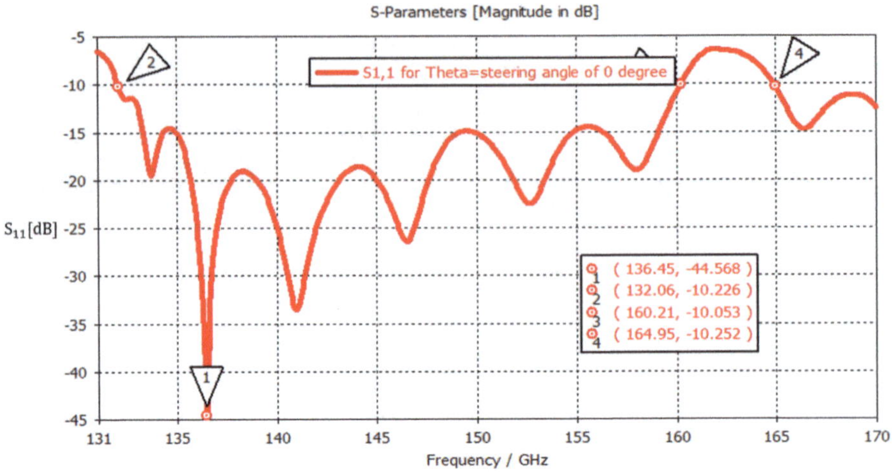

Fig. 8.12 Simulation results of the S_{11} for the second THz beamforming microstrip array antenna

Figures 8.13a–c show the simulation results of the 3D far-field radiation pattern, gain, and directivity at the E-plane and H-plane of the designed first THz beamforming microstrip array antenna at a frequency of 165 GHz at the polar plots, respectively. From Fig. 8.13a it has been shown that the directivity is 26.8 dBi, and the total efficiency is −1.24 dB for 165 GHz. From Fig. 8.13b, it has been shown that the gain is 26 dB in the E-plane with the direction of the main lobe at 0°, while the –3 dB angular width is 4.1° and the side lobe level (SLL) is −13.4 dB. The gain is 26.4 dB in the H-plane with the direction of the main lobe at 2°, while the –3 dB angular width is 8.1°, and the SLL is −7.4 dB. From Fig. 8.13c, it has been shown that the directivity is 26.8 dBi in the E-plane with the direction of the main lobe at 0°, while the –3 dB angular width is 4.1° and the SLL is −13.4 dB. The directivity is 27.1 dBi in the H-plane with the direction of the main lobe at 2°, while the –3 dB angular width is 8.1° and the SLL is −7.4 dB.

The simulation results of the E-field depending on different steering angles at frequencies of 152.5, 157.5, and 167.5 GHz of the offered second THz beamforming microstrip array antenna at 2D Cartesian plot results are shown in Fig. 8.14a–c. This figure exhibits that the E-field is 17.07–25.66 dB, and the steering angles are between −13.5° and 13.0°.

Figures 8.15a, b show the simulation results from the equalization of S_{11} and the gain of the first and second antennae. From Fig. 8.15a, it has been shown that 8.5 GHz reduces the antenna's BW by inserting the FSSs. While from Fig. 8.15b, it has been shown that the gain with FSSs was more significant by over than 4 dB at 157–165 GHz vs. not including FSSs, and the peak change was 10.3 dB at 160 GHz.

Figures. 8.16a, b show the simulation results of the S_{11} and the gain for the first antenna with or not including etching accuracy ±20 μm [51]. This deviation is supposed to be when this antenna is manufactured. These figures show that the first antenna's BW and gain may change by about 6.31 GHz, 1.5–10.43 dB when this first

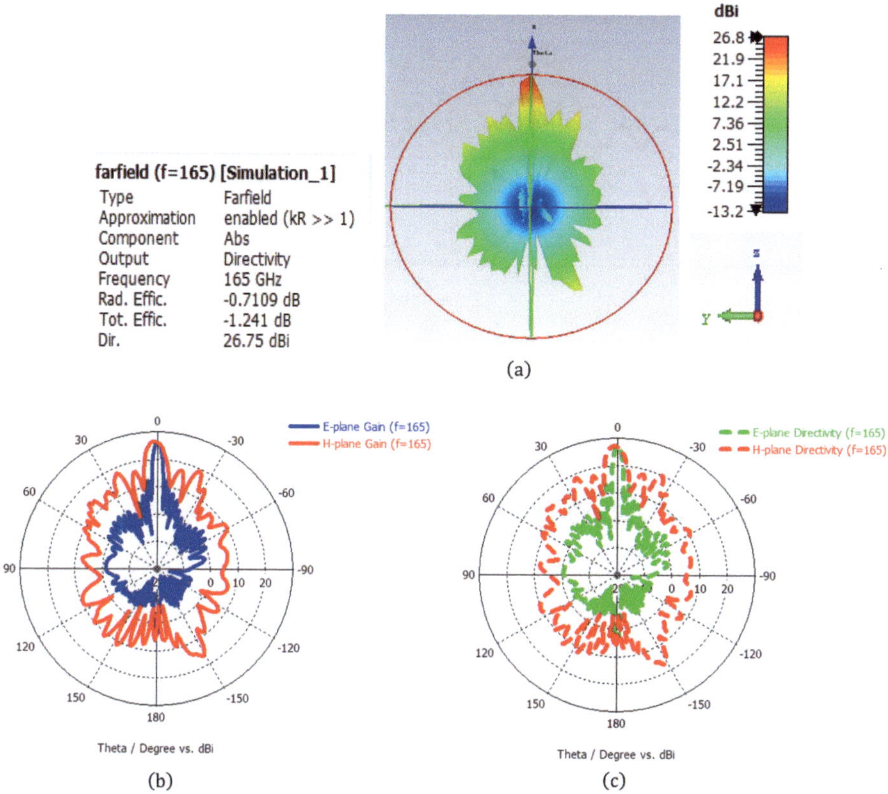

Fig. 8.13 Simulation results of directivity, E-plane, and H-plane of the second THz beamforming microstrip array antenna, (**a**) 3D far-field radiation pattern, (**b**) Gain, and (**c**) Directivity

antenna is fabricated. Furthermore, from these simulation results, it may have been shown that the first antenna design was not optimized well because of the fluctuations in the gain and at S_{11} when this first antenna is fabricated.

Figures 8.17a, b show the simulation results of the S_{11} and the gain for the first proposed THz beamforming microstrip array antenna with a deviation of $\varepsilon_r = 2.82 \pm 0.02$. This parameter is frequency-dependent and may change at the THz frequencies. These figures show that the first antenna's BW and gain may change by about 0.7 GHz, 6.5 dB when this first antenna is measured for experimental verification, and due to that ε_r is frequency-dependent and may change at the THz frequencies.

Figures 8.18a, b show the simulation results of the S_{11} and the gain for the first proposed THz beamforming microstrip array antenna with a deviation of tanδ. This parameter is frequency-dependent and may change at the THz frequencies. These figures show that the first antenna's BW and gain may change by about 0.6 GHz, 6.2 dB when this first antenna is measured for experimental verification, and due to

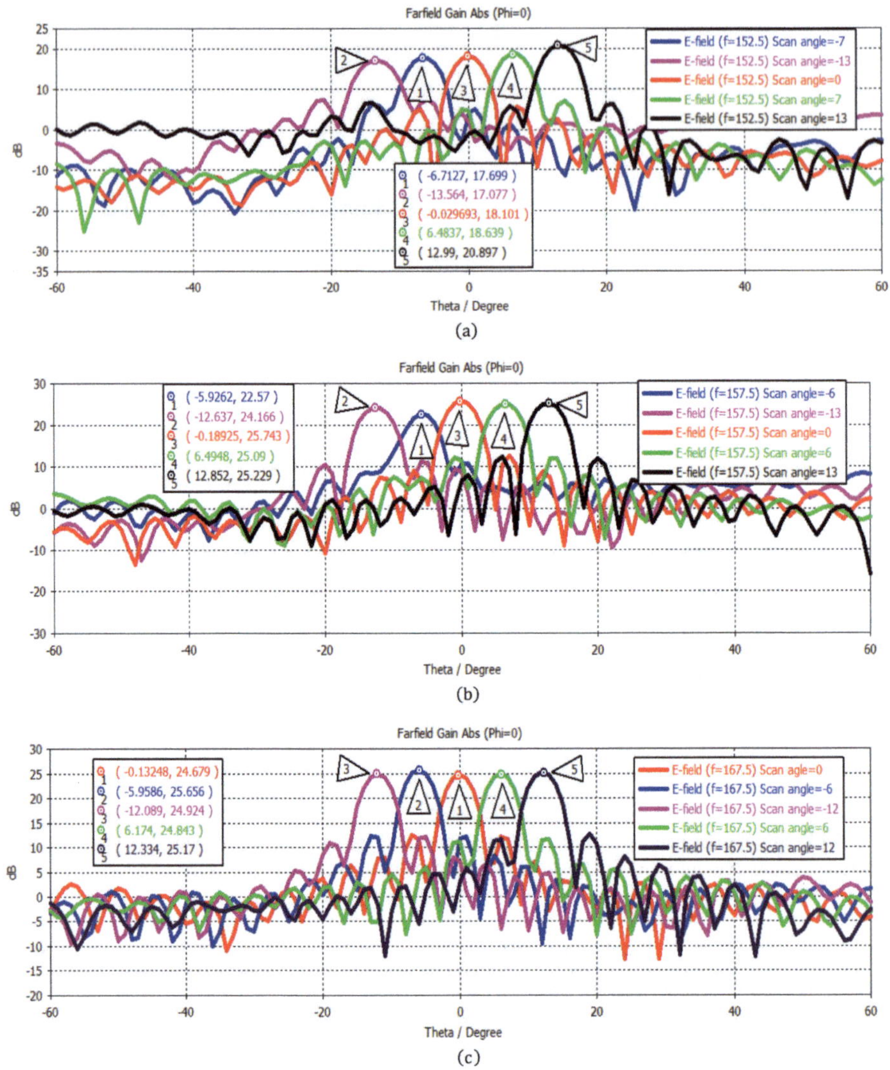

Fig. 8.14 Simulation results of the E-field of the second THz beamforming microstrip array antenna, (**a**) For 152.5 GHz, (**b**) For 157.5 GHz, (**c**) For 167.5 GHz

that tanδ is frequency-dependent and may change at the THz frequencies, and this parameter shows the possible losses of the proposed first antenna due to the high frequency.

Figures 8.19a, b show the equalization of simulation results of the S_{11} and the gain of the first antenna for substrate parameters at 125 GHz and 20 GHz, respectively, made with CST MWS and Ansys HFSS simulators. From Fig. 8.19a, it has been shown that a good agreement between the simulation results is achieved

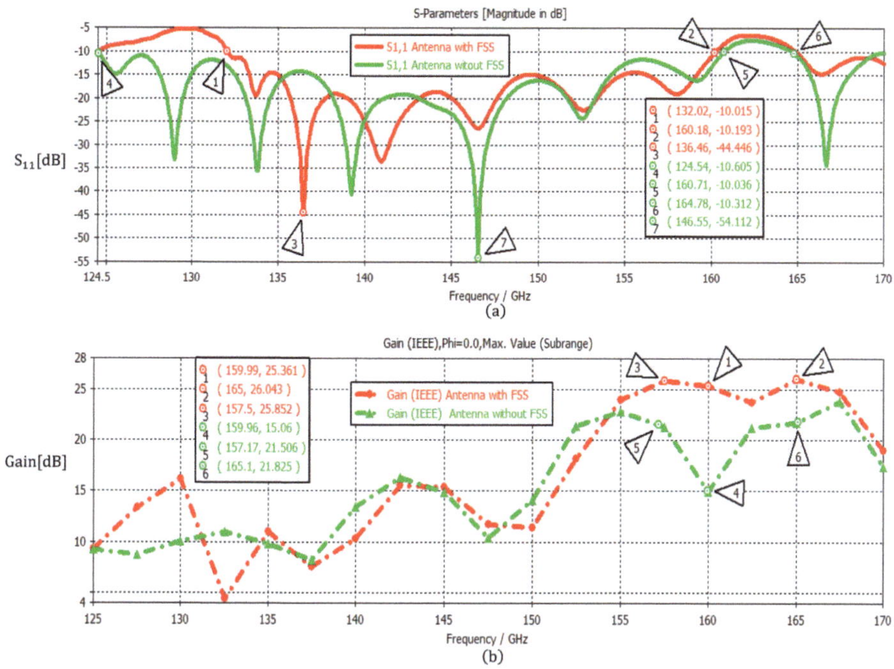

Fig. 8.15 Equalization of simulation results between the first and second THz beamforming antenna, (**a**) S_{11}, and (**b**) Gain

regarding the first THz beamforming antenna in the frequency range of 160–170 GHz. When the tanδ and ε_r of the Isola Astra® MT77 substrate were measured at 125 GHz, from Fig. 8.19b, it has been shown that a good agreement between the simulation results is achieved regarding the first THz beamforming antenna in the frequency range of 155–165 GHz when the tanδ and ε_r of the Isola-Astra® MT77 was measured at 20 GHz.

8.9.2 THz Beamforming Rotman Lens Microstrip Array Antenna

The THz Rotman lens was made with Isola Astra MT77 microstrip laminate with the copper thickness (t), substrate height (h), ε_r, and tanδ of 17.5 μm, 127 μm, 2.82, 0.001 @ 125 GHz, respectively. Figure 8.20 shows the suggested D-band Rotman lens. Every dummy port terminates with a discrete port condition with an impedance of 50 Ω.

The parameters of the Rotman lens, inserted into RLD software, are presented in Table 8.2.

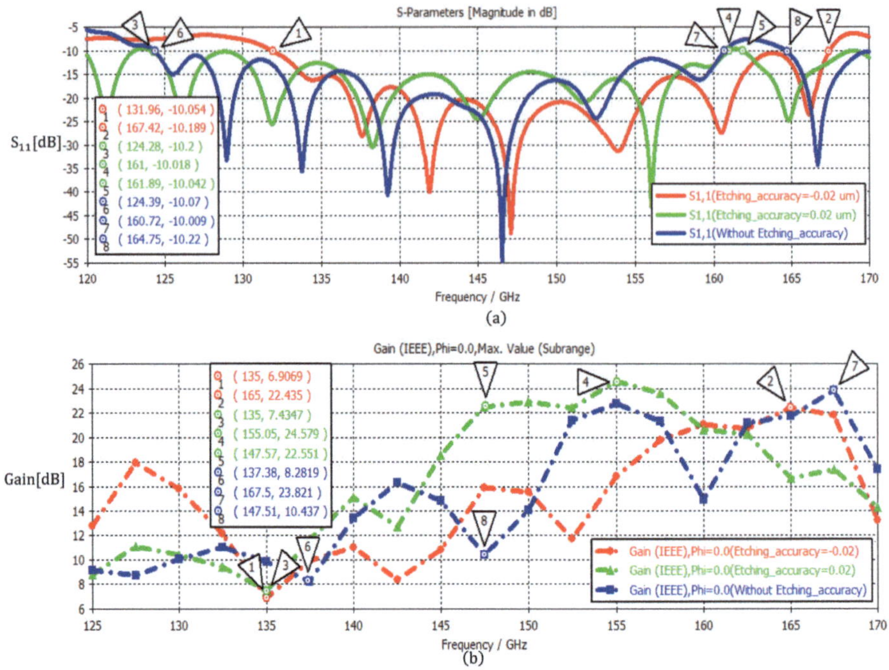

Fig. 8.16 Equalization of simulation results between the first THz beamforming microstrip array antenna with or not including the declared etching accuracy, (**a**) S_{11}, and (**b**) Gain

The THz beamforming Rotman lens microstrip array antenna is based on the same laminate and parameters as the Rotman lens laminate. This THz beamforming microstrip antenna is based on corporate fed and nonstandard patches for gain and BW enhancement. Figure 8.21 shows the suggested THz beamforming Rotman lens microstrip array antenna. The dimensions of the substrate were $37.92 \times 30 \times 0.127\ mm^3$, which equals $15.8\lambda_0 \times 12.5\lambda_0 \times 0.53\lambda_0$ @ 125 GHz, where λ_0 is the free-space wavelength.

Figure 8.22 illustrates the S-Parameters ($S_{11} - S_{13,\ 1}$) simulation results of the suggested THz Rotman lens. It has been shown that the BW ($[S_{11}$ & S_{21} & S_{31} & S_{41} & $S_{51}] \leq -15$dB) of the THz Rotman lens is: >50 GHz (>40%) while the resonances frequencies are between 100 GHz and 150 GHz, and the insertion losses (S_{61} & S_{71} & S_{81} & S_{91} & $S_{10,\ 1}$ & $S_{11,\ 1}$ & $S_{12,\ 1}$ & $S_{13,\ 1}$) are between -10.278 up to -19.927 dB.

Figure 8.23a, b illustrates the S-parameters phases ($\measuredangle S_{6,\ 1} - \measuredangle S_{13,\ 1}$) in [°] vs. frequency at [GHz] of the suggested THz Rotman lens. From Fig. 8.23a, it has been shown that the S-parameters phases ($\measuredangle S_{6,\ 1} - \measuredangle S_{13,\ 1}$) are between $-180°$ and 180°. Furthermore, from Fig. 8.23b, it has been shown that the average changes of the S-parameters phases ($\measuredangle S_{6,\ 1} - \measuredangle S_{13,\ 1}$) between the array port and subsequent array port of the suggested THz Rotman lens is 11.1°.

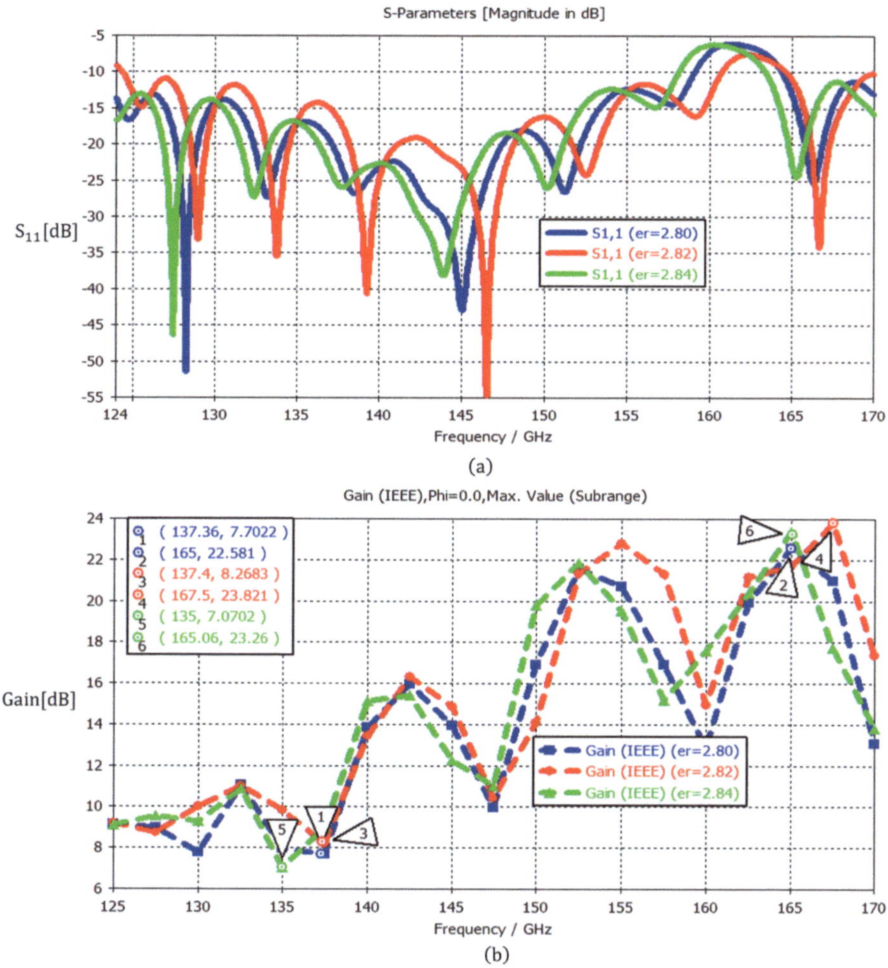

Fig. 8.17 Equalization of simulation results between the first THz beamforming microstrip array antenna with a deviation of ε_r, (**a**) S_{11}, and (**b**) Gain

Figure 8.24a–e illustrates the suggested THz Rotman lens's surface current when Port 1 up to Port 5 is excited separately. From Fig. 8.24a–e, it has been shown that the surface current of the suggested THz Rotman lens is 300–409 A/m.

Figures 8.25a–c illustrates the simulation result of return losses (S_{11}, S_{22}, S_{33}, S_{44}, S_{55}), realized gain, and the suggested THz beamforming Rotman lens microstrip array antenna's total efficiency. From Fig. 8.25a, it has been shown that the BW(S_{11}, S_{22}, S_{33}, S_{44}, $S_{55} \leq -10$ dB) of the antenna is 31.02 GHz(24.3%) while the resonance frequencies are 117.95–143.24 GHz. From Fig. 8.25b, it has been shown that the realized gain of the antenna is 4.69–14.09 dB. From Fig. 8.25c, it has been shown that the total efficiency of the antenna is 50.75–75.34% at 110–145 GHz.

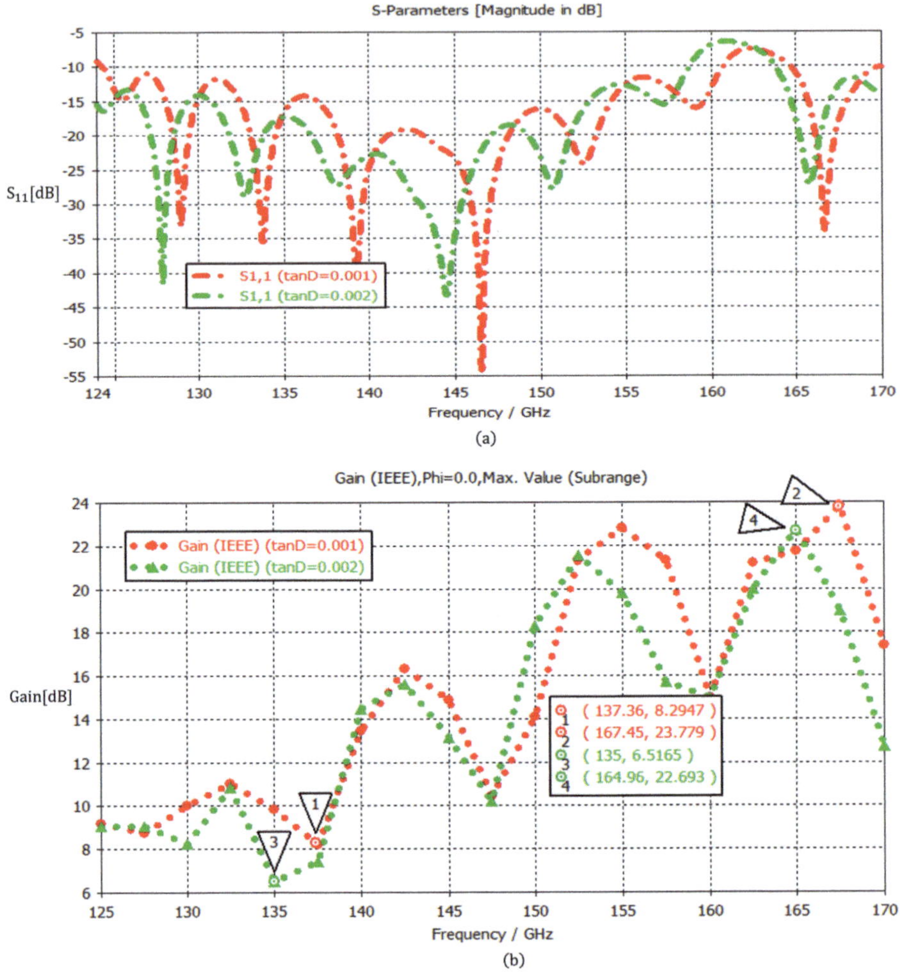

Fig. 8.18 Equalization of simulation results between the first THz beamforming microstrip array antenna with a deviation of tanδ, (**a**) S_{11}, and (**b**) Gain

Figure 8.26 illustrates the simulation result of the directivity radiation pattern of the suggested THz beamforming Rotman lens microstrip array antenna for 117.5 GHz. From Fig. 8.26, it has been shown that the directivity, radiation efficiency, and total efficiency of the suggested THz beamforming microstrip array antenna for 117.5 GHz is 15.2 dBi, −1.28 dB, and –1.57 dB, respectively.

Figure 8.27a–f illustrates the simulation results of the E-field of the suggested THz beamforming Rotman lens microstrip array antenna depending on the frequencies of 115, 125, and 135 GHz and the beamforming angles when each Port 1 up to 5 is excited separately. From Fig. 8.27a–f, it has been shown that the E-field levels of the antenna are 9.6–13.92 dB. At the same time, the beamforming angles are between −23.35° and 23.76°.

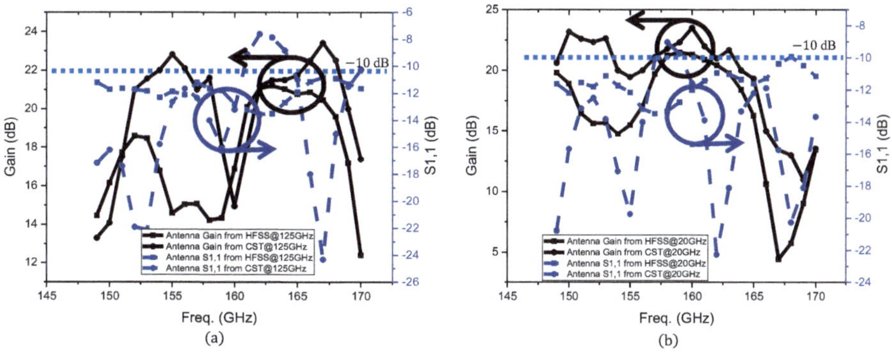

Fig. 8.19 Equalization of simulation results in the gain and the S_{11} for the first THz beamforming microstrip array antenna made with Ansys HFSS and CST MWS simulators, (**a**) Substrate parameters at 125GHz, and (**b**) Substrate parameters at 20GHz

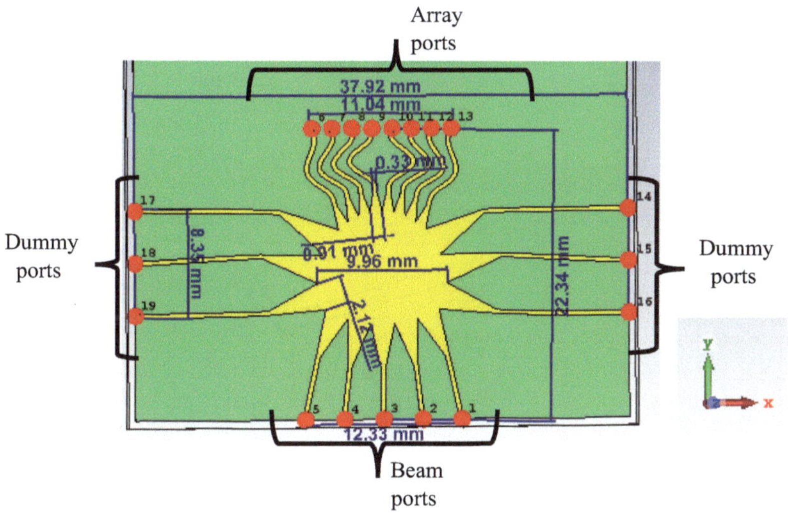

Fig. 8.20 The suggested THz Rotman lens

The antenna, a resonance component, and dimensions depend on the wavelength. At higher frequencies, the wavelength is smaller, so the antenna dimensions are also smaller, so when we fabricate the proposed THz beamforming Rotman lens microstrip array antenna, the minimal PCB fabrication etching was around 50 μm, while the declared PCB etching precision was about ±10 μm [51]. Thus, the minimal PCB etching at the proposed THz beamforming antenna design dimensions was not below 50 μm.

Simulations were also executed on the possibility of declared etching precision, i.e., for each eight radiator's sizes, to illustrate the possible errors of future proposed prototype THz beamforming Rotman lens microstrip array antenna, fabrication for

Table 8.2 The design parameters defined in RLD software

Parameter	Value at RLD
Copper conductor thickness (t)	17.5 μm
Substrate thickness (h)	127 μm
Substrate permittivity (ε_r)	2.82 @ 125 GHz
Substrate loss tangent (tanδ)	0.001 @ 125 GHz
Length × width × height	$37.92 \times 22.34 \times 0.127$ mm^3
Center frequency	125 GHz
Bandwidth	±20 GHz → 125 GHz ± 20 GHz
Beamwidth	30°
Max. scan angle	±30°
Focal angle (Flare angle)	12°
Array number output ports	8
Number of beams	5
Dummy ports	6
Z system	50 Ω
Port width	334 μm

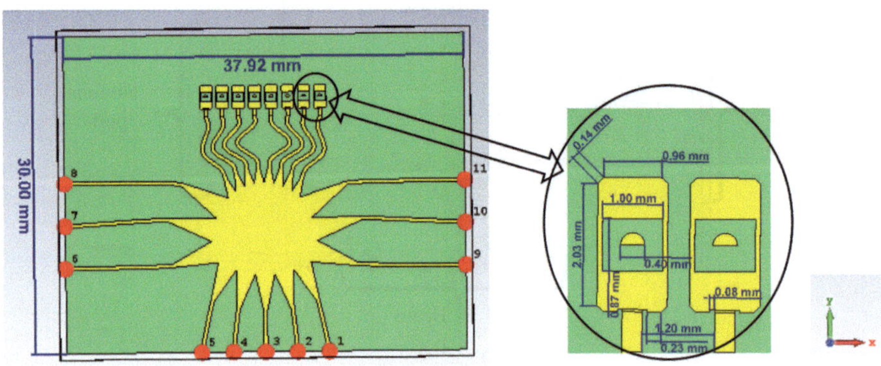

(a) (b)

Fig. 8.21 The suggested THz beamforming Rotman lens microstrip array antenna, (**a**) Front size of the suggested antenna, and (**b**) Zoom partial size of radiators and their dimensions

experimental verification. The possible fabrication declared PCB etching precision was subtracted/added in the CST MWS simulator, and two more simulations were performed with the new values. Therefore, three more graphs were attained for the realized gain, S_{11}, and beam steering angle when Port 1 is excited, which may show the realized gain, S_{11} and beam steering angle, which can be attained in the prototype fabricated D-band Rotman lens beam-steering microstrip array antenna.

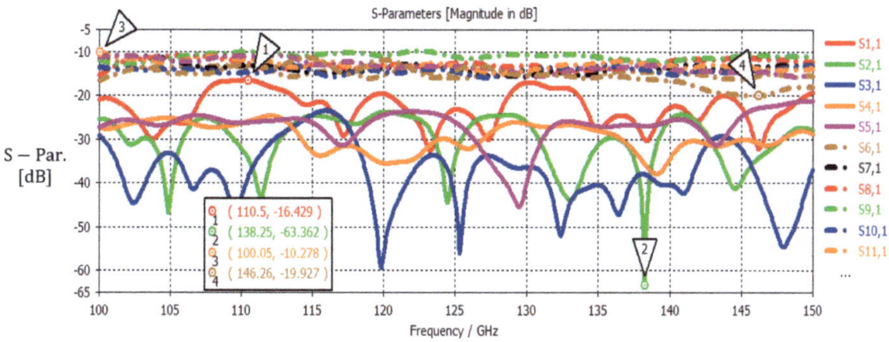

Fig. 8.22 The simulation results of S-Parameters ($S_{11} - S_{13,\ 1}$) of the suggested THz Rotman lens

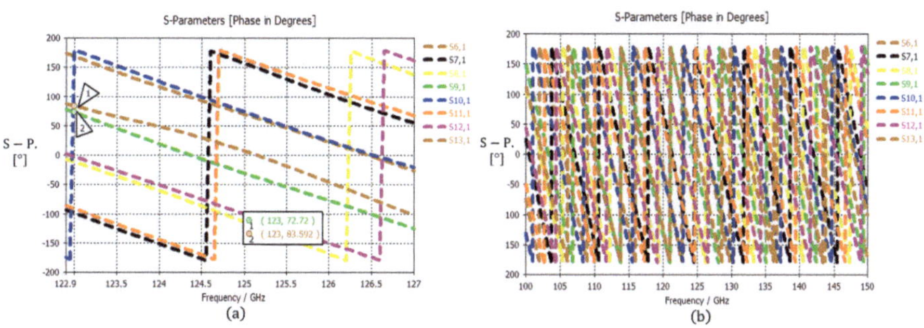

Fig. 8.23 (**a, b**) Shows the simulation results of the S-parameters phases ($\measuredangle S_{6,\ 1} - \measuredangle S_{13,\ 1}$) of the suggested THz Rotman lens

Figures 8.28a–c illustrate the comparison of the simulation results of the S_{11}, the realized gain and E-field for 110 GHz and 135 GHz of the suggested THz beamforming Rotman lens microstrip array antenna, which Port 1 excited with or without future fabrication declared PCB etching precision ±10 μm [51, 52]. From Fig. 8.28a, it may have been shown that the $BW_{S_{11}<-15\ dB}$ of the future fabricated THz beamforming Rotman lens microstrip array antenna will not be changed and will be 31.02 GHz due to declared PCB etching precision – the same as the simulated proposed THz beamforming Rotman lens microstrip array antenna. From Fig. 8.28b, the realized gain of the future fabricated THz beamforming Rotman lens microstrip array antenna may have been shown by 0.1 dB from the simulated proposed THz beamforming Rotman lens microstrip array antenna due to declared PCB etching precision. Figure 8.28c shows that the beam-steering angle will change by 1° from the simulated proposed THz beamforming Rotman lens microstrip array antenna due to declared PCB etching precision.

Figure 8.29a, b illustrates the comparison of the simulation results of the return loss and the realized gain of the suggested THz beamforming Rotman lens

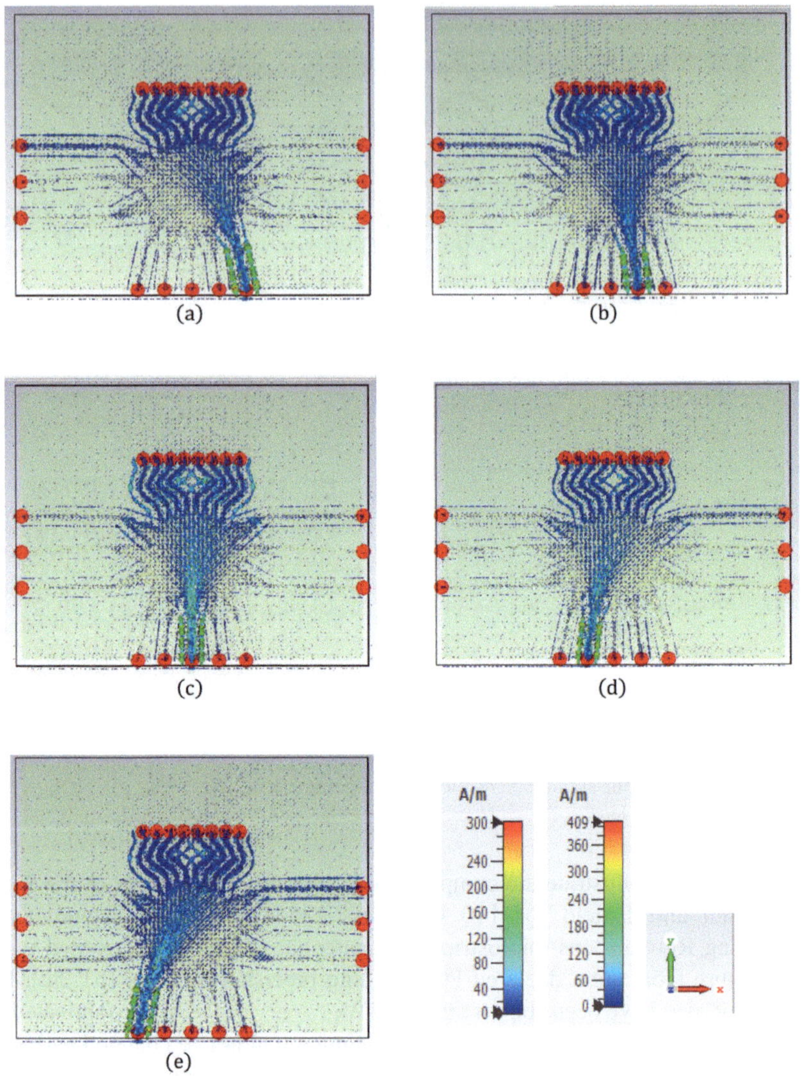

Fig. 8.24 A surface current of the proposed optimized THz Rotman lens with (**a**) Port 1 excited, (**b**) Port 2 excited, (**c**) Port 3 excited, (**d**) Port 4 excited, and (**e**) Port 5 excited

microstrip array antenna, which Port 1 excited and done with FIT CST MWS and FEM CST MWS simulator. From Fig. 8.29a, it has been shown that the BW of the suggested THz beamforming Rotman lens microstrip array antenna is the same at both solvers. From Fig. 8.29b it has been shown realized gain of the suggested THz beamforming Rotman lens microstrip array antenna from the FEM CST MWS simulator is 7.46–14.16 dB, and the realized gain of the antenna from the FIT CST

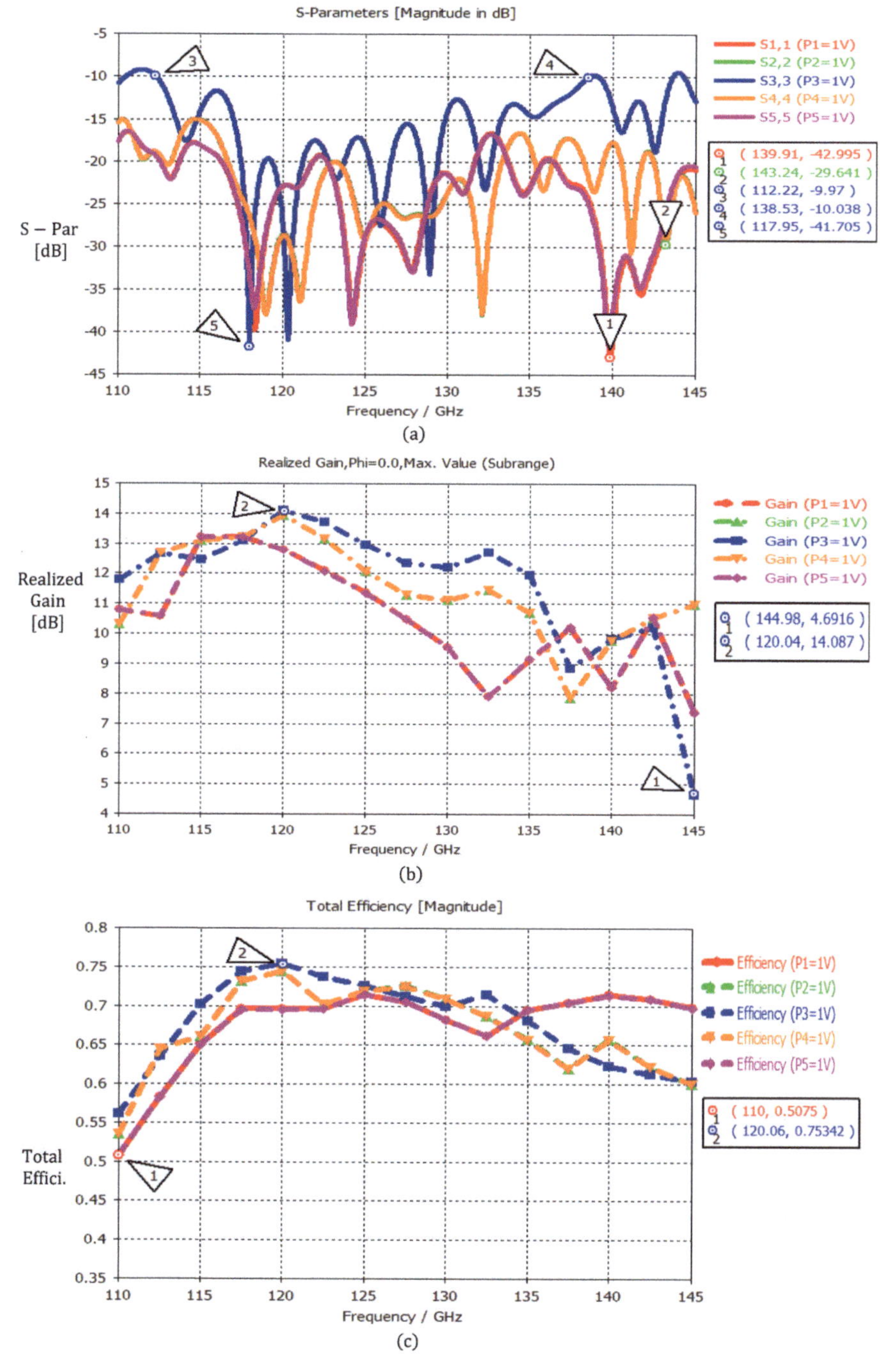

Fig. 8.25 Simulation result for the suggested THz beamforming Rotman lens microstrip array antenna, (**a**) Return loss, (**b**) Gain, and (**c**) Total efficiency

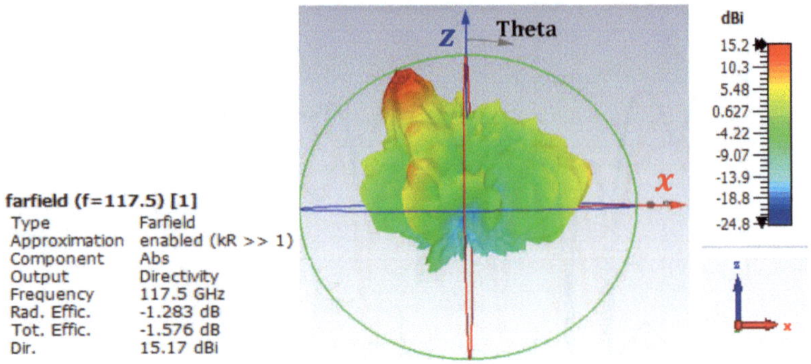

Fig. 8.26 Simulation result of the directivity radiation pattern of the suggested THz beamforming Rotman lens microstrip array antenna for 117.5 GHz

MWS simulator is 4.67–13.2 dB. So it may have been shown from Fig. 8.29a, b that a good agreement between the simulation results is achieved concerning the suggested THz beamforming Rotman lens microstrip array antenna, which validates the proposed design equations, and designed THz beamforming Rotman lens microstrip array antenna.

8.9.3 THz 4-Ports MIMO UWB Antipodal Vivaldi Microstrip Antenna

The THz 4-port MIMO UWB antipodal Vivaldi microstrip antenna is based on Rogers RO3003 microstrip substrate with the copper thickness (t), substrate height (h), dielectric constant (ε_r), and loss tangent (tanδ) of 17.5 μm, 127 μm, 3.02, 0.0019 @ 100 GHz, respectively. Figure 8.30 shows the suggested THz 4-port MIMO UWB antipodal Vivaldi microstrip antenna. Figure 8.30a shows the THz 4-port MIMO UWB antipodal Vivaldi microstrip antenna in a front view, while Fig. 8.30b shows the THz 4-port MIMO UWB antipodal Vivaldi microstrip antenna in a front view with a back view without mentioning the microstrip laminate. The dimensions of the substrate were 8 × 2.2 × 0.127 mm^3, which equals $44.2\lambda_0 \times 12.15\lambda_0 \times 0.701\lambda_0$ @ 1656.1 GHz, where λ_0 is the free-space wavelength.

Figure 8.31a–c illustrates the simulation result of return losses (S_{11}), isolation, and the maximum gain at a solid angle of the THz 4-port MIMO UWB antipodal Vivaldi microstrip antenna. From Fig. 8.31a it has been shown that the BW ($S_{11},S_{22},S_{33},S_{44} \leq -10$ dB) of this THz MIMO antenna is >1846.7 GHz, while the resonance frequency is 1656.1 GHz. From Fig. 8.31b, it has been shown that the isolation is 23.48–89.26 dB of this THz MIMO antenna. From Fig. 8.31c, it has been shown that the maximum gain at a solid angle of this THz MIMO antenna is 5.19–8.66 dB.

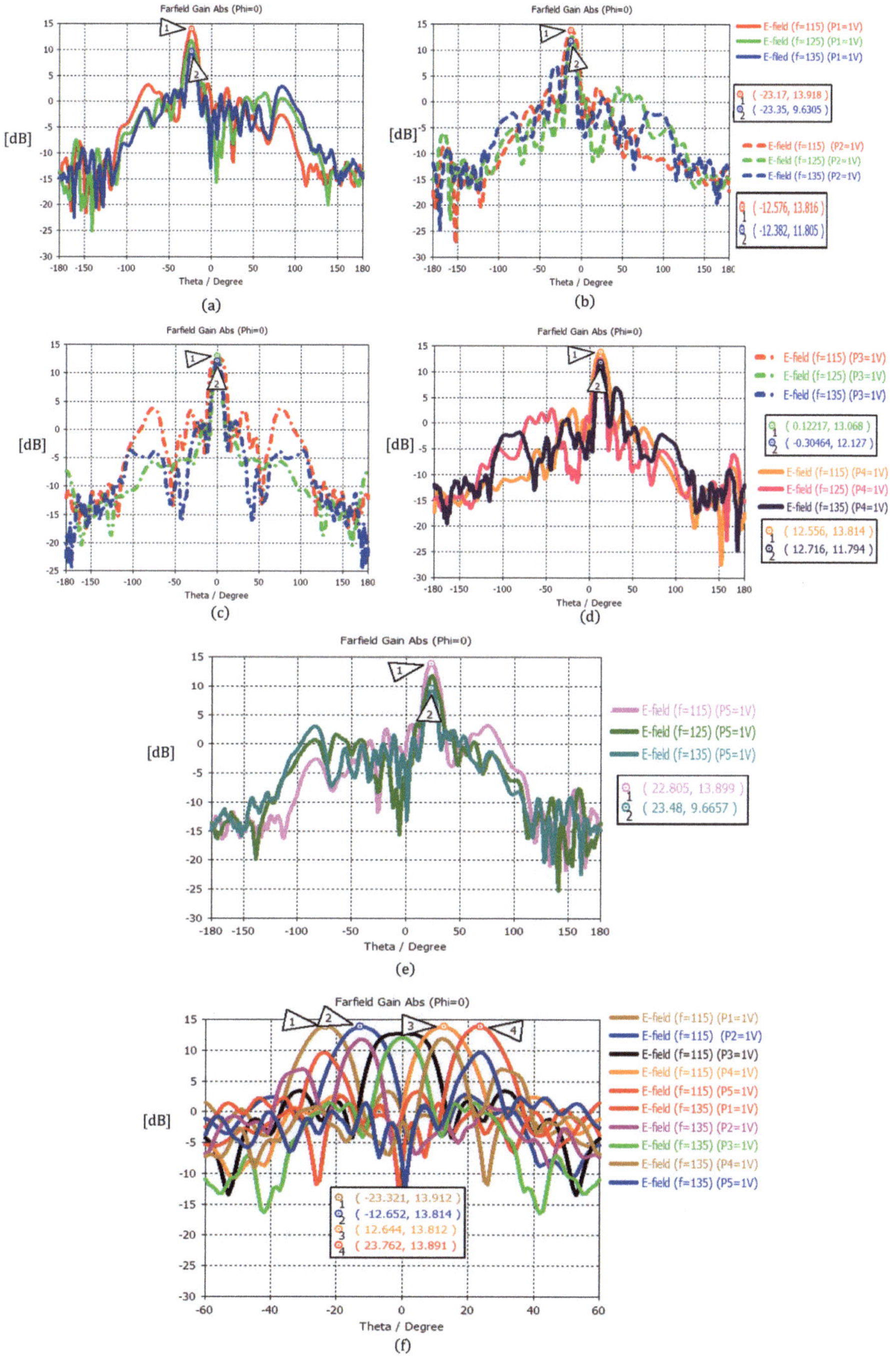

Fig. 8.27 The simulation results of the E-field for the suggested THz beamforming Rotman lens microstrip array antenna with (**a**) Port 1 excited, (**b**) Port 2 excited, (**c**) Port 3 excited, (**d**) Port 4 excited, (**e**) Port 5 excited, and (**f**) Combining all simulation results

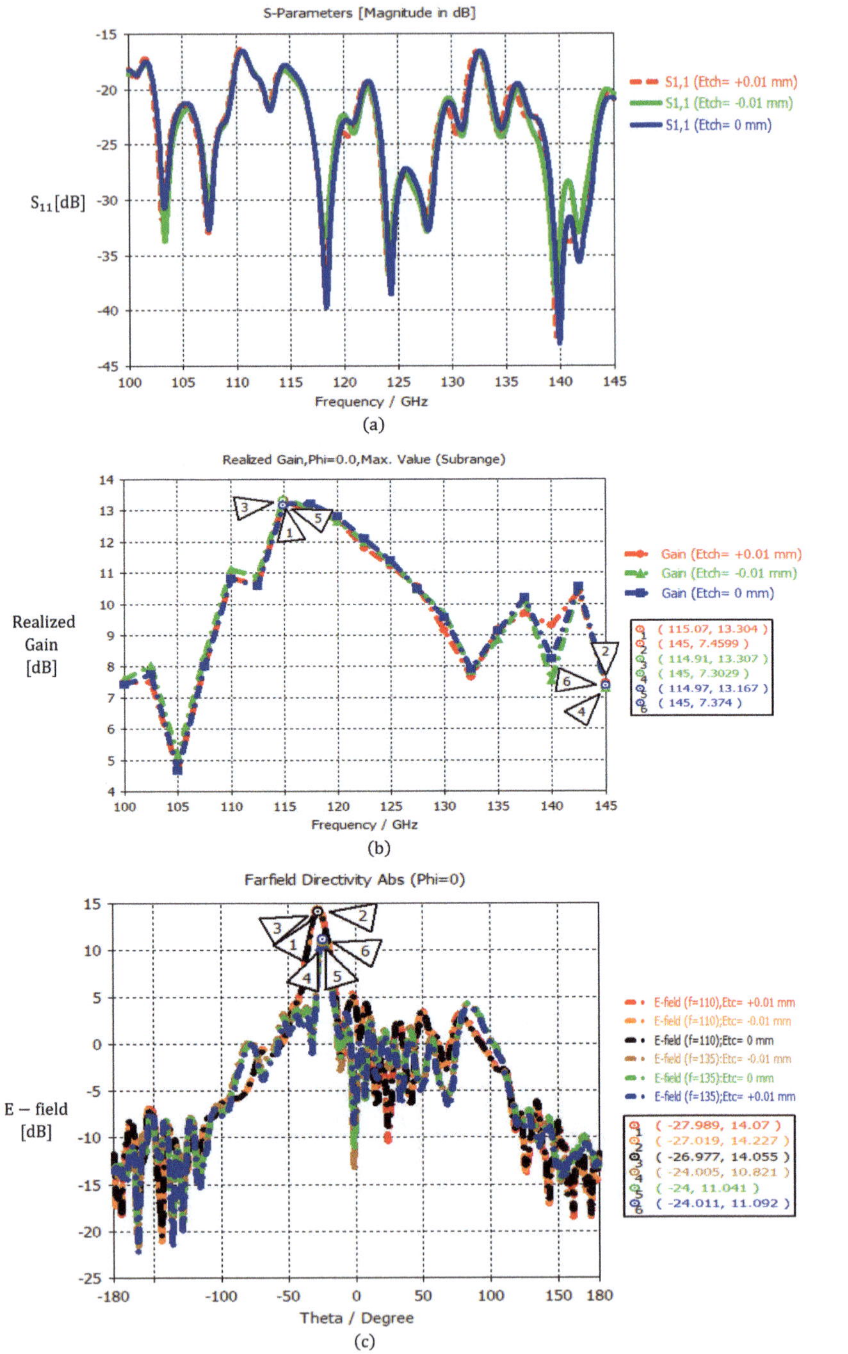

Fig. 8.28 Simulation results in comparison of the suggested THz beamforming Rotman lens microstrip array antenna with the declared etching precision, (**a**) S_{11}, (**b**) Realized gain, and (**c**) E-field for 110 GHz and 135 GHz

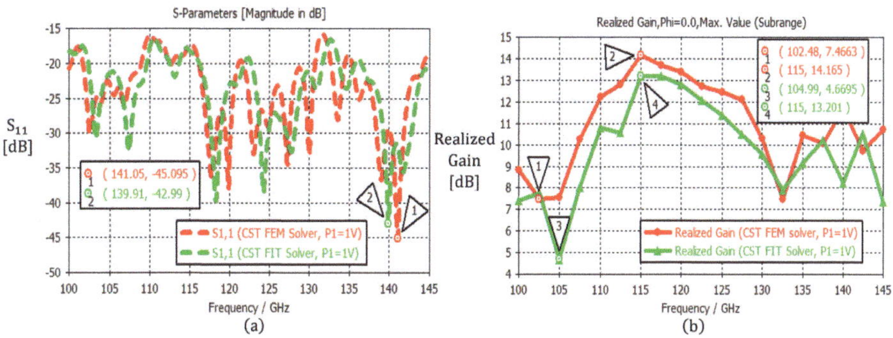

Fig. 8.29 Comparison of the simulation results of the suggested THz beamforming Rotman lens microstrip array antenna made with FIT CST MWS and FEM CST MWS simulator, (**a**) The S_{11}, and (**b**) the realized gain

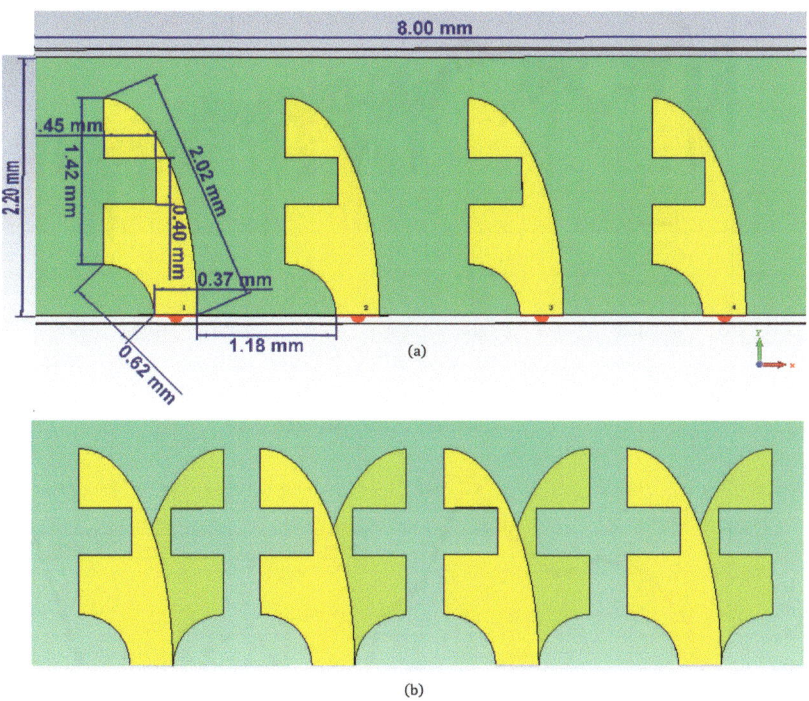

Fig. 8.30 The suggested THz 4-port MIMO UWB antipodal Vivaldi microstrip antenna, (**a**) Front view, and (**b**) Front view with back view

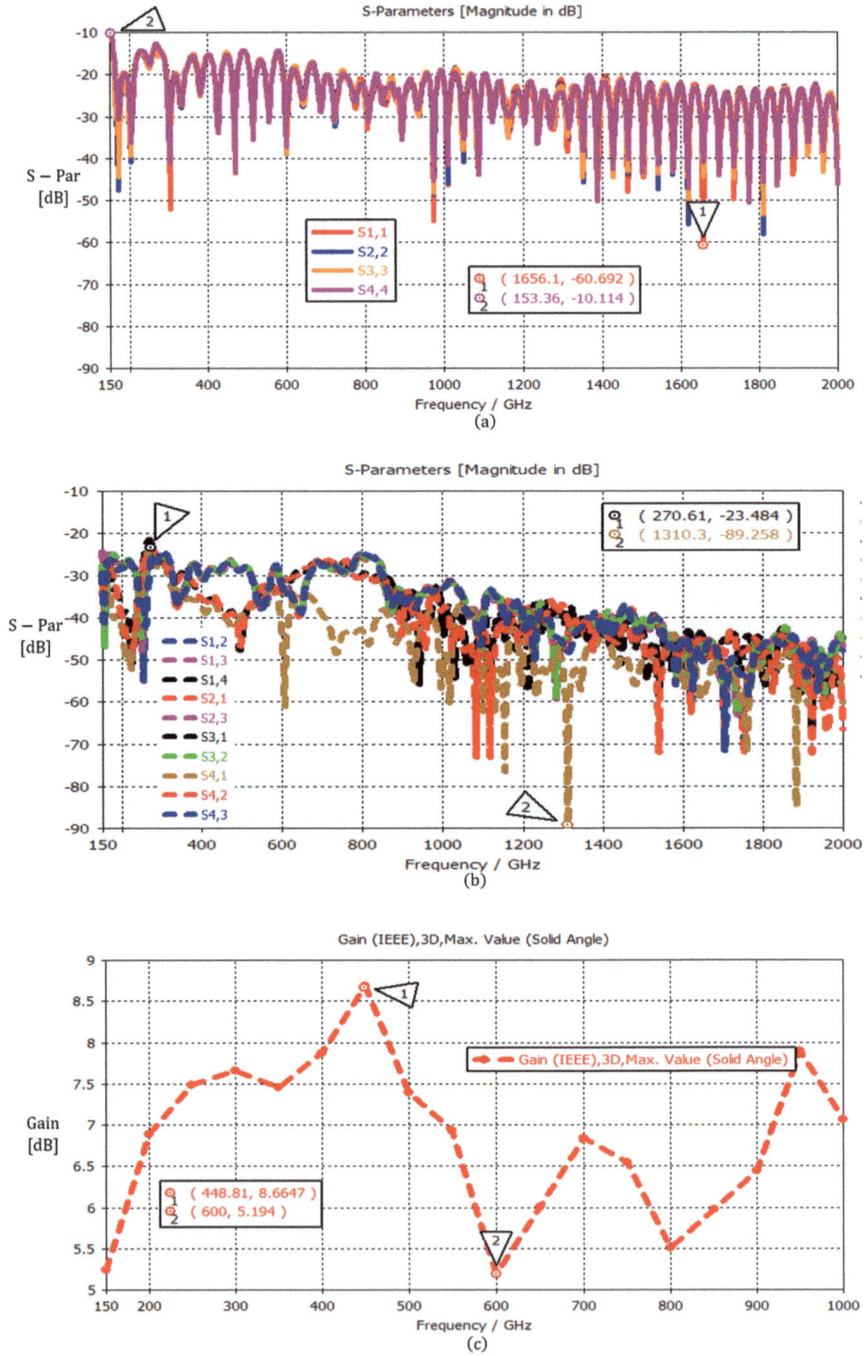

Fig. 8.31 Simulation result for the THz MIMO antenna, (**a**) S_{11}, (**b**) Isolation, and (**c**) Maximum gain (solid angle)

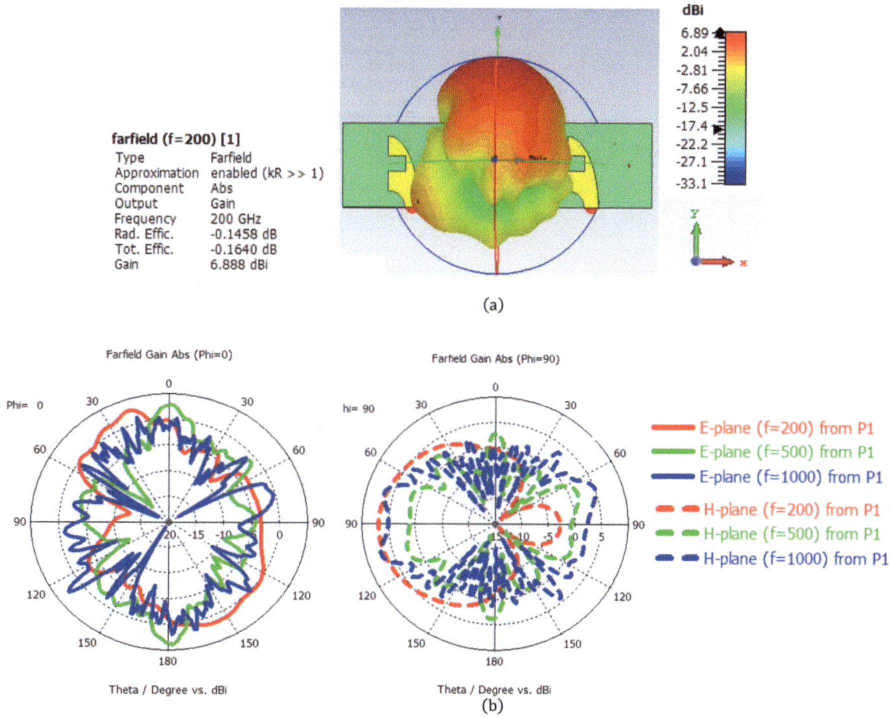

Fig. 8.32 Simulation result of the 3D far-field pattern and E-plane with H-plane of the THz MIMO antenna, (**a**) 3D far-field pattern at 200 GHz, and (**b**) E-plane and H-plane for 200, 500, and 1000 GHz

Figure 8.32a–c illustrates the simulation result of the 3D far-field pattern for 200 GHz and E-plane and H-plane for 200 GHz, 500 GHz, and 1000 GHz of the THz MIMO UWB antipodal Vivaldi microstrip antenna from port 1. From Fig. 8.32a, it has been shown that the gain of this THz MIMO antenna is 6.88 dB @200 GHz, while the total efficiency of this THz MIMO antenna is −0.164 dB @200 GHz. From Fig. 8.32b, it has been shown that the E-plane is 2.54–3.69 dB @200–1000 GHz, while the H-plane is 3.74–7.1 dB @200–1000 GHz of this THz MIMO antenna.

Figures 8.33a, b illustrates the ECC and DG simulation result from the THz MIMO UWB antipodal Vivaldi S-parameter microstrip antenna. From Fig. 8.32a, it has been shown that the maximum ECC is 0.000916 < 0.2 for the frequency range of 150–2000 GHz of this THz MIMO antenna, while from Fig. 8.32b, it has been shown that the DG is 9.9965–10 dB for the frequency range of 150–2000 GHz of this THz MIMO antenna.

Figure 8.34a, b shows the simulation results of the S_{11},S_{22},S_{33}, and S_{44} and the maximum gain at a solid angle for the THz MIMO antenna with or without future fabrication etching tolerance accuracy of ±10 μm [51]. This deviation is supposed to

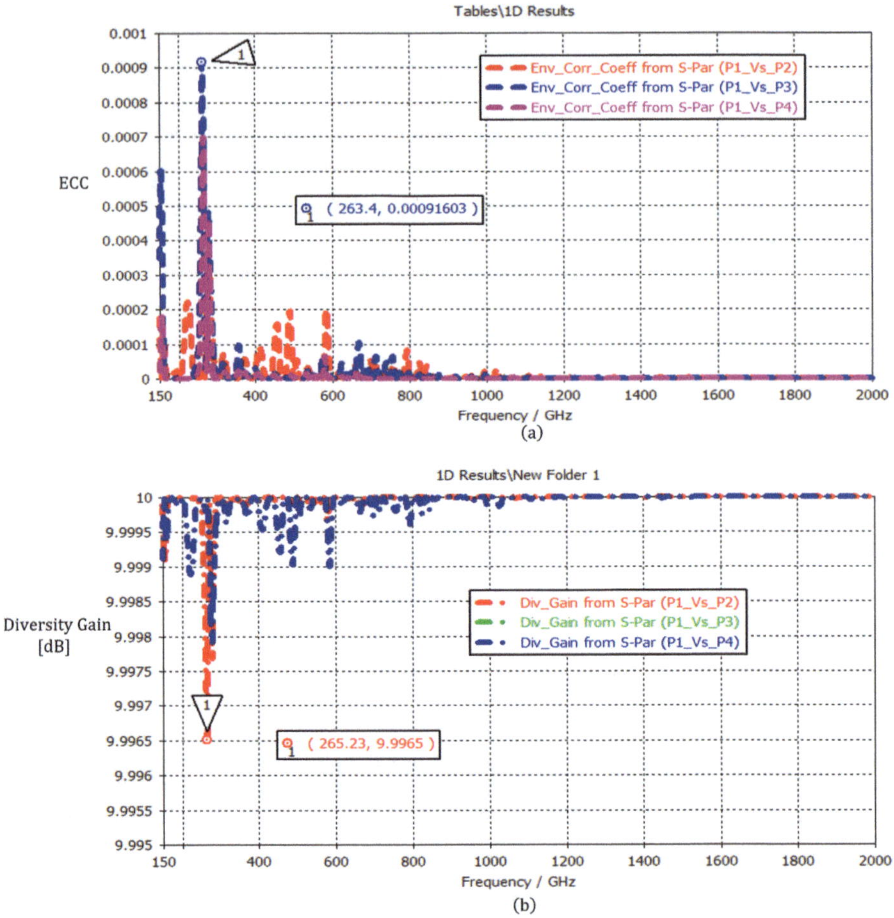

Fig. 8.33 Simulation result of ECC and DG from the S-parameter of the THz MIMO UWB antipodal Vivaldi microstrip antenna, (**a**) ECC, and (**b**) DG

be when this THz MIMO antenna is fabricated for experimental verification. These figures show that the THz MIMO antenna's BW and maximum gain may change by about 0.4–0.7 GHz, 0.04–0.21 dB when this THz MIMO antenna is fabricated due to the etching tolerance as described.

Figure 8.35a, b shows the simulation results of the S_{11},S_{22},$S_{33,}$ and S_{44} and the maximum gain at a solid angle for the THz MIMO when the ε_r and tanδ of Rogers RO3003 microstrip substrate are set at the CST MWS simulator for a frequency of 100 or 2000 GHz, whereas for 100 GHz the $\varepsilon_r = 3.02$ and $\tan\delta = 0.0019$ and for 2000 GHz $\varepsilon_r = 3.02 \pm 0.1$ and $\tan\delta = 0.0019$. This change at $\varepsilon_r = 3.02$ of ± 0.1 for a frequency of 2000 GHz can be chosen as a possible deviation because the ε_r is frequency dependent [48–50]. Figure 8.35c, d shows the simulation results of the S_{11},S_{22},$S_{33,}$ and S_{44} and the maximum gain at a solid angle for the THz MIMO when

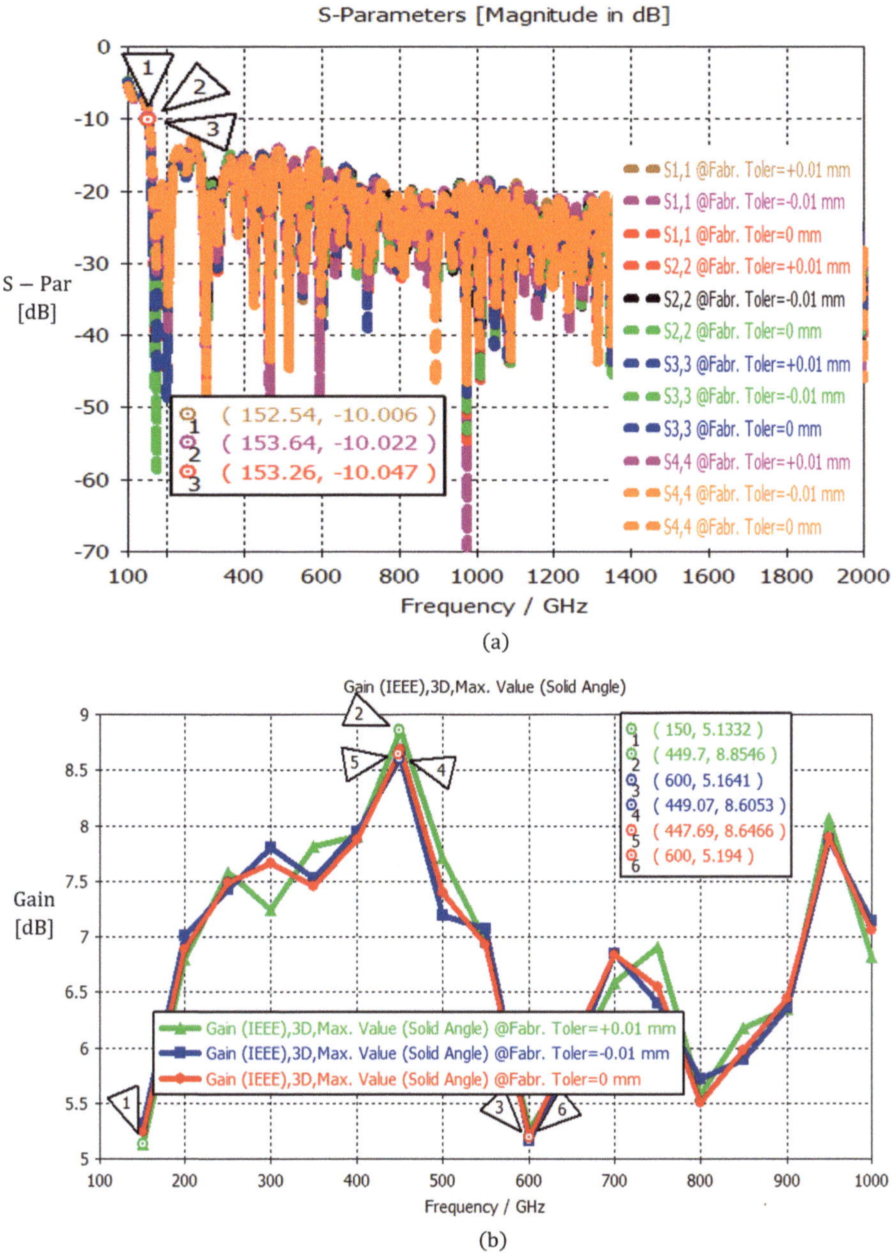

Fig. 8.34 Simulation results from the comparison of the gain and BW of the THz MIMO antenna due to fabrication tolerance error, (**a**) Comparison of simulation results of S_{11},S_{22},S_{33}, and S_{44}, (**b**) Comparison of simulation results of maximum gain at a solid angle

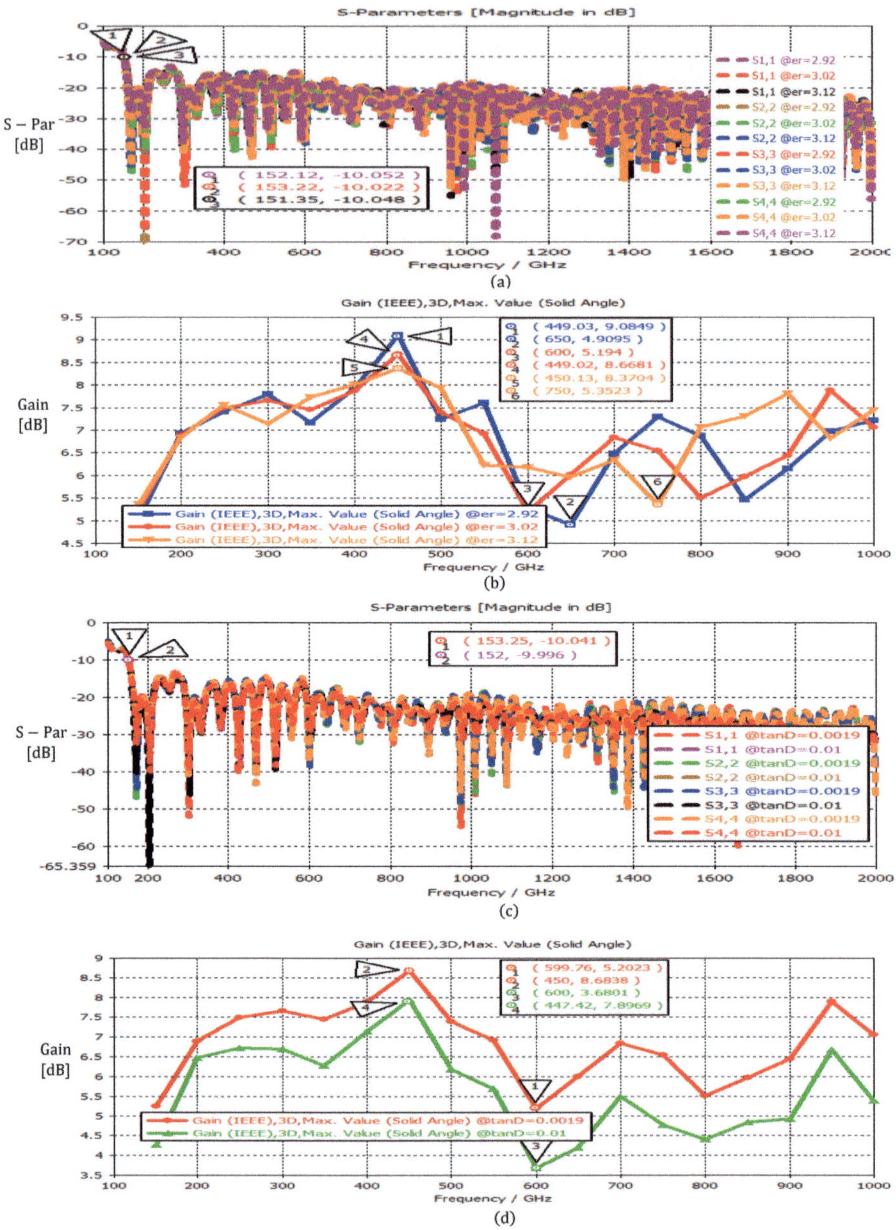

Fig. 8.35 Simulation results from the comparison of the gain and BW of the THz MIMO antenna due to changes at ε_r and tanδ, (**a**, **b**) Comparison of simulation results of S_{11},S_{22},S_{33}, S_{44} and maximum gain at a solid angle due to changes at ε_r, (**c**, **d**) Comparison of simulation results of S_{11},S_{22},S_{33}, S_{44} and maximum gain at a solid angle due to changes at tanδ

the ε_r and tanδ of Rogers RO3003 microstrip substrate are set at the CST MWS simulator for a frequency of 100 GHz or frequency of 2000 GHz, whereas for 100 GHz the $\varepsilon_r = 3.02$ and $\tan\delta = 0.0019$ and for 2000 GHz $\varepsilon_r = 3.02$ and$\tan\delta = 0.01$. This change at tanδ from 0.0019 to 0.01 for a frequency of 2000 GHz can be chosen as a possible deviation because the tanδ is frequency dependent. Moreover, when the frequency increases, the tan δ increases [48–50]. These deviations, shown in Fig. 8.35a–d, should be when this THz MIMO antenna is fabricated for experimental verification. These figures show that the THz MIMO antenna's BW and maximum gain at a solid angle may change by about 1.25–2 GHz, 0.5–0.9 dB when this THz MIMO antenna is fabricated due to the deviation and frequencies-dependent of the ε_r and tanδ of the Rogers RO3003 microstrip substrate.

The performance summary of these designed and simulate THz antennas presented above in this chapter is presented in Table 8.3.

8.9.4 Discussion

The THz regime is a pioneer in developing applications such as high data rate 6G wireless communication networks. Therefore, using multi-beam UM-MIMO and beamforming array antennas is necessary to enable these technologies, as described in detail in this chapter [1–3]. This chapter shows the design and simulation of the THz beamforming 30 × 16 E-shaped radiators microstrip array antenna, THz beamforming 30 × 16 E-shaped radiators microstrip array with bandpass and bandstop FSSs, THz beamforming Rotman lens 1 × 8 microstrip antenna, and THz 4-ports MIMO UWB antipodal Vivaldi microstrip antenna. The performance summary is shown in Table 8.3. This table shows that the gain enhancement due to adding FSSs is up to about 3.2 dB.

Furthermore, the maximum steering angle of the proposed THz beamforming antennas is up to ±23.8° and may be increased up to ±60°. Under the components that exist in contemporary technology, the central problem of the beamformers at the THz band is the limited phase angle range up to about ± (0–40°) [18], so by analog beamforming within the Rotman lens, we can enable high data rate 6G wireless communication networks at a distance of a 100 m. From error analysis, the proposed THz beamforming antenna peak gain and impedance BW may alter by up to 2.1 dB and 6.31 GHz because of the predictable fabrication etching tolerance accuracy ±10 μm [51, 52], the microstrip laminates frequency-pendent dielectric properties (ε_r) and the frequency-pendent losses (tanδ) may change by about 1.1 dB, 0.6 GHz, because of the antennas' high frequency working beyond 100 GHz. The primary models' simulation results were verified with another antenna solver [53–57], and a good agreement was achieved.

In addition, from error analysis, the proposed THz MIMO antenna maximum gain at a solid angle and impedance BW may alter by up to 0.04–0.21 dB, 0.4–0.7 GHz because of the predictable fabrication etching tolerance accuracy ±10 μm [51, 52], the microstrip laminates frequency-pendent dielectric properties (ε_r) and the

Table 8.3 The performance summary of the designed and simulated THz antenna in this chapter

THz antennas technology	Gain and BW enhancement techniques	Freq. nom. (GHz)	BW (GHz)	Maximum gain (dB)	Steering angle (°)	Isolation (dB)
Beamforming 30 × 16 E-shaped radiators microstrip array	Hybrid-fed and proximity-coupled fed	146.55[a]	>41.88[a]	23.8[a]	±13.7[a]	n.a.
Beamforming 30 × 16 E-shaped radiators microstrip array and bandpass and bandstop FSSs	Hybrid-fed, proximity-coupled fed, and FSSs	136.45	>33.3	26.04	±13.5	n.a.
Beamforming Rotman lens 1 × 8 microstrip array	Corporate-fed and non-standard patches	118–143.2[a]	31.02[a]	Realized gain: 14.09[a]	±23.8[a]	n.a.
THz 4-ports MIMO UWB antipodal Vivaldi microstrip antenna	Nonstandard patches (antipodal Vivaldi) with slots	1656.1	> 1846.7	8.66	n.a.	≤89.26

[a]Simulated and compared to a different type of antenna solver/simulator

frequency-pendent losses (tanδ) may change by about 0.5–0.9 dB, 1.25–2 GHz, because of the antennas' high frequency working beyond 100 GHz. Moreover, when the frequency increases, the tan δ and the proposed THz MIMO antenna and THz beam-steering antenna losses increase [48–50].

8.10 Open Research Challenges

Based on this chapter, multi-beam and beamforming THz array antenna for 6G have several open research challenges:

1. Further investigation and improvements on high gain and wide beam steering angle beyond ±40° THz beamforming microstrip array antenna.
2. Additional analysis and evolutions on THz UM-MIMO microstrip array antenna with high number ports, gain, and wide bandwidth.

8.11 Summary

This chapter discusses THz multi-beam UM-MIMO and beamforming THz array antenna for 6G. Four models of a THz antenna, when three models of THz beamforming microstrip array antennas and 1 THz UWB MIMO antenna model have been designed, and the simulation results are presented in this chapter. The first model was a THz beamforming 30 × 16 E-shaped radiator microstrip array antenna for the resonance frequency of 146.55 GHz. The achieved gain, impedance BW, and steering angles were 23.8 dB, >41.88 GHz, and ±13.7°, respectively. The second model was THz beamforming 30 × 16 E-shaped radiators microstrip array and bandstop and passband FSSs antenna for the resonance frequency of 136.45 GHz, and the achieved gain, impedance BW, and steering angles were 26.04 dB, >33.3 GHz, and ±13.5°, respectively. The third model was THz beamforming Rotman lens microstrip array antenna for the resonance frequency of 118–143.2 GHz, and the achieved realized gain, impedance BW, and steering angles were 14.09 dB, 31.02 GHz, and ±23.8°, respectively. From error analysis, the proposed THz beamforming microstrip array antenna, gain, and the BW because of the predictable fabrication etching tolerance accuracy ±10 μm, the microstrip laminates frequency-pendent dielectric properties (ε_r), and the frequency-pendent losses (tanδ) may change by a maximum of about 2.1 dB and 6.31 GHz because of the antennas' high frequency working beyond 100 GHz. The simulation results of the primary models were verified with another kind of antenna simulator, and a good agreement was achieved. The designed THz UWB 4-ports MIMO microstrip antenna included four antipodal Vivaldi microstrip radiators. Furthermore, the simulated results of the proposed THz MIMO antenna showed an impedance BW, maximum gain at a solid angle, minimum ECC, and minimum DG of >1846.7 GHz, 8.66 dB, 23.48 dB, 0.000916, and 9.9965 dB, respectively. From state-of-the-art simulation results with future fabrication etching tolerance accuracy, microstrip substrate ε_r, and microstrip substrate tan δ deviations, which can be when an experimental verification with a prototype fabricated THz MIMO antenna, the impedance BW, and maximum gain can be altered by a maximum of about 1.25–2 GHz, 0.5–0.9 dB from the simulated results. This chapter's proposed antennas can be a base for next-generation THz wireless communication 6G antennas.

References

1. E. C. Strinati, S. Barbarossa, J. L. G. Jimenez, D. Kténas, N. Cassiau, and C. Dehos, "6G: The next frontier," [Online] arXiv, pp. 1–16, May 2019, https://arxiv.org/abs/1901.03239.
2. D. Headland, Y. Monnai, D. Abbott, C. Fumeaux, and W. Withayachumnankul, "Tutorial: Terahertz beamforming, from concepts to realizations," *APL Photonics*, vol. 3, pp. 051101/1–32, Jan. 2018.
3. M. Bharathi, A. Amsaveni, R. Priyadharshini, "Study of MIMO antenna at terahertz frequency," *Proceedings of IEEE International Conference on Advancements in Electrical, Electronics, Communication, Computing, and Automation (ICAECA)*, Coimbatore, India, 08–09 Oct. 2021, pp. 1–3.
4. Y. Monnai, X. Lu, and K. Sengupta, "Terahertz beam steering: from fundamentals to applications," *Journal of Infrared, Millimeter, and Terahertz Waves*, pp. 1–43, Feb. 2023.
5. C. A. Balanis, "Microstrip and Mobile Communications Antennas," In book: Antenna Theory Analysis and Design, 4th Edition, NJ: John Wiley & Sons, pp. 783–874, 2016.
6. M. H. Rahaman, A. Bandyopadhyay, S. Pal, and K. P. Ray, "Reviewing the scope of THz communication and a technology roadmap for implementation," *IETE Technical Review,* vol. 38, no. 5, pp. 465–478, June 2020.
7. K. Sengupta, T. Nagatsuma, and D. M. Mittleman, "Terahertz integrated electronic and hybrid electronic–photonic systems," *Nature Electronics*, vol. 1, pp. 622–635, 2018.
8. M. A. B. Othman, "Sub-Terahertz: Generation and Detection," *Ph.D. Thesis, Electrical and Electronic Engineering*, University of Nottingham, UK, June 2013.
9. J. S. Rieh, *Introduction to Terahertz Electronics*. Springer, Switzerland, pp. 81–82, 2021.
10. D. T. McGrath and V. P. Pyati, "Phased array antenna analysis with hybrid finite element method," *IEEE Transaction Antennas Propagation*, vol. 42, no. 12, pp. 1625–1630, Dec. 1994.
11. N. Amitay, V. Galindo, and C. P. Wu, "Theory and Analysis of Phased Array Antennas," John Wiley and Sons: New York, pp. 1–443, 1972.
12. X. Fu, F. Yang, C. Liu, X. Wu, and T. J. Cui, "Terahertz beam steering technologies: from phased arrays to field-programmable metasurfaces," *Advanced Optical Materials*, vol. 8, no. 3, pp. 1–22, Aug. 2019.
13. Y. Yang, O. D. Gurbuz, and G. M. Rebeiz, "An eight-element 370–410 GHz phased-array transmitter in 45-nm CMOS SOI with peak EIRP of 8–8.5 dBm," *IEEE Transaction on Microwave Theory and Techniques*, vol. 64, no. 12, pp. 4241–4249, 2016.
14. S. K. Koul, C. Mahajan, A. K. Poddar, and U. L. Rohde, "A Microelectromechanical switch with metamaterial contacts, part I: Concepts and technology," *Microwave Journal*, pp. 82–108, May 2020.
15. I. Hertl, M. Vavrda, P. Hanulak, "Rotman lens design for millimeter-wave sensor application," *IEEE Proceedings of 21st International Conference Radioelektronika*, Brno, Czech Republic, 19–20 April 2011, pp. 1–4.
16. I. I. Idrus., T. A. Latef, N. K. Aridas., M. S. A. Talip., Y. Yamada, T. A. Rahman, I. Adam., M. N. M. Yasin, "A low-loss and compact single-layer butler matrix for a 5G base station antenna," *PloS One Journal*, vol. 14, no. 12, pp. e0226499/1–23, Nov. 2019.
17. D. Muller, S. Diebold, S. Reiss, H. Massler, A. Tessmann, A. Leuther, T. Zwick and I. Kallfass, "D-band digital phase shifters for phased-array applications," *Proc. of the IEEE German Microwave Conference (GeMiC 2015),* Nuremberg, Germany, 16–18 March 2015, pp. 205–208.

18. Z. S. He, S. An, J. Liu, and C. Jin, "Variable high precision wide D-band phase shifter," *IEEE Access*, vol. 8, pp. 140438– 140444, July 2020.
19. Q. H. Abbasi, S. F. Jilani, A. Alomainy, M. A. Imran, *Antennas and Propagation for 5G and Beyond*. IET Telecommunication: London, UK, pp. 267–301, 2020.
20. E. H. Mujammami, I. Afifi, and A. B. Sebak, "Optimum wideband high gain analog beamforming network for 5G applications," *IEEE Access*, vol. 7, pp. 52226–52237, April 2019.
21. S. E. Ershadi, A. Keshtkar, A. Bayat, A. H. Abdelrahman, and H. Xin, "Rotman lens design and optimization for 5G applications," *International Journal of Microwave and Wireless Technologies*, vol. 10, no. 9, pp. 1048–1057, May 2018.
22. W. Rotman and R. Turner, "Wide-angle microwave lens for line source applications," *IEEE Transactions on Antennas and Propagation*, vol. 11, no. 6, pp. 623–632, Nov. 1963.
23. R. C. Hansen, "Design trades for Rotman lenses," *IEEE Transactions on Antennas and Propagation*, vol. 39, no. 4, pp. 464–472, April 1991.
24. D. M. Pozar, *Microwave Engineering*. John Wiley and Sons: 2nd edition, New York, pp. 148–150, 1998.
25. C. A. Balanis, *Antenna Theory: Analysis and Design*. John Wiley & Sons, 4th edition, New Jersey, pp. 931– 980, 2016.
26. S. Darzi, T. S. Kiong, M. T. Islam, H. R. Soleymanpour, and S. Kibria, "A memory-based gravitational search algorithm for enhancing minimum variance distortionless response beamforming," *Applied Soft Computing*, vol. 47, pp. 103–118, 2016.
27. M. Behjati, M. Ismail, and R. Nordin, "Limited feedback MU-MIMO in LTE-A system: Concepts, performance, and future works," *Wireless Personal Communication*, vol. 84, no. 2, pp. 935–957, 2015.
28. W. Hong, Z. H. Jiang, C. Yu, D. Hou, H. Wang, C. Guo, Y. Hu, L. Kuai, Y. Yu, Z. Jiang, Z. Chen, J. Chen, Z. Yu, J. Zhai, N. Zhang, L. Tian, F. Wu, G. Yang, Z.C. Hao, J. Y. Zhou, "The role of millimeter-wave technologies in 5G/6G wireless communications," *IEEE Journal of Microwaves*, vol. 1, no. 1, pp. 101–122, Jan. 2021.
29. S. Noh, M. D. Zoltowski, and D. J. Love, "Training sequence design for feedback assisted hybrid beamforming in massive MIMO systems," *IEEE Transactions on Communications*, vol. 64, no. 1, pp. 187–200, Nov. 2015.
30. S. Hur, T. Kim, D. J. Love, J. V. Krogmeier, T. A. Thomas, and A. Ghosh, "Millimeter wave beamforming for wireless backhaul and access in small cell networks," *IEEE Transactions on Communications*, vol. 61, no. 10, pp. 4391–4403, June 2013.
31. M. A. Jamshed, A. Nauman, M. A. Babar Abbasi, and S. W. Kim, "Antenna selection and designing for THz applications: suitability and performance evaluation: a survey," *IEEE Access*, vol. 8, pp. 113246–113261, June 2020.
32. H. Sarieddeen, M. S. Alouini, and T. Y. Al-Naffouri, "Terahertz-band ultra-massive spatial modulation MIMO," *IEEE Journal On Areas In Communication*, vol. 37, no. 9, pp. 1–19, June 2019.
33. U. Nissanov and G. Singh, "Beamforming D-band phased array microstrip antennas," *Sensors International*, vol. 3, pp. 100196/1–14, July 2022.
34. R. Chataut and R. Akl, "Massive MIMO systems for 5G and beyond networks–overview, recent trends, challenges, and future research direction," *Sensors*, vol. 20, no. 10, pp. 2753/1–35, May 2020.
35. B. K. Kanaujia, S. K. Gupta, J. Kishor, and D. Gangwar, "Printed Antennas-Theory and Design," CRC Press: Boca Raton, pp. 328–329, 2021.
36. A. Singh, M. Andrello, N. Thawdar, and J. M. Jornet, "Design and operation of a graphene-based Plasmonic nano-antenna array for communication in the terahertz band," *IEEE Journal on Selected Areas in Communications*, vol. 38, no. 9, pp. 2104–2117, Sept. 2020.
37. I. F. Akyildiz, C. Han, and S. Nie, "Combating the distance problem in the millimeter wave and terahertz frequency bands," *IEEE Communications Magazine*, vol. 56, no. 6, pp. 102–108, June 2018.
38. L. Yan, C. Han, and J. Yuan, "A dynamic array-of-subarrays architecture and hybrid precoding algorithms for terahertz wireless communications," *IEEE Journal on Selected Areas in Communications*, vol. 38, no. 9, pp. 2041–2056, Sept. 2020.

39. A. Shafie, G. N. Yang, C. Han, J. M. Jornet, M. Juntti, and T. Kurner, "Terahertz communications for 6G and beyond wireless networks: challenges, key advancements, and opportunities," *IEEE Network*, pp. 1–8, Sept. 2022.
40. K. Sengupta and A. Hajimiri, "A 0.28 THz power-generation and beam-steering array in CMOS based on distributed active radiators," *IEEE Journal of Solid-State Circuits*, vol. 47, no. 12, pp. 3013–3031, Dec. 2012.
41. N. Buadana, S. Jameson and E. Socher, "A 280-GHz digitally controlled four port chip-scale dielectric resonator antenna transmitter with DiCAD true time delay," *IEEE Solid-State Circuits Letters*, vol. 3, pp. 454–457, 2020.
42. A. Tang, N. Chahat, Y. Zhao, G. Virb, C. Lee, F. Hsiao, L. Du, Y. C. Kuan, M. C. F. Chan, G. Chattopadhyay, and I. Mehdi, "A 65nm CMOS 140 GHz 27.3 dBm EIRP transmit array with membrane antenna for highly scalable multi-chip phase arrays," *Proceedings of IEEE MTT-S International Microwave Symposium (IMS2014)*, Tampa, FL, 2014, pp. 1–3.
43. X. Liu, L. Samfass, K. Kolpatzeck, L. Häring, J. C. Balzer, M. Hoffmann, and A. Czylwik, "Terahertz beam steering concept based on a MEMS-reconfigurable reflection grating," *Sensors*, vol. 20, no. 10, pp. 2874/1–13, May 2020.
44. M. F. Ali, R. K. Singh, and R. Bhattacharya, "Reconfigurable graphene-based two port dual-band and MIMO antenna for THz applications," *Proceedings of IEEE Students Conference on Engineering & Systems (SCES)*, Prayagraj, India, 10–12 July 2020, pp. 1–5.
45. S. Das, D. Mitra, and S. R. B. Chaudhuri, "Fractal loaded planar super wideband MIMO antenna in THz frequency range," *Proceedings of IEEE Asia-Pacific Microwave Conference (APMC)*, Hong-Kong, 08–11 Dec. 2020, pp. 703–705.
46. A. Mohanty and S. Sahu, "Micro 4-port THz MIMO antenna for nano communication networks," *Photonics and Nanostructures - Fundamentals and Applications*, vol. 53, pp. 101092/1–14, Nov. 2022.
47. C. Vamsi, A. K. Dwivedi, G. Bharti, V. R. Verma, and A Sharma, "Efficient graphene-based circularly polarized MIMO antenna for THz applications," *Applied Optics*, vol. 61, no. 28, pp. 8155–8161, Oct. 2022.
48. J. D. Preez and S. Sinha, *Millimetre-Wave Antennas Configurations and Application*. Springer Signals and Communication Technology, Switzerland, pp. 39–60, 2016.
49. G. Kouemou, *Radar Technology*. IntechOpen, Croatia, pp. 1–430, 2009.
50. L. K. S. Granados, "Antennas for Millimeter-Wave Applications," Ph.D. Thesis, Electrical and Electronic Engineering, Santa María Tonantzintla University, Puebla, Mexico, pp. 1–89, 2016.
51. U. Nissanov, G. Singh, and A. Segun, "Sixth-generation (6G) microstrip antenna with high-gain," *International Journal on Communications Antenna and Propagation*, vol. 11, no. 4, pp. 279–287, Aug. 2021.
52. J. Francey and T. Bateman, "PCB technology requirements for millimeter-wave interconnect and antenna," *The PCB Magazine*, pp. 36–42, Apr. 2017.
53. U. Nissanov, G. Singh, P. Kumar, and N. Kumar, "High-gain terahertz microstrip array antenna for future generation cellular communication," *Proceedings of IEEE International Conference on Artificial Intelligence, Big Data, Computing and Data Communication (icABCD 2020)*, Durban, South Africa, 6–7 Aug. 2020, pp. 1–6.
54. U. Nissanov, G Singh, and A Akinola, "Low sidelobe levels (SLLs) microstrip antennas for THz bio-sensors and THz communications," *Sensors International*, vol. 2, pp. 100097/1–10, April 2021.
55. U. Nissanov and G. Singh, "MmWave/THz reconfigurable ultra-wideband (UWB) microstrip antenna," *Progress in Electromagnetic Research C*, vol. 111, pp. 207–224, April 2021.
56. U. Nissanov, G. Singh, and N. Kumar, "High gain microstrip array antenna with SIW and FSS for beyond 5G at THz band," *OPTIK: International Journal for Light and Electron Optics*, vol. 236C, pp. 166568/1–13, March 2021.
57. U. Nissanov, G. Singh, E. Gelbart, and N. Kumar, "Highly directive microstrip array antenna with FSS for future generation cellular communication at THz band," *Wireless Personal Communications*, vol. 118, no. 1, pp. 599–617, Jan. 2021.

Chapter 9
Reconfigurable Terahertz Microstrip Antenna for 6G Communication

9.1 Introduction

Reconfigurable intelligent surfaces (RISs), or intelligent reconfigurable surfaces (IRSs), comprised of approximately passive, low-cost, and reconfigurable metamaterials (MtMs), can counterfeit design the propagation surroundings by offering independent and controllable phase shifts on electromagnetic (EM) waves. Furthermore, integrating the RISs within 6G wireless networks engineers to extraordinary energy focusing on receivers enhances direct communications [1–2].

9.2 6G Reconfigurable Intelligent Surfaces / Intelligent Reconfigurable Surfaces

RISs/IRSs are a crucial future technology for the 6G wireless communication systems primarily because of their advantages in tuning wireless signals, consequently smartly controlling propagation surroundings. So, RISs/IRSs can smartly deflect the beams of incident RF waves in diverse paths by fine-tuning the phase shifts using an intelligent processor. This process acquiesces RIS to alter an LoS THz channel within a multipath software-detailed channel within simulated LoS links. Therefore, RIS can be adequately used to facilitate THz signals' blockage susceptibility, improving THz 6G wireless communication networks' reliability. Both active and passive RIS/IRS can be used at the THz band. The active RIS has the superiority of delivering distributed THz access points (APs) or signal repeaters to deflect THz signals facing definite directions by showing random phase shifts. In addition, RIS can be used to design virtual multi-connectivity (MC) links when the number of APs in the structure or proximity is confined [3–5]. Finally, RISs/IRSs can supply a propagation path without LoS [6]. Figure 9.1 shows using the RISs/IRSs technologies to enable 6G communication links despite obstacles.

U. Nissanov, G. Singh, *Antenna Technology for Terahertz Wireless Communication*,
https://doi.org/10.1007/978-3-031-35900-2_9

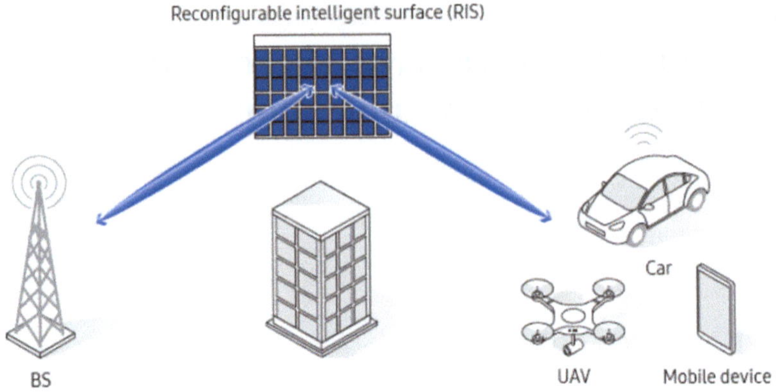

Fig. 9.1 RISs/IRSs technologies concept [6]

9.3 Terahertz Reconfigurable Microstrip Antenna

As the contemporary period of wireless communications began, the spreading psychic has been noticeable, like a casual behaving unity through the receiver and the transmitter that usually reduces the received signal quality because of the not controllable reciprocal actions of the transmitted RF within the environment substances. Still, this common perception has been altered by the recent revelation of RISs, intelligent mirrors, hypersurfaces (HSFs), and programmable wireless environments (PWEs), which enable the proactive control of the refraction, scattering, and reflection characteristics of the RF waves by conquering the harmful effects of natural environment wireless spreading [7].

As the 5G cellular communication networks are deployed, scientists are investigating technologies for the 6G future-generation cellular communications to obtain faster and extra dependable data transmissions. These telecommunication systems include UM-MIMO, reconfigurable, and beamforming THz microstrip antennas have given these wanted parameters [8]. THz wireless communication has captured the heed of numerous antenna scientists, particularly since the emergence of graphene as an antenna design material. The graphene can enable surfaces' Plasmon polaritons (SPPs) modes at THz frequencies and scale down different communication systems for communication and sensing goals. The design of antennas by utilizing graphene supplies proper matching, high miniaturization, decent total efficiency, and reconfiguration abilities [9–11].

The growing demand for 6G communications has motivated recent research on state-of-the-art transceiver hardware architectures and relevant communication algorithms. Such hardware architectures contain vast numbers of EM radiators, motivating the way for UM-MIMO communications [12–13]. The 6G cellular communications need to be reconfigurable to mitigate this issue. They will own the intelligence and capacity to encourage the most appropriate communications tactics founded on feedback from channel activities sensing and signal quality estimation by RIS/IRS. 6G reconfigurable THz antennas suggest a remedy that allows the integration of multiple radios into SoC design. Integrating numerous reconfigurable radiators has become a desired and fundamental property of modern RF systems to the 6G THz next-generation cellular communication, satellite communications, sensing, and imaging. There is a movement toward incorporating RF, intelligent, and cognitive devices to sense the surrounding RF environment and communicate simultaneously in crowded surroundings. Several new required abilities encompass software-defined radios (SDRs), frequency-agile, and cognitive-radios (CRs) to sustain the reconfigurable multiband, multiservice operation, and power management efficient spectrum. The reconfigurable antenna's capability to change their capacity supports them in changing their impedance BW, working frequency, polarization, and primary beam-lobe orientation or integration. Such antennas can supply more extensive covering by beamforming the angle of the primary lobe, refraining from noise sources, and upgrading the beamforming ability. These ideas can notably decrease hardware intricacy, the component's number, and expense equalized to current RF technology that depends on communications design without flexible hardware. A reconfigurable antenna may be used to refrain interference, increase security, and mitigate signal-quality reduction generated by multipath fading employing UM-MIMO and CRs communications methods. A UM-MIMO system with many radiators can efficiently supply spectral efficiency within simple signal processing algorithms. Therefore, UM-MIMO's potential has motivated the incorporation of this technology in the 5G new radio (5G-NR) interface. Furthermore, these UM-MIMO transceivers within incredibly massive radiators are critical technologies for 6G communications [13].

Reconfigurable antennas may be graded upon four diverse kinds [13–14] (a) Frequency reconfigurable antenna: a radiating element, which may change its operating frequency with hoping between unequal frequency BWs. Frequency reconfigurable is apperceived by making some notches or tuning in the antenna reflection coefficient. (b) Pattern reconfigurable antenna: the radiator element, which may adjust its radiation pattern. The radiation pattern antenna changes this kind's directivity, shape, or direction. (c) Polarization reconfigurable antenna: a radiating element that may alter its polarization, like left-hand or right-hand circular polarized (LHCP, RHCP), horizontal/vertical. In this type, the antenna can alter, for example, from horizontal to RHCP. (d) This type is an integration of the three types above. For instance, an antenna that can achieve a polarization diversity with a frequency reconfigurable antenna simultaneously. Reconfigurable antennas with adaptive radiation feathers are needed as alternative antennas, and their configurations are equaled to certain dements to obtain this pliancy.

Four primary methods of reconfiguration techniques are used to facilitate reconfigurable antennas [13–14]. (a) Antennas are established upon selectively switching in or out portions of the antenna pattern within RF MEMS or PIN diode and varactor diodes. The PIN diode is a p-type semiconductor with an un-doped intrinsic and n-type semiconductor to alter their surface currents that are electrically reconfigurable. (b) Antennas support photoconductive changing components, which are named optically reconfigurable antennas. (c) Physically reconfigurable antennas can be apperceived via altering the antenna structure shape. (d) Reconfigurable antennas can be fulfilled through intelligent components, within which the feathers of the antenna's laminate can be altered, like liquid crystals and ferrites.

9.4 PIN Diode Fundamentals

The PIN diode is an RF current-controlled resistor. The n-type and p-type sections are generally deeply doped because they are used for ohmic connections. The resistance level of the PIN diode is derived via a forward-biased direct current (DC). At attenuator and switch implementations, the PIN diode must optimally master the level of RF signal without any signal distortion, changing the form of the RF signal. A significant characteristic of the PIN diode is the inherent capability to manage significant RF signals when applying significantly lower levels of DC biasing. PIN diode switches are frequently used as superior mmWave switches because of superior cut-off frequencies (up to 1500 GHz), supreme third-order-intercept-point (IP3), and high power handling ability [15]. A PIN diode, applied as a switch, has a defined pair of electrical equivalents: ON and OFF states. The ON state is often depicted per a parasitic inductor L, generally less than 0.15 nH, in serial within a small ohmic resistor R_{ON}. Moreover, the following equation can be used [16–17]:

$$R_{ON} = \frac{T_i^2\left(\mu_n + \mu_p\right)}{4\left(\mu_n \cdot \mu_p\right)Q} \tag{9.1}$$

where μ_n is the mobility of an electron, T_i is the intrinsic "I" region thickness, μ_p is the mobility of a hole and $Q = I_{PN} \cdot \tau_c$ is the "I" region charge, where τ_c is the average carrier lifetime in the intrinsic region, while I_{PN} is the DC bias. R_{ON} typically have a resistance of about 10 Ω. A parasitic capacitor represents the OFF state (C_{OFF}) in serial within the exact parasitic inductor L as the ON state [16–17]:

$$C_{OFF} = \frac{\varepsilon_0 \varepsilon_i A}{T_i} \tag{9.2}$$

where A is the p-type semiconductor area, ε_i is the intrinsic "I" region, relative ε_r, and ε_0 is the vacuum permittivity. C_{OFF} is generally less than 50 fF (fF $= 10^{-15}$F). Figure 9.2 shows the PIN diode equivalent electrical states, Fig. 9.2a shows the OFF state, and Fig. 9.2b shows the ON state.

The isolation and insertion loss are the electrical feathers connected with the design and fabrication quality. The insertion loss depends on matching the signal

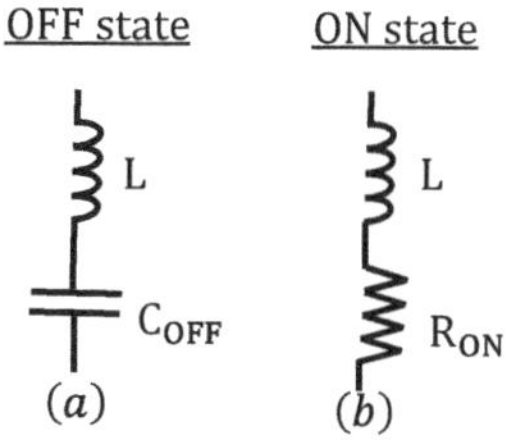

Fig. 9.2 Electrical equivalent models of the PIN diode (**a**) OFF state and (**b**) ON state [16]

paths and the conductor/transmission-line interconnect within the PIN diode switch. The insertion loss of the PIN diode switch can be reduced by reducing the intrinsic layer thickness T_i, the growing bias current, carrier mobility, and lifetime. While the isolation depends on the series or shunt connection of the PIN diode switch.

9.5 Problem Formulation

RISs/IRSs, determined by the reflective radio, have recently attracted significant attention from the pair industry and academia because of their ability to intelligently reconfigure 6G wireless communication environments within commercial reflecting elements (REs) [18–20]. Based on the EM scattering fundamentals, each RE can reflect the incident RF signal within an adjustable reflecting factor (phase and amplitude). Within the appropriate design of the reflecting factors at each RE, RIS/IRS can increase the signal receiving at the wanted destinations and mitigate the interference to accidental users [21], consequently artificially executing a favorable propagation situation and boosting the existing 6G wireless communication not needing any sizeable RF chain [22]. Therefore, equalized to standard wireless communication within complicated RF chain components, RIS has developed as a critical enabler in achieving additional spectrum and energy-efficient wireless communications.

9.6 Related Work

A couple of reconfigurable 0.3 THz and 1.2 THz graphene-brassy Vivaldi and graphene Vivaldi antennas have been designed with the CST MWS. The reconfigurable feature is performed by altering the bias of the electrical field ($\vec{E}$) of the graphene stratum. The simulation results of a graphene Vivaldi antenna and the graphene-brassy Vivaldi of the maximal gain and working frequency band are 1 dB, 0.28–1.4 THz, 7 dB, 0.21–0.67 THz, appropriately [23]. A 2.5 THz hybrid DRA coupled with a graphene Plasmonic dipole antenna has been designed with

CST MWS. The reconfigurable was done by altering the bias of the $\vec{E}$ of the graphene stratum. The maximal simulation results of the radiation efficiency and gain are 70%, 7 dB [24]. A dual-band of 4–5 THz and 6.5–7.5 THz reconfigurable patch antenna using a graphene backing cavity was designed with the CST MWS. The reconfigurable was done by altering the bias of the $\vec{E}$ of the graphene stratum. The bandwidth, primary-lobe angles, and maximal directivity simulation results are 1 THz, −15–25°, 7.2 dBi, appropriately [25]. Based on the simple patch radiator, a 0.22 THz 15 × 15 reconfigurable 15 × 15 reflectarray antenna (RRA) has been designed with the Ansys HFSS. A PIN diode was reconfigurable. The simulation results of the aperture efficiency, maximal directivity, and range of beam-steering angle are 43.7%, 21 dBi, 0–50°, appropriately [26]. A couple of uniform graphenes and PBG graphene, dual-band of 0.85–1.04 THz antennas have been designed with the CST MWS. The reconfigurable was done by altering the bias of the $\vec{E}$ of the graphene stratum. The simulation results of the maximal gain, directivity, BW, and radiation efficiency achieved for the PBG graphene and uniform graphene are 16.4 dB, 17 dBi, 33.34 GHz, 86.67%, 15.89 dB, 16.8 dBi, 28.13 GHz, and 87%, appropriately [27]. A 0.6 THz with couple-radiators patch reconfigurable was designed with the CST MWS. The reconfigurable was done with a couple of PIN diodes. The simulation results of the primary-lobe angles, efficiency, BW, maximal gain, and directivity are −26°, 26°, 97%, 54.66 GHz, 7 dB, and 7.1 dBi, appropriately [28]. A 1.78 THz with three graphene formations with a reconfigurable antenna was designed with the CST MWS. The reconfigurable was done by altering the bias of the $\vec{E}$ of the graphene stratum. The simulation results of the main-lobe angles and the maximal directivity for the three antennas are −30 up to −70°, 5.8 dBi, appropriately [29]. A 0.5 THz beam reconfigurable quasi-Yagi-Uda microstrip with graphene-patch directors antenna working at 0.495–0.51 THz has been designed with the Ansys HFSS. The reconfigurable was done by altering the bias of the $\vec{E}$ of the graphene stratum. The simulation results of the primary-lobe angles, BW, maximal gain, and radiation efficiency are 30° up to 150°, 15 GHz, 7.5 dB, and 96%, appropriately [30]. A 4.85–4.88 THz bandwidth reconfigurable based on a homocentric ring slot graphene antenna, while the operation frequency was between 4.667–5.073 THz, has been designed with the CST MWS. The reconfigurable was done by altering the bias of the $\vec{E}$ of the graphene stratum. The BW and maximal gain simulation results are appropriately 0.255–0.406 THz, 8.6 dB [31]. A 0.99–1.24 THz reconfigurable graphene-based microstrip antenna and an operating frequency were between 0.7 and 1.505 THz designed with the Ansys HFSS. The reconfigurable was done by altering the bias of the $\vec{E}$ of the graphene stratum. The maximal bandwidth and radiation efficiency simulation results are appropriately 0.805 THz, 69% [32]. The performance summary of these THz antennas discussed above is presented in Table 9.1 according to the related work.

Table 9.1 THz antennas classification and performance according to related work in Chap. 9

Refs.	Antenna techno.	Reconfig. technique	Resona. freq. (THz)	Working freq. (THz)	Peak. Dir./ gain (dBi/dB)	Primary-lobe angles (°)
[23]	Graphene Vivaldi	$\vec{E}$ bias of the graphene	≈1.2	0.28–1.4	1 dB	n.a
	Graphene-metal Vivaldi		≈0.3	0.21–0.67	7 dB	n.a
[24]	Hybrid DRA with a graphene dipole antenna	$\vec{E}$ bias of the graphene	2.5	1–4	7 dB	n.a
[25]	Microstrip graphene backing-cavity	$\vec{E}$ bias of the graphene	4.5, 7	4–5, 6.5–7.5	7.2 dBi	−15–25
[26]	15 × 15 microstrip patch RRA	PIN diode	0.22	0.215–0.225	21 dBi	0–50
[27]	Graphene PBG graphene	$\vec{E}$ bias of the graphene	0.904 0.991	0.89–0.91 0.97–1.16	15.9 dB/ 16.9 dBi	n.a
[28]	A couple of radiators patch	2 PIN diodes	0.6	0.573–0.627	16.4 dB/ 17 dBi	−26,26
[29]	Graphene antenna	$\vec{E}$ bias of the graphene	1.78	n.a	5.8 dBi	−30, −70
[30]	Graphene microstrip quasi–Yagi Uda	$\vec{E}$ bias of the graphene	0.5	0.495–0.51	7.5 dB	30–150
[31]	Graphene	$\vec{E}$ bias of the graphene	4.754–4.88	4.667–5.073	8.6	n.a
[32]	Graphene	$\vec{E}$ bias of the graphene	0.99–1.24	0.8	n.a	n.a

9.7 Simulation Results and Discussion

9.7.1 THz Ultra-Wideband (UWB) Reconfigurable Microstrip Antenna

The authors chose that the antenna will be resonated at 100–303 GHz for this design. Moreover, the chosen substrate of BCB from the CYCLOTENE series 3000 polymer is a dielectric laminate, in which (tanδ and ε_r) are measured up to 65 GHz, while these specifications virtually steady up to 1.5 THz [33], and this laminate ε_r is virtually steady when the surroundings alter the temperature until 150 °C [34–36], and the chosen BCB series 3000 polymer parameters of ε_r, tan δ, laminate height (h), gold thickness (t), (2.6, 0.0009) at 65 GHz 30 μm, 2 μm, appropriately, are

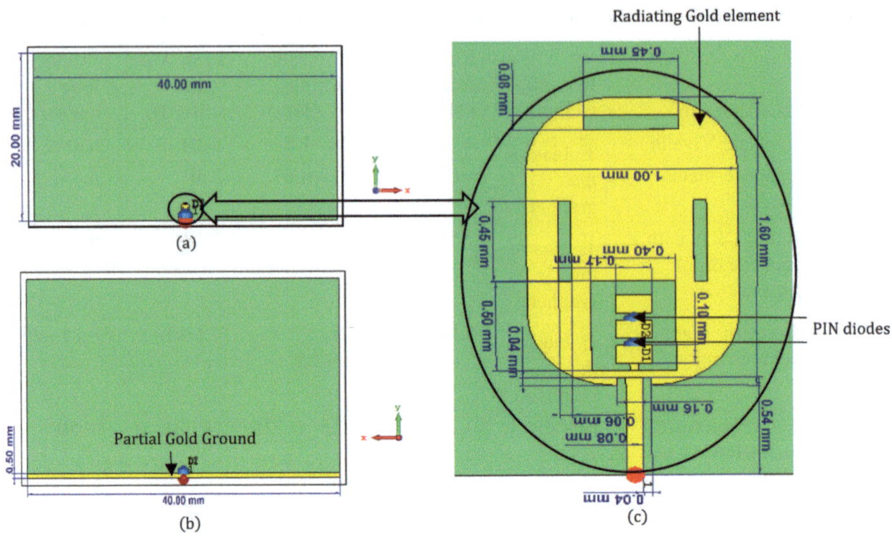

Fig. 9.3 Configuration of the THz UWB reconfigurable microstrip antenna, (**a**) Front view, (**b**) Back view, and (**c**) Zoomed view of the gold radiator

Table 9.2 The different modes of the THz UWB reconfigurable microstrip antenna

Mode	PIN diode D1 level	PIN diode D2 level
4	ON	ON
3	ON	OFF
2	OFF	ON
1	OFF	OFF

($h = 30\mu m \leq 0.125\lambda_{high}$),($t = 2\ \mu m \leq 0.125\lambda_{high}$, $\varepsilon_r = 2.6 < 3.5$, $\tan\delta = 0.0009 < 0.0095$), and they will guarantee that the conduction losses, the substrate losses, and the surface radiation losses are reduced to a minimum, which is the main concern at the THz band [37–39], and the simulation results should be as much as can be close with empirical validation. This antenna included a state-of-the-art gold radiator based within BCB polymer and two PIN diode switches (D1, D2) presented in Fig. 9.3a–c. The magnitudes of the substrate were $40 \times 20\ mm^2$.

The PIN diode switches (D1, D2) were simulated via CST MWS as a lumped-components (C, R, and L) following the PIN diode equivalent electrical models at OFF and ON levels, while the fair values of C, R, and L of the PIN diodes were set up following the optimization via CST MWS simulator as 4 fF, 10 Ω, and 0.1 nH, appropriately. The variant modes are shown in Table 9.2.

The simulated result of the S_{11} of the THz UWB reconfigurable microstrip antenna is shown in Fig. 9.4a–d, which depends on modes 1–4. From these figures it has been shown that the BWs of mode 1–mode 4 are: BW($S_{11} \leq -10dB$) > 159.98 GHz(>80.54%), BW($S_{11} \leq -10dB$) > 201.77 GHz(>100.44%), BW($S_{11} \leq -10dB$) > 159.96 GHz(>88.88%), BW($S_{11} \leq -10dB$) > 168.05 GHz (>91.31%), appropriately, while the resonance frequencies of mode 1–4 are: 0.202 THz, 0.179 THz, 0.212 THz, and 0.128 THz, appropriately.

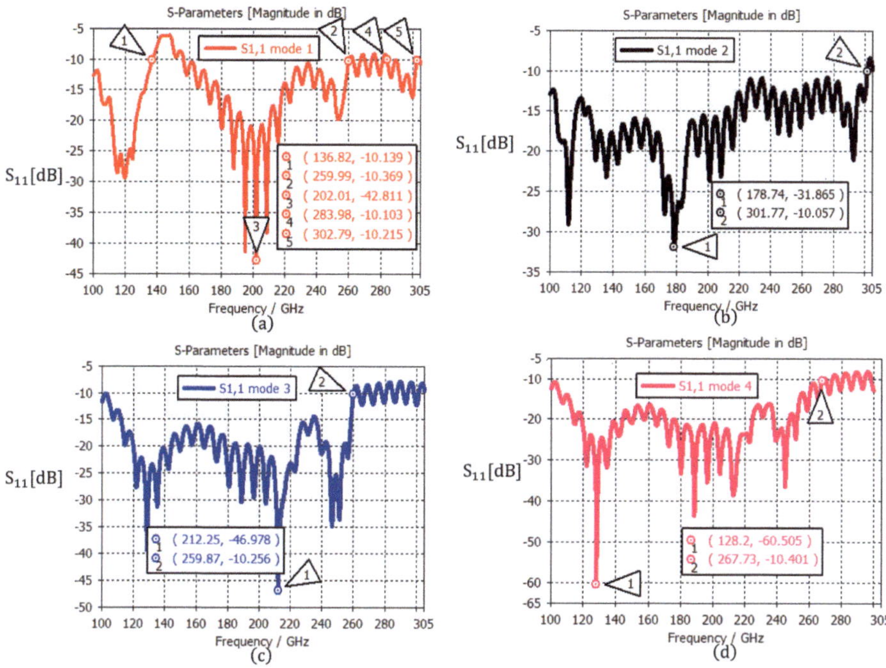

Fig. 9.4 Simulation results of the S_{11} for the THz UWB reconfigurable microstrip antenna, (**a**) For mode 1, (**b**) For mode 2, (**c**) For mode 3, and (**d**) For mode 4

The simulated result of the gain at Phi = 44° for the THz UWB reconfigurable microstrip antenna is shown in Figs. 9.5a–d, which depends on modes 1–4. From these figures it has been shown that the maximal gain of the modes 1–4 are: 8 dB, 8.47 dB, 8.46 dB, and 8.6 dB, appropriately, at 100–305 GHz.

Figure 9.6a, b show the simulation results of the gain and directivity at the E-plane and H-plane of the offered THz UWB reconfigurable microstrip antenna at a frequency of 0.16 THz at the polar plots for modes 1–4. In addition, the apperceived simulation results are shown in the following Table 9.3.

Figures 9.7a–d shows the simulation results of the directivity in (dBi) at the 3D result of the proposed THz UWB reconfigurable microstrip (modes 1–4) @160 GHz. It has been shown that the directivity at modes 1–4 was 8.32 dBi, 8.82 dBi, 8.76 dBi, and 8.9 dBi @ 160 GHz, respectively.

The simulated result of radiating and the total efficiency for the THz UWB reconfigurable microstrip antenna for modes 1–4 is shown in Fig. 9.8a–d. This figure shows that the maximal radiating and total efficiency for modes 1–4 at 100–305 GHz are 94.4% and 93.44%, respectively.

The simulated result of the E-field for the THz UWB reconfigurable microstrip antenna for modes 1–4 at 0.1 THz, 0.18 THz, and 0.3 THz is shown in Fig. 9.9a–d. From these Figures, it has been shown that the E-field of the THz UWB

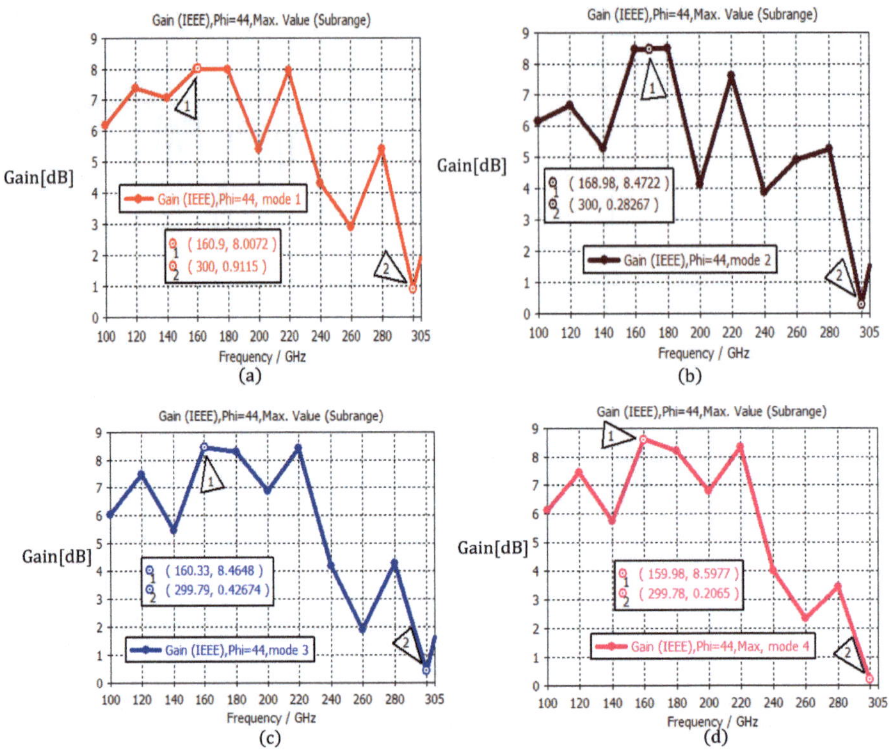

Fig. 9.5 Simulation results of the gain for the THz UWB reconfigurable microstrip antenna, (**a**) For mode 1, (**b**) For mode 2, (**c**) For mode 3, and (**d**) For mode 4

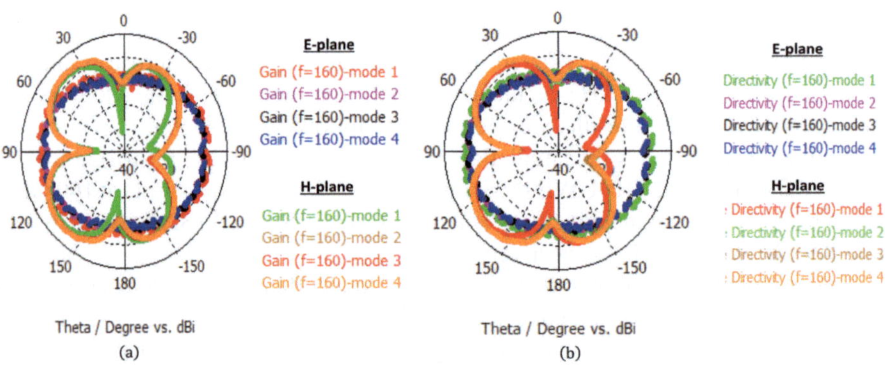

Fig. 9.6 Simulation results of E-plane and H-plane of the THz UWB reconfigurable microstrip antenna for modes 1–4, (**a**) Gain, and (**b**) Directivity

Table 9.3 The simulation results of the gain and directivity at the E-plane and H-plane of the THz UWB reconfigurable microstrip antenna

Name	Gain/directivity [dB/dBi]	Main lobe direction [°]	Angular width (–3 dB) [°]	SLL [dB]
E-plane gain mode 1	1.82	115	70.4	−1.3
E-plane gain mode 2	−0.4	107	73.7	−0.5
E-plane gain mode 3	−0.4	107	73.8	−0.9
E-plane gain mode 4	−0.44	−107	69.1	−0.6
H-plane gain mode 1	2.84	138	37.6	−0.5
H-plane gain mode 2	4.02	138	40.2	−0.6
H-plane gain mode 3	4.05	139	40.3	−0.6
H-plane gain mode 4	4.16	138	40.6	−0.6
E-plane directivity mode 1	2.21	115	70.4	−1.3
E-plane directivity mode 2	0.061	107	73.7	−0.5
E-plane directivity mode 3	−0.015	107	73.8	−0.9
E-plane directivity mode 4	−0.04	−107	69.1	−0.6
H-plane directivity mode 1	3.24	138	37.6	−0.5
H-plane directivity mode 2	4.47	138	40.2	−0.6
H-plane directivity mode 3	4.43	139	40.3	−0.6
H-plane directivity mode 4	4.56	138	40.6	−0.6

reconfigurable antenna is between −0.46 and 6.1 dB at the frequency of 0.1 THz, 0.18 THz, and 0.3 THz. Moreover, from these figures, it also has been shown that the main lobes' angles are: 103°, 97°, 96°, 90°, 77°, and −90° when these angles depend on the PIN diodes D1 and D2 modes.

The simulated result of surface current scattering for the THz UWB reconfigurable microstrip antenna at the frequency of 0.1 THz and 0.3 THz of the THz UWB reconfigurable antenna for modes 1–4 are shown in Fig. 9.10a–h. From these figures, it has been shown that the THz UWB reconfigurable antenna surface-current scattering is up to 1531 A/m at the frequency of 0.1 THz and 0.3 THz. Moreover, from these fFigures. it has been shown that the surface current scattering is centralized close to the gold radiator edges and at the feeding line due to the skin depth effect. Moreover, the surface current scattering is oppositely connected to the frequency, i.e., at a low frequency, the surface current scattering is upper, and on the contrary, the surface current scattering is virtually autonomous of the PIN diodes D1 and D2 modes.

The simulation results of S_{11} and the gain from comparisons between mode 1 and mode 4 of THz UWB reconfigurable microstrip antenna that depends on the couple PIN diodes modes are shown in Fig. 9.11a, b. From Fig. 9.11a it has been shown that the THz UWB reconfigurable microstrip antenna has six variant operation bands, which are 0.1–0.136 THz, 0.156–0.26 THz, 0.283–0.302 THz, 0.1–0.301 THz, 0.1–0.26 THz, and 0.1–0.268 THz. From Fig. 9.11b it has been shown that the gain of THz UWB reconfigurable antenna is between 0.17 and 8.59 dB at 0.1–0.3 THz.

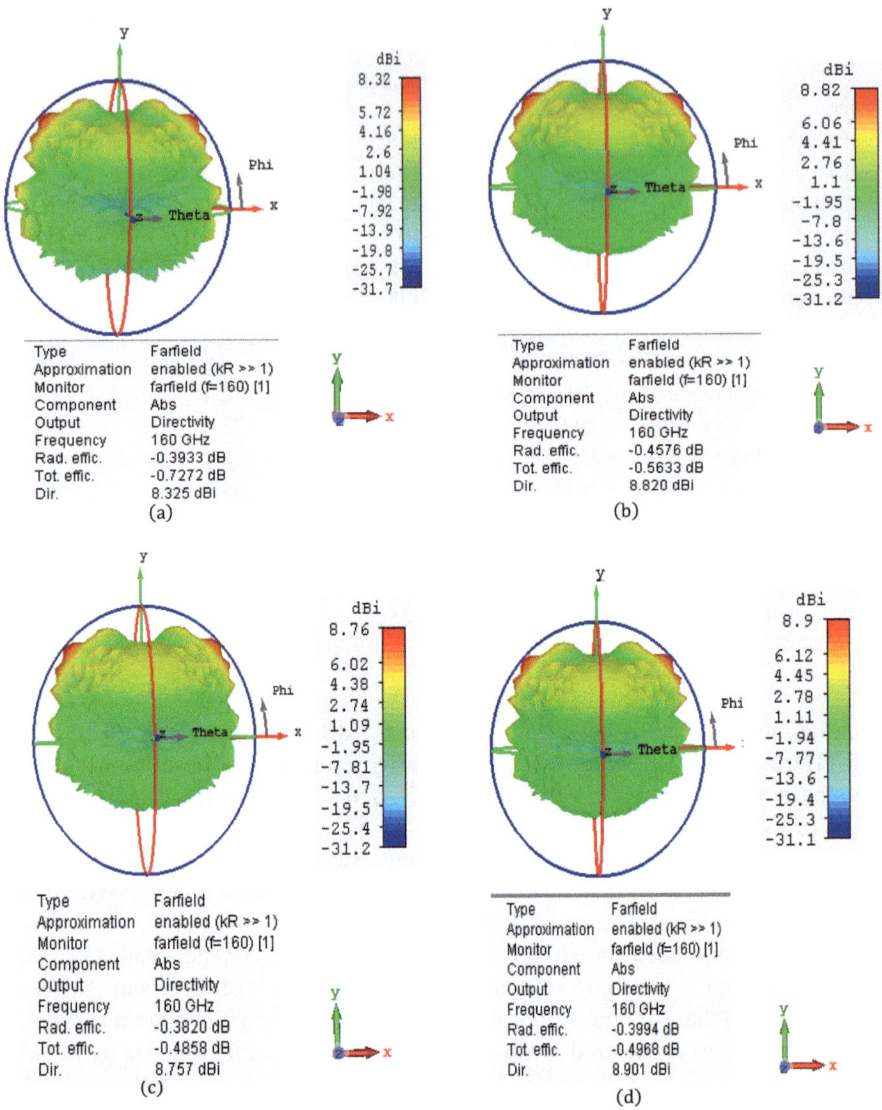

Fig. 9.7 Simulation results of the directivity at the 3D result of the THz UWB reconfigurable microstrip antenna, (**a**) For mode 1, (**b**) For mode 2, (**c**) For mode 3, and (**d**) For mode 4

Figure 9.12a–d exhibits the simulation results of the S_{11} for the proposed THz UWB reconfigurable microstrip antenna (modes 1–4) with or without etching accuracy ±5 μm; half of the declared etching accuracy [40] is supposed to be when this antenna is made. This figure may show that the frequency working band may change at a frequency range of about 8 GHz when this reconfigurable antenna is made.

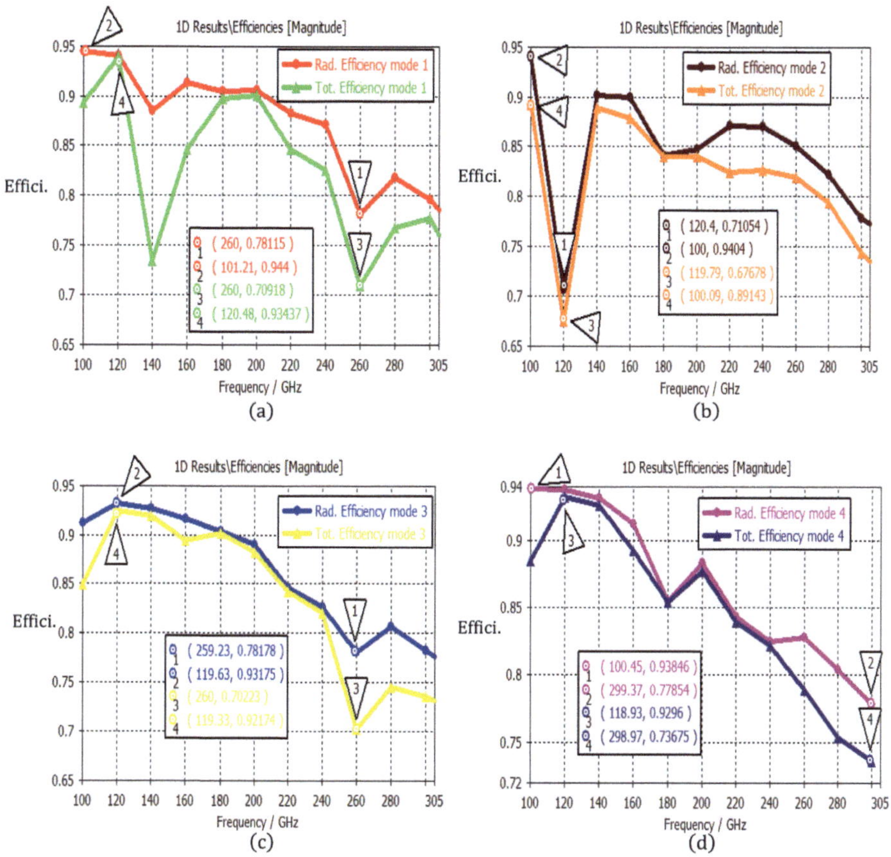

Fig. 9.8 Simulation results of the radiation and the total efficiency of the THz UWB reconfigurable microstrip antenna, (**a**) For mode 1, (**b**) For mode 2, (**c**) For mode 3, and (**d**) For mode 4

Figure 9.13a, b exhibits the simulation results from a comparison of the S_{11} and the gain of the offered reconfigurable antenna that depends on a couple of PIN diode modes (modes 1–4), depending on the frequency-domain (FEM) and the time-domain (FIM) solvers at the CST MWS. It has been shown that a good agreement between simulation results is achieved regarding the offered reconfigurable antenna that verified the offered reconfigurable antenna design.

The performance summary of these designed and simulate THz antennas presented above in Chap. 9 is presented in Table 9.4.

From Table 9.4, it has been shown that the maximal BW enhancement between typical Unit-cell THz planar rectangular microstrip antenna and the THz UWB reconfigurable microstrip antenna was 192.88%.

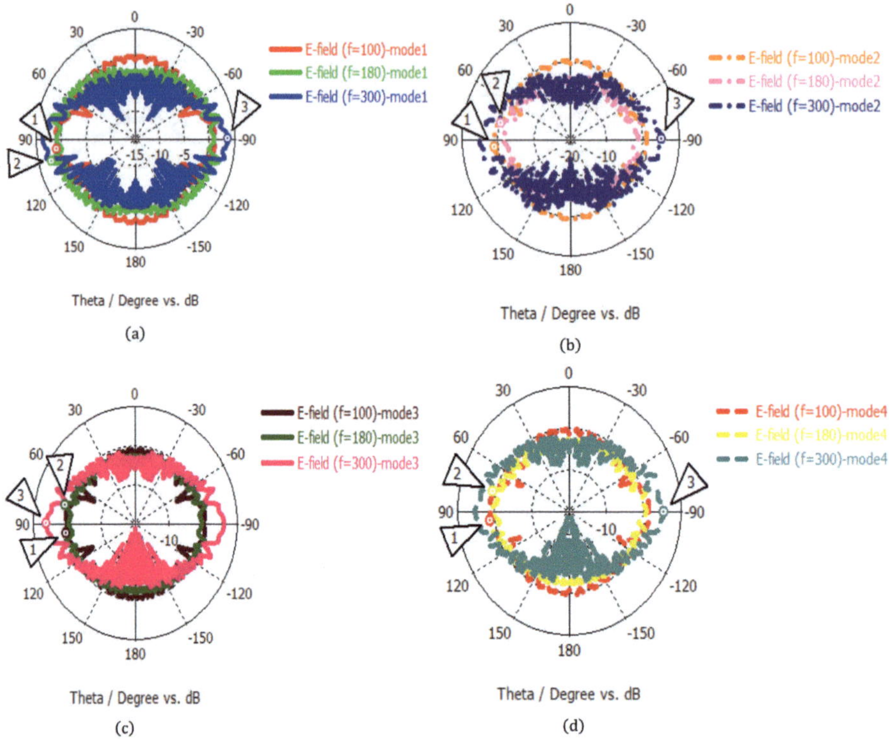

Fig. 9.9 Simulation results of the E-field for the THz UWB reconfigurable microstrip antenna, (**a**) For mode 1, (**b**) For mode 2, (**c**) For mode 3, and (**d**) For mode 4

9.7.2 Discussion

RISs/IRSs can smartly deflect the beams of incident RF waves in diverse paths by fine-tuning the phase shifts using an intelligent processor. This process acquiesces RIS to alter an LoS THz channel within a multipath software-detailed channel within simulated LoS links. The rapid growth of cellular communications systems is blocked because the EM range has become highly congested. Therefore, 6G cellular communications must be reconfigurable at THz to mitigate this problem. Moreover, they need the intelligence and capacity to sustain the appropriate communications tactics established upon the signal feedback estimation and feedback from sensing the channel performances. 6G THz reconfigurable antennas will allow supplying a remedy, which allows the integration of numerous RF into SoC design. Integrating multiple radiators with reconfigurable antennas has become a desired feature and a critical novel RF system for 6G wireless communication networks. So, RIS can be adequately used to facilitate THz signals' blockage susceptibility, improving THz 6G wireless communication networks' reliability [3–5].

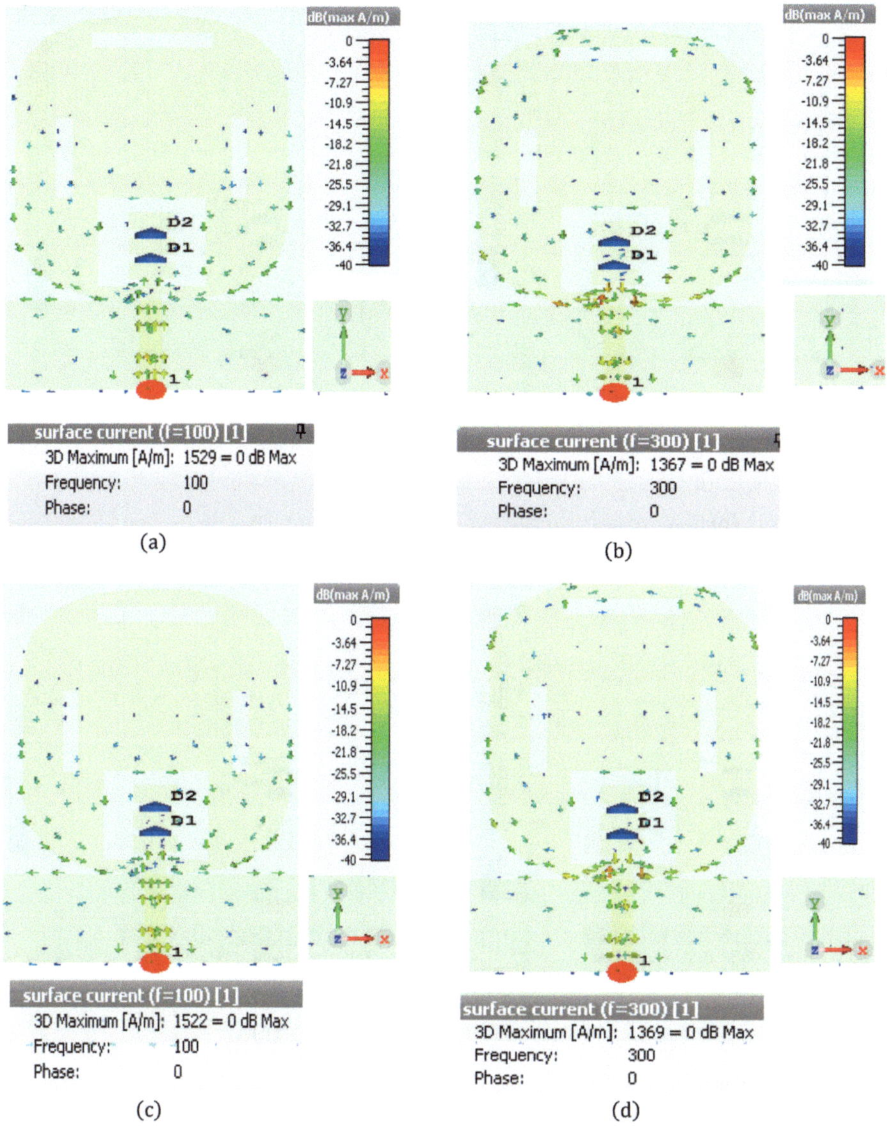

Fig. 9.10 Simulation results of the surface current scattering of the THz UWB reconfigurable microstrip antenna, (**a**)–(**b**) for mode 1, (**c**)–(**d**) for mode 2, (**e**)–(**f**) for mode 3, (**g**)–(**h**) for mode 4

This chapter shows the design and simulation of the THz UWB reconfigurable microstrip antenna. The offered THz UWB reconfigurable microstrip antenna is established on the BCB from the series 3000 polymer as a dielectric laminate, in which the (tanδ and ε_r) are measured up to 65 GHz, while these specifications are virtually steady up to 1.5 THz [33], and this laminate ε_r is virtually steady when the

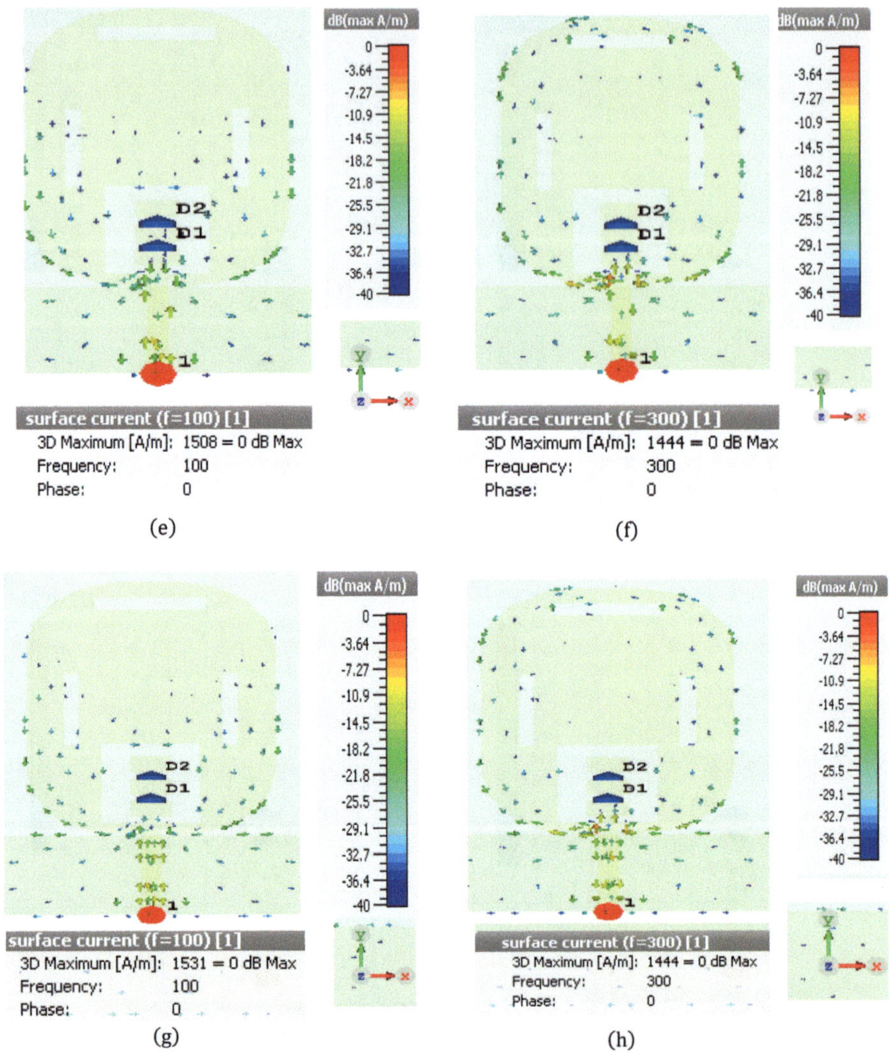

Fig. 9.10 (continued)

surroundings alter the temperature until 150 °C [34–36], and the chosen BCB series 3000 polymer parameters of ε_r, tan δ, laminate height (h), gold thickness (t), (2.6, 0.0009) at 65 GHz 30 μm, 2 μm, appropriately, which are ($h = 30\mu m \leq 0.125\lambda_{high}$), ($t = 2\ \mu m \leq 0.125\lambda_{high}, \varepsilon_r = 2.6 < 3.5, \tan\delta = 0.0009 < 0.0095$), and they will guarantee that the conduction losses, the substrate losses, and the surface radiation losses are reduced to a minimum, which is the main concern at the THz band [37–39], and the simulation results should be as much as can be close with empirical validation.

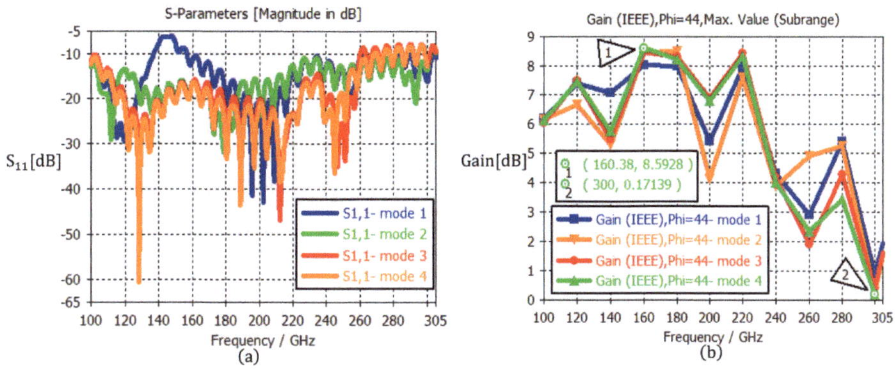

Fig. 9.11 The simulation results from comparison for the THz UWB reconfigurable microstrip antenna for modes 1–4, (**a**) S_{11},(**b**) Gain

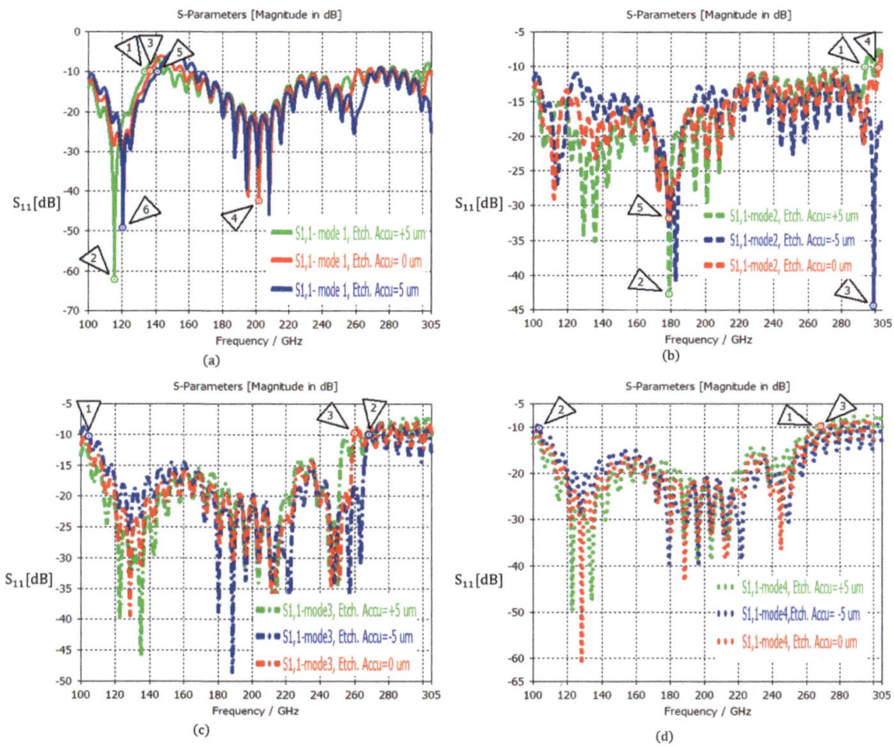

Fig. 9.12 Simulation results of the S_{11} of the proposed THz UWB reconfigurable microstrip antenna with or without the etching accuracy, (**a**) for mode 1, (**b**) for mode 2, (**c**) for mode 3, (**d**) for mode 4

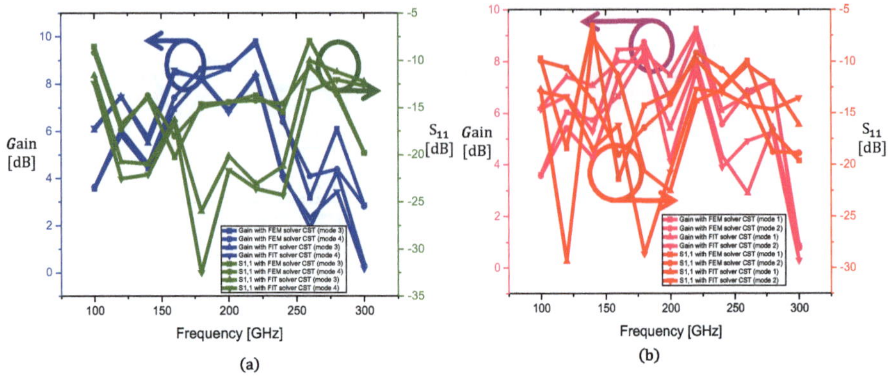

Fig. 9.13 Simulation results from a comparison of the gain and the $S_{11}$for the THz UWB reconfigurable microstrip antenna made with the frequency-domain (FEM) solver and time-domain (FIM) solver at the CST MWS simulator, (**a**) for mode 1–2, (**b**) for mode 3–4

Table 9.4 The performance summary of the designed and simulated THz antenna in Chap. 9

Offered THz antenna	Resonance freq. (GHz)	BW (GHz/%)	BW enhance. vs. *typical UCell (%)
Reconfigurable microstrip antenna at mode 1	202.01	>159.98 GHz (>80.54%)	>151.08
Reconfigurable microstrip antenna at mode 2	178.74	>201.77 GHz (>100.44%)	>192.88
Reconfigurable microstrip antenna at mode 3	212.25	>159.96 GHz (>88.88%)	>167.76
Reconfigurable microstrip antenna at mode 4	128.2	>168.05 GHz (>91.31%)	>172.62
*Typical unit-cell microstrip antenna	–	5%	–

The technique to reconfigure the offered microstrip antenna is electrical switching with two PIN diodes [16–17]. The offered reconfigurable microstrip antenna may alter its main-lobes angle, impedance BW, or each within a single antenna [16–17]. The identical resistor, inductor, and capacitor for the OFF and ON levels of these PIN diodes must be 10 ohms, 0.1 nH, and four fF, appropriately. Following the simulation within the FIT solver at the CST MWS simulator, the offered THz UWB reconfigurable microstrip antenna has six operation bands. The bands are 0.1–0.136 THz, 0.156–0.26 THz, 0.284–0.302 THz, 0.1–0.301 THz, 0.1–0.26 THz, and 0.1–0.268 THz that depends upon a couple of PIN diodes modes, while the main-lobe angles and the maximal gain were 103°,97°, 96°, 90°, 77°, –90°, and 8.59 dB, appropriately and depends on the PIN diodes modes.

A state-of-the-art comparison was made to the offered THz UWB reconfigurable microstrip antenna, and it may have been shown that the working frequency band of the offered reconfigurable microstrip antenna can be changed by about 8 GHz when this antenna is made due to etching accuracy, which is ±5 μm, half of the declared etching accuracy [40].

A simulator validation/comparison was chosen [41–46]. The simulation results of the BW and gain of the offered THz UWB reconfigurable microstrip antenna were done via the FIM solver at the CST MWS simulator compared with the simulation via the FEM solver at the same simulator. This comparison is as close as can be to an empirical verification with the manufacturing of an archetype THz antenna because at both simulators, the tan δ, $\varepsilon_r$was applied at a frequency of 65 GHz. At the same time, these specifications are virtually steady up to 1.5 THz [33]. Therefore, a fair concurrence is achieved, supporting this design.

9.8 Open Research Challenges

Based on this chapter, the THz reconfigurable microstrip antenna for 6G has several open research challenges:

1. Further investigation and improvements on high gain and broadband beyond 100.5% of THz reconfigurable microstrip antenna.
2. Additional analysis and evolutions on THz PIN diodes and graphene for THz reconfigurable microstrip antennas.

9.9 Summary

This chapter discusses THz's reconfigurable microstrip antenna for 6G. The rapid growth of cellular communications systems is blocked because the EM range has become highly congested. Therefore, 6G wireless communications networks must be at THz and reconfigurable to mitigate this problem. Moreover, they need the intelligence and capacity to sustain the appropriate communications tactics established upon the signal feedback estimation and feedback from sensing the channel performances using the RISs/IRSs. 6G THz reconfigurable antennas will allow supplying a remedy, which allows the integration of numerous RF into SoC design. Integrating multiple radiators with reconfigurable antennas has become a desired feature and a critical novel RF system for B5G or 6G cellular communication.

The idea of a THz UWB reconfigurable planar microstrip antenna for the frequency range 0.1–0.33 THz within a new-fangled gold radiator patch within a couple of PIN diodes mounted on a BCB polymer is shown. The identical resistor, inductor, and capacitor for the OFF and ON levels of these PIN diodes must be 10 ohms, 0.1 nH, and four fF, appropriately. Furthermore, these PIN diodes should operate at a frequency range of 0.1–0.31 THz. Unfortunately, to our best knowledge, these PIN diodes are not available at the present THz/ mmWave technology. Therefore, the reconfigurable kinds of the offered antenna are beams, BWs, and frequency reconfigurations. This reconfigurable microstrip antenna was designed via the FIT solver within the CST MWS when the equalization was via the FEM solver within the CST MWS. The simulation results apperceived from both solvers were in decent agreement, approving the recommended antenna design.

A state-of-the-art comparison was made to the offered THz UWB reconfigurable microstrip antenna. It may have been shown that the reconfigurable microstrip antenna's working frequency band can be changed by about 8 GHz when this antenna is made due to etching accuracy, which is ±5 μm.

Following the simulation within the FIT solver at the CST MWS simulator, the offered THz UWB reconfigurable microstrip antenna has six operation bands. The bands are 0.1–0.136 THz, 0.156–0.26 THz, 0.284–0.302 THz, 0.1–0.301 THz, 0.1–0.26 THz, and 0.1–0.268 THz that depends upon a couple of PIN diodes levels, while the main-lobe angles and the maximal gain were 103°, 97°, 96°, 90°, 77°, –90°, and 8.59 dB, appropriately and depends on the PIN diodes levels.

Moreover, the simulation results of the maximal BW and total efficiency apperceived of the offered antenna were >201.77 GHz (>100.44%), 93.4%, appropriately.

References

1. Y. Wu, S. Singh, T. Taleb, A. Roy, H. S. Dhillon, M. R. Kanagarathinam, and A. De, *6G Mobile Wireless Networks.* Springer Nature, Switzerland, pp. 53–82, 2021.
2. R. Chen, M. Liu, Y. Hui, N. Cheng, and J. Li, "Reconfigurable intelligent surfaces for 6G IoT wireless positioning: A contemporary survey," *IEEE Internet of Things Journal*, vol. 9, no. 23, pp. 2357–23582, Dec. 2022.
3. M. Polese, J. M. Jornet, T. Melodia, and M. Zorzi, "Toward end-to-end, full-stack 6G terahertz networks," *IEEE Communications Magazine*, vol. 58, no. 11, pp. 48–54, Nov. 2020.
4. Ian F. Akyildiz, C. Han, and S. Nie, "Combating the distance problem in the millimeter wave and terahertz frequency bands," *IEEE Communications Magazine*, vol. 56, no. 6, pp. 102–108, June 2018.
5. A. Shafie, G. N. Yang, C. Han, J. M. Jornet, M. Juntti, and T. Kurner, "Terahertz communications for 6G and beyond wireless networks: challenges, key advancements, and opportunities," *IEEE Network*, pp. 1–8, Sep. 2022.
6. C. Huang, A. Zappone, G. C. Alexandropoulos, M. Debbah, and C. Yuen, "Reconfigurable intelligent surfaces for energy efficiency in wireless communication," *IEEE Transactions on Wireless Communications*, vol. 18, no. 8, pp. 4157–4170, Aug. 2019.
7. W. Saad, M. Bennis, and M. Chen, "A vision of 6G wireless systems: applications, trends, technologies, and open research problems," *IEEE Network*, vol. 34, no. 3, pp. 134–142, May 2020.
8. J. Zhao, "A survey of reconfigurable intelligent surfaces: towards 6G wireless communication networks with massive MIMO 2.0," [Online] arXiv, pp. 1–7, July 2019, https://arxiv.org/abs/1907.04789.
9. Y. Luo, Q. Zeng, X. Yan, T. Jiang, R. Yang, J. Wang, Y. Wu, Q. Lu, and X. Zhang, "A graphene-established tunable negative refractive index metamaterial and its application in dynamic beam-tilting terahertz antenna," *Microwave and Optical Technology Letters*, vol. 61, no. 12, pp. 2266–2672, Dec. 2019.

10. G. Bansal, A. Marwaha, and A. Singh, "A graphene-established multiband antipodal Vivaldi nanoantenna for UWB applications," *Journal of Computational Electronics*, vol. 19, pp. 709–718, Feb. 2020.
11. C. L. Wang, Y. Q. Wang, H. Hu, D. J. Liu, D. L. Gao, and L. Gao, "Reconfigurable sensor and nanoantenna by graphene-tuned Fano resonance," *Optics Express*, vol. 27, no. 24 / 25, pp. 35925–35935, Nov. 2019.
12. N. Schlesinger, G. C. Alexandropoulos, M. F. Imani, Y. C. Eldar, and R. Smith, "Dynamic metasurface antennas for 6G extreme massive MIMO communications," [Online] arXiv pp. 1–7, June 2020, https://arxiv.org/abs/2006.07838.
13. C. G. Christodoulou, Y. Tawk, S. A. Lane, and S. R. Erwin, "Reconfigurable antennas for wireless and space applications," *Proceedings of the IEEE*, vol. 100, no. 7, pp. 2250–2261, July 2012.
14. N. O. Parchin, H. J. Basherlou, Yasir I. A. Al-Yasir, A. , M. Abdulkhaleq, and R. A. Abd-Alhameed, "Reconfigurable antennas: switching techniques-a survey," *Electronics*, vol. 9, no.2, pp. 336/1–14, May 2020.
15. D. Liu, B. Gaucher, U. Pfeiffer, and J. Grzyb, *Advanced Millimeter-Wave Technologies: Antennas, Packaging, and Circuits.* John Wiley and Sons, UK, pp. 467–468, 2009.
16. Y. Tanaka, H. Uda, H. Hayashi, H. Ueda, and M. Usui, "A 76–77GHz high isolation GaAs PIN-diode switch MMIC," R&D Review of Toyota CRDL, vol. 37, no. 2, pp. 19–26, May 2002.
17. A. Bondarik and D. Sjöberg, "Pattern reconfigurable wideband stacked microstrip patch antenna for 60GHz," *International Journal of Antennas and Propagation*, vol. 2016, pp. 5961309/1–11, May 2016.
18. Y. C. Liang, R. Long, Q. Zhang, J. Chen, H. V. Cheng, and H. Guo, "Large intelligent surface/ antennas LISA: Making reflective radios smart," *Journal of Communications and Information Networks*, vol. 4, no. 2, pp. 40–50, Jun. 2019.
19. Y. C. Liang, Q. Zhang, E. G. Larsson, and G. Y. Li, "Symbiotic radio, cognitive backscattering communications for future wireless networks," *IEEE Transactions on Cognitive Communications and Networking*, vol. 6, no. 4, pp. 1242–1255, Sep. 2020.
20. M. A. ElMossallamy, H. Zhang, L. Song, K. G. Seddik, Z. Han, and G. Y. Li, "Reconfigurable intelligent surfaces for wireless communications: Principles, challenges, and opportunities," *IEEE Transactions on Cognitive Communications and Networking*, vol. 6, no. 3, pp. 990–1002, May 2020.
21. M. Di Renzo, A. Zappone, M. Debbah, M. S. Alouini, C. Yuen, J. de Rosny, and S. Tretyakov, "Smart radio environments empowered by reconfigurable intelligent surfaces: How it works, state of research, and the road ahead," *IEEE Journal on Selected Areas in Communications*, vol. 38, no. 11, pp. 2450–2525, July 2020.
22. M. Di Renzo, M. Debbah, D. T. Phan-Huy, A. Zappone, M. S. Alouini, C. Yuen, V. Sciancalepore, G. C. Alexandropoulos, J. Hoydis, H. Gacanin J. de Rosny, A. Bouncer, G. Lerosey, and M. Fink, "Smart radio environments empowered by reconfigurable AI meta-surfaces: an idea whose time has come," *EURASIP Journal on Wireless Communications and Networking*, vol. 2019, no. 1, pp. 129/1–20, May 2019.
23. J. Jin, Z. Cheng, J. Chen, T. Zhou, C. Wu, and C. Xu, "Reconfigurable terahertz Vivaldi antenna established on a hybrid graphene-metal configuration," *International Journal of RF and Microwave Computer-Aided Engineering*, vol. 30, no. 5, pp. e22175/1–8, Jan 2020.
24. S. E. Hosseininejad, M. Neshat, R. Faraji-Dana, S. Abadal, M. C. Lemme, P. H. Bolivar, E. Alarcon, and A. Cabellos-Aparicio, "Terahertz dielectric resonator antenna coupled to graphene plasmonic dipole," *Proceeding of the IET 12th European Conference on Antennas and Propagation* (EuCAP 2018), 2018, pp. 1–5.
25. Y. Dong, P. Liu, D. Yu, G. Li, and F. Tao, "Dual-band reconfigurable terahertz patch antenna with a graphene-stack-established backing cavity," *IEEE Antennas and Wireless Propagation Letters*, vol. 15, pp. 1541–1544, Feb. 2016.
26. L. Sun, B. Li, M. Wu, and X. Lv, "A 1-bit 220 GHz reconfigurable reflectarray," *Proceedings of IEEE International Conference on Microwave and Millimeter Wave Technology* (ICMMT 2019), Guangzhou, China, 19–22 May 2019, pp. 1–2.

27. R. K. Kushwaha and P. Karuppanan, "Parasitic-coupled high-gain graphene antenna employed on PBG dielectric grating substrate for THz applications," *Microwave Optical Technology Letter*, vol. 62, no. 1, pp. 439–447, Jan. 2020.
28. M. Usman, S. Tanoli, F. Khan, Wasi-Ur-Rehman Khan, S. M. Umar, and S. Ullah, "Pattern reconfigurable two elements printed patch antenna for THz wireless applications," *Proceedings of IEEE 3rd International Conference on Computing, Mathematics and Engineering Technologies (iCoMET2020)*, Sukkur, Pakistan, 29–30 Jan. 2020, pp. 1–7.
29. G. Varshney, "Reconfigurable graphene antenna for THz applications: A mode conversion approach," *Nanotechnology*, vol. 31, no. 13, pp. 135208/1–16, Jan. 2020.
30. Y. Wen-Long, G. Xu-Guang, Z. Yi-Ming, and L. Ping, "Terahertz beam reconfigurable microstrip quasi-Yagi-Uda antenna established on monolayer graphene," *Journal of Infrared, Millimeter, and Terahertz Waves*, vol. 39, no. 1, pp. 39–46, Feb. 2020.
31. H. B. Krid, Z. Houaneb, and H. Zairi, "Reconfigurable graphene annular ring antenna for medical and imaging applications," *Progress in Electromagnetics Research M*, vol. 89, pp. 53–62, 2020.
32. S. Dash and A. Patnaik, "Behavior of graphene-established planar antenna at microwave and terahertz frequency," *Photonics and Nanoconfigurations - Fundamentals and Applications*, vol. 40, pp. 100800/1–13, July 2020.
33. A. Borgia, "New Materials and Technologies for Compact Antennas and Circuits at Millimeter Frequencies," Ph.D. Thesis, University Della Calabria, Italy, 2012.
34. M. Woehrmann and M. Toepper, "Polymerization of Thin-Film Polymers," INTECH-Open, Croatia, pp.113–138, 2012.
35. T. Su, W. Men, Z. Wang, L. Xuan, and W. Zhao, "POSS benzocyclobutene (POSS-BCB) resin: A hybrid thermosetting material with high thermal stability and a low dielectric constant," *High-Performance Polymers*, vol. 30, no. 9, pp. 1123–1129, 2018.
36. C. Huang, L. Pan, R. Liu, and Z. Wang, "Thermal and electrical properties of BCB-liner through-silicon vias," *IEEE Transactions on Components, Packaging, and Manufacturing Technology*, vol. 4, no. 12, pp. 1936–1946, Dec. 2014.
37. J. D. Preez and S. Sinha, *Millimetre-Wave Antennas Configurations and Application*. Springer, Switzerland, 2016.
38. G. Kouemou, *Radar Technology*. INTECH-Open, Croatia, pp. 297–317, 2009.
39. D. Liu, B. Gaucher, U. Pfeiffer, and J. Grzyb, *Advanced Millimeter-Wave Technologies Antennas, Packaging, and Circuits*. Wiley, UK, pp. 49–80, 2009.
40. J. Francey and T. Bateman, "PCB technology requirements for millimeter-wave interconnect and antenna," *The PCB Magazine*, vol. 7, no, 4, pp. 36–42, 2017.
41. U. Nissanov, G. Singh, P. Kumar, and N. Kumar, "High-gain terahertz microstrip array antenna for future generation cellular communication," *Proceedings of IEEE International Conference on Artificial Intelligence, Big Data, Computing and Data Communication (icABCD 2020)*, Durban, South Africa, 6–7 Aug. 2020, pp. 1–6.
42. U. Nissanov, G. Singh, and A. Segun, "Sixth-generation (6G) microstrip antenna with high-gain," *International Journal on Communications Antenna and Propagation*, vol. 11, no. 4, pp. 279–287, August 2021.
43. U. Nissanov, G. Singh, and A. Akinola, "Low sidelobe levels (SLLs) microstrip antennas for THz bio-sensors and THz communications," *Sensors International*, vol. 2, pp. 100097/1–10, April 2021.
44. U. Nissanov and G. Singh, "MmWave/THz reconfigurable ultra-wideband (UWB) microstrip antenna," *Progress in Electromagnetic Research C*, vol. 111, pp. 207–224, April 2021.
45. U. Nissanov, G. Singh, and N. Kumar, "High gain microstrip array antenna with SIW and FSS for beyond 5G at THz band," Optik: *International Journal for Light and Electron Optics*, vol. 236C, pp. 166568/1–13, March 2021.
46. U. Nissanov, G. Singh, E. Gelbart, and N. Kumar, "Highly directive microstrip array antenna with FSS for future generation cellular communication at THz band," *Wireless Personal Communications*, vol. 118, no. 1, pp. 599–617, Jan. 2021.

Chapter 10
Machine Learning in Terahertz Communication

10.1 Introduction

To design an efficacious communication network at terahertz (THz) frequency for 6G wireless communication is more complicated than that of those at a lower frequency, as discussed in the former chapters, because the frequency of THz wave is 1 ~ 4 times higher than that of the microwave communication, which exhibits a larger communication capacity. In addition, the narrow beam and significantly higher directivity of the THz wave lead to intense anti-interference ability and high security. Machine learning (ML)/deep learning (DL)/artificial intelligence (AI) can be used as scalable and agile tools to improve resources for accomplishing network-level and user-level performance objectives while promising near-real-time function. The 6G wireless networks will own the intelligence and capacity to encourage the most appropriate communications tactics based on channel activities sensing and signal quality estimation feedback by ML, DL, and AI technologies [1–2], which could be exploited in 6G wireless for automation. It will be mixed in the resource setting and network selection to upgrade the performance, particularly in slow-down-sensitive technologies and applications [3–5].

In addition, to increase the bandwidth and boost the data rate to new dimensions, we need to use THz communication to induce 6G wireless applications such as holographic communication and digital twinning. Moreover, the THz frequencies supply access to wider bandwidths. These wider bandwidths allow the users to change how user interact with their devices by improving aspects like gesture recognition to assist extended reality (XR)-based applications, for example, the Metaverse. Extended reality is a term including augmented reality (AR), virtual reality (VR), and mixed reality (MR), where VR is a computer-simulated realism experience using a headset that generates images and sounds and creates a fanciful world [6].

U. Nissanov, G. Singh, *Antenna Technology for Terahertz Wireless Communication*,
https://doi.org/10.1007/978-3-031-35900-2_10

10.2 Machine Learning

The terahertz beamforming design is central to the UM-MIMO and multi-user MIMO (MU-MIMO) communication networks. Nevertheless, the traditional beamforming schemes are firmly based on definite channel evaluation; consequently, their complication is comparably high. Therefore, much research is required to consider ML-based approaches to facilitate the THz beamforming design process [7]. However, the fundamental drawback of the ML architectures shown by T. Peken et al. [8] is that they demand many evaluations of channel matrices for practice that can generate an unlikely latency in fast-changing surroundings. To solve this problem, Mo et al. [9] have suggested using an unsupervised K-means algorithm that exploits each antenna radiator's electric-field feedback concerning design beam codebooks, which uses users' average received power gain equipment (UEs) that are located through a cluster. Despite not needing significant training periods, this technique is conscious of major changes in the wireless surroundings that can emerge because of users' or scatters' motions. Sun et al. [10] have developed an interference guiding riddle using coordinated beamforming in extremely dense 6G wireless networks, designed at "almost real-time" sum-rate exaggeration. Further, a Q-learning-based ML method has been interpolated to mitigate the problem, which exclusively needs significantly scaled channel fading criteria and accomplishes identical results to the comparable analytical technique, which needs complete channel state information data. To summarize this section, and as shown in Fig. 10.1, the ML can be sorted into four examples: unsupervised, supervised, transfer learning, and reinforcement [7–10].

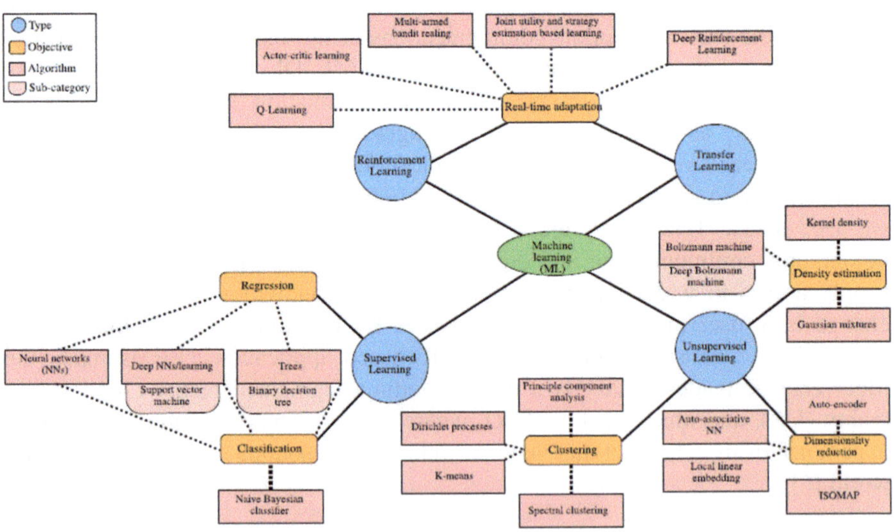

Fig. 10.1 Classification of ML [7]

10.3 Deep Learning

Today's commercial growth is deeply established on novelty in digital technology. However, present-day society does not use 5G, but the demand for 6G wireless communication is standard. DL is an ML subdivision within numerous stacked processing layers, which facilitate classifications, predictions, or other determinations by learning data depictions. DL is well exploited in state-of-the-art communication foundations to download the data, security, and resource reserves. The Internet of Things (IoT) in every innovative application like health care, vehicles, home, and cities requires DL to accurately predict and classify incoming data. Tasks analyzing, collecting, and interpreting extensive data need DL. It causes computation rapidly, comparable to human thinking. The dynamic application needs an automated decision-making structure that can use DL methods. Next-generation ultra-high wireless networking implementation needs demanding technology to activate with high credibility. The 6G period within DL can communicate in a wireless network with ultra-high credibility, ultra-high latency, and moderate energy utilization. In addition, the 6G wireless network must be adequate to treat big data within feasible networking [11–12].

Consequently, big data classification, analytics, and prediction using the DL scheme give imperishable data handling. Intelligent 6G applications like innovative waste handling, smart women safety things, and intelligent environment monitoring demand DL algorithms. Diseases analysis, automatic vehicle acquisition, and security applications are computed effectively using DL algorithms. The 6G wireless networking employing edge computing, cloud computing, fog computing, and inescapable computing techniques must be approached by applying DL algorithms [13–14].

DL in the 6G physical layer (PHY) will be at four sections: (1) UM-MIMO categories. Signal detection, channel estimation, intelligent transmitter, and end-to-end systems. (2) Advance waveforms categories. OFDM transmitter design. (3) RIS-based system categories. Signal detection, channel estimation, passive beamforming, and resource allocation. (4) PHY security. Anti-jamming, anti-spoofing, and anti-eavesdropping [12–13]. Figure 10.2 shows four leading research directions in 6G DL.

While the three typical DL architectures are (1) Deep neural network (DNN), (2) Convolutional neural network (CNN), and (3) recurrent neural network (RNN) [15].

10.4 Artificial Intelligence

The astounding advancement of wireless data drives the boundaries of communications system achievement. However, this wireless data improvement poses a challenge, and next-generation 6G wireless communications networks will demand state-of-the-art algorithms for various services. Therefore, AI applications in the

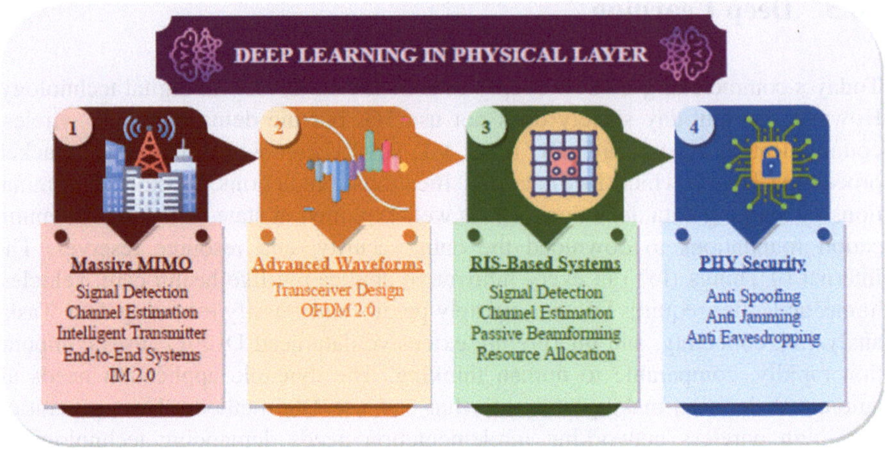

Fig. 10.2 Four leading research directions in 6G PHY with DL [12]

6G wireless communication networks are receiving much attention. For example, the use of ML and DL to implement AI is presently being researched for 6G radio channel propagation prediction [16–18].

AI-enabled 6G channel modeling demands enormous data as input to the algorithms. This data is used to exercise these algorithms and make forecasts. In such a manner, the calculation intricacy can be significantly reduced because the forecasts are precise, and the algorithms are powerful for modeling various schemes. However, the drawbacks are caused by hybrid deterministic-stochastic models regarding limited channel sounding and simulation time and inhibit large-scale characterization. At THz, small-scale synergy is mainly applicable because of the short wavelength above 100 GHz and cannot be modeled effectively over large BWs within the current techniques. In other words, the massive number of data, which would have to be handled as a result of the enormous number of diffuse and specular reflections and the resulting clustering demeanor, which are unavoidable at THz, would aim at enormous calculation resource demands, which can demand the use of AI techniques to facilitate efficient and precise characterization of the 6G wireless propagation channel [19]. Figure 10.3 shows the relationship between AI, ML, and DL.

10.5 Extended Reality

Extended reality (XR) is an umbrella concept for all-enveloping technologies like AR, VR, and MR. Multisensory XR combines the five ordinary senses: touch, sight, hearing, taste, and smell. XR is the medium that combines avatars in the Metaverse and users in reality [20]. The Metaverse will employ the subsequent software and hardware components: hand and non-hand-based input apparatus, object, scene, speech recognition, motion rendering, synthesis, and head-mounted devices

Fig. 10.3 The relationship between AI, ML, and DL [12]

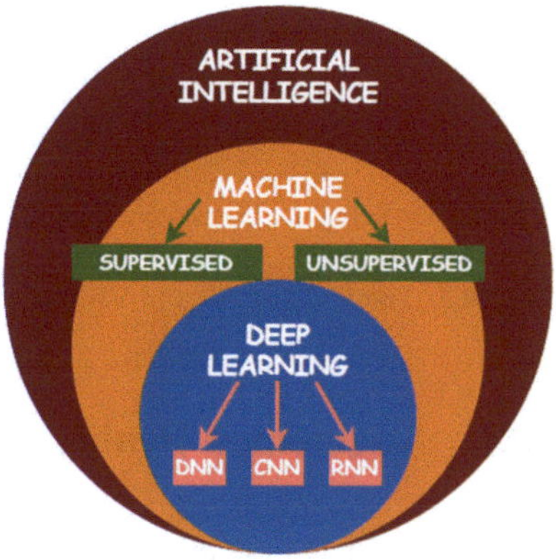

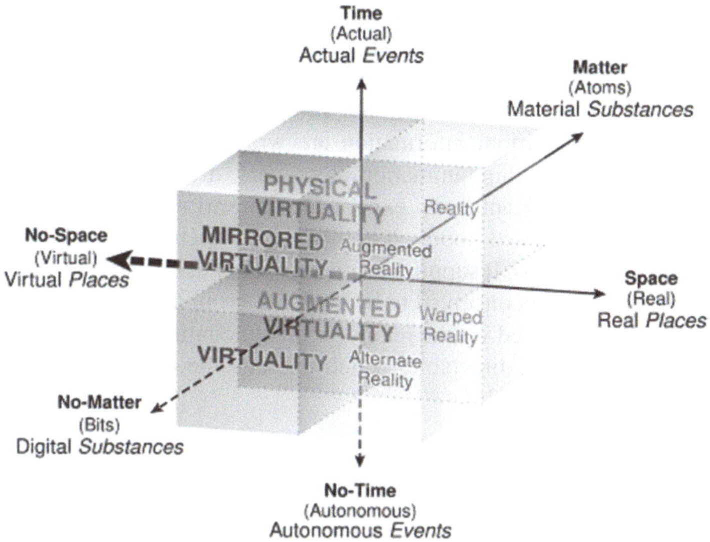

Fig. 10.4 The Multiverse as an architecture of state-of-the-art XR experiences [20]

(HMDs). Additionally, the Metaverse is assumed to supply gamified experiences over emerging Web 3.0 technologies and is predictable to be the harbinger of the so-called Multiverse that will officiate as an architecture of the advanced XR experience domain. Furthermore, while the Metaverse mainly centers on AR and VR, the Multiverse suggests 3D, eight diverse characters, and six variables of time, matter, space, non-space, non-time, and non-matter of state-of-the-art XR experience domains, along with AR and VR, as illustrated in Fig. 10.4 [21–22].

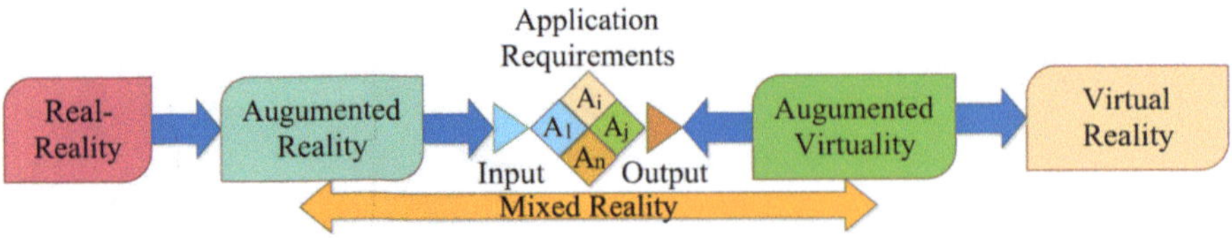

Fig. 10.5 The RVC of AR and VR [24]

10.6 Augmented Reality

6G wireless networks assist ameliorated holographic projection via THz bandwidth, massive apparatus connectedness, and ultra-low latency. Nevertheless, the data is transmitted via autonomous networks above unreliable channels. Consequently, blockchain (BC) unlocks modern dimensions of intelligent resource handling, hearing, user access domination, and chronology in reserved transactions to provide data security, privacy, and stakeholder trust. So, the 6G and blockchain combination in future AR and VR implementations is an approaching research topic [23].

In AR, objects and virtual information are wrapped in reality. This sensation intensifies the reality within animation, text, and digital images. AR enables enveloping, colorful, and authentic experiences in various utilizations. AR comprises the virtuality of actual goods; it combines multisensory capabilities such as cognitive, video, and audio. It supplies a concurrent 3D demonstration of virtual and actual objects. AR can be conceived to be a middle-ground within VR and actuality space. AR can be interpreted via the reality-virtuality continuum (RVC) term, where Fig. 10.5 shows the RVC timetable, which exhibits the change from reality to a simulated 3D environment [24].

10.7 Virtual Reality

VR is conceivably one of the highest predictable technologies of the forthcoming decade [25]. Nonetheless, relying on wired VR systems necessarily confines the VR application realm. In lieu, deploying wireless VR on top of 6G wireless networks can allow its actual potential. Therefore, securing ultra-low latency communications (URLLC) to incorporate VR utilities on top of 6G wireless networks is vital. Furthermore, to obtain good user practice, the end-to-end (E2E) delay for 6G wireless VR must be very low, about tens of milliseconds. In a VR experience, users are entirely engaged in the simulated digital surrounding. Users essentially put on an HMD or VR headset to achieve a 360° scene of an artificial world [26].

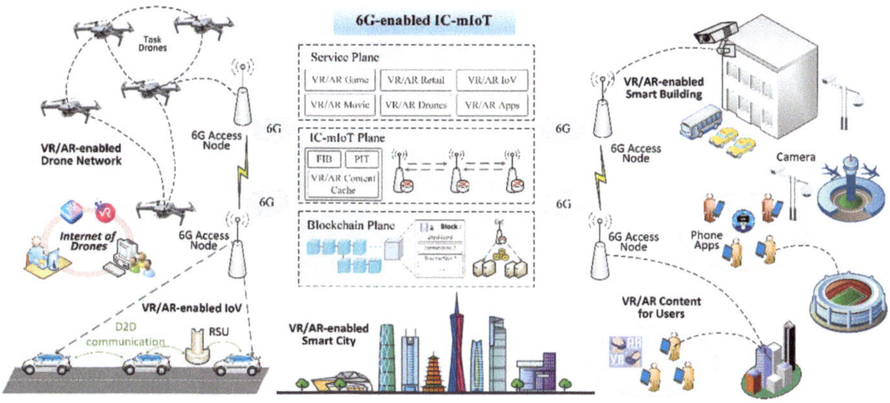

Fig. 10.6 The main field of 6G-enabled information massive IoT for omnipresent VR [23]

Furthermore, within URLLC, wireless VR services will also need high data rates to provide 360° substance to their users. High data rates within URLLC require a significant exit from archetypal URLLC utilities confined to low-rate sensors [27]. One can investigate the high BW affordable at the THz frequency bands, allowing high-rate wireless access for VR to solve this dare. Still, the fidelity of the THz channel can be inhibited by its sensitivity to molecular absorption, blockage, and communication propagation range [28]. As shown in Fig.10.6, VR substance utilization in 6G will enter different scenarios involving the integrated sky, earth, sea, and tactile wireless network.

10.8 Mixed Reality

MR is a term used to incorporate real-world surroundings with computer-generated ones. Virtual and physical commodities can coincide in MR surroundings and communicate in real time. Therefore, MR is mainly compatible with AR. However, unlike AR content that can be pulled off by applying a mobile device, an HMD needs to be used to involve MR, as shown in Fig. 10.7. MR incorporating haptics has occasionally been assigned to visual-haptic-mixed-reality (VHMR) [29–30].

MR may permit more intercommunication within digital content as users can go hands-free. Nonetheless, it also needs to be more approachable to the commercial. However, there is no apprehension that MR has enormous potential in 6G wireless communication networks. Many software and hardware developments are essential to implement MR within 6G wireless communication networks [31].

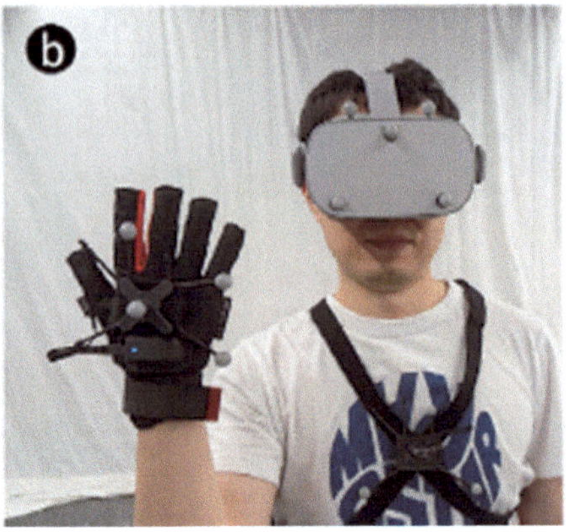

Fig. 10.7 Using an HMD to involve MR [30]

10.9 Open Research Challenges

This section presents the future open research challenges in 6G communication in AI/ML domains. Based on this chapter, the ML, AI, and XR in Terahertz for 6G has several open research challenges regarding the Collection and Availability of Rear-World Datasets, Holographic Communication, Extended Reality (XR), Multi-Sense Experience, Digital Twin, Pervasive Intelligence, Computation-intensive AI/ML technologies, and Privacy and Security [32–34]:

1. Collection and Availability of Real-World Datasets: The massive future wireless communication applications will be based on AI/ML technology, which requires an enormous quantity of datasets during the training phase, which is still unavailable publicly. Also, the availability of such data sets freely/publically may give rise to issues like users' data protection and privacy.
2. Holographic Communication: Holographic communication provides a real-time, detailed capture of human emotions and expressions transmitted through the communication networks, which discusses real-time capturing, encoding, transporting, and rendering of 3D representations. There have been considerable advancements in this technology with the developments in artificial intelligence. Enhanced proficiencies will enable networks to adapt, reliant on user behavior and routines, while pre-empting what will be required from the network to enhance the overall quality and reliability of the experience. However, holographic communication requires broader bandwidth in the order of terabits per second [32]. In addition, it also needs ultra-low latency and high-precision synchronization.
3. Extended Reality (XR): In the 5G communication technology era, extended reality combines augmented and virtual reality and is still in its teenage stage. However, to attain a similar visual quality to 2D video streaming, extended reality devices with a 360° field vision require a substantially higher data rate communication capacity.

4. Multi-Sense Experience: The communication network applications, which imply not only the optical and acoustic senses but also involve touch (tactile), senses of smell, and taste to create fully immersive experiences, will explore new and emerging application areas [33]. Using smell and taste to create a genuinely immersive experience might lead to new products in the food and textile sectors. However, real breakthroughs in actuators that convey a digital sense of smell and taste seem at this point far away. Assuming those problems can be overcome, there will be interesting potential networked applications such as remote learning solutions as well as digital advertising, which may exploit the fact that memory retention could be improved by association with smells and tastes. Further, digital experiences can be enhanced, particularly as smells and tastes can evoke or amplify emotions.
5. Digital Twin: It creates an exact virtual reproduction of a physical thing. The software clone involves many main features, information, and properties. Thus, the twin is used to duplicate an object automatically and intelligently. Early digital twin functioning took the curiosity of several vertical businesses and manufacturers. However, with the introduction of 6G networks, it is expected to be ultimately deployed. 6G communication with AI/ML supports emerging use cases that will drive the digital society toward 2030 and beyond, including holographic media, digital twins, and many more, with suitable application-network interaction allowing for specialized service delivery to the needs of the applications. Thus, the network 2030 will provide a rich set of mechanisms to allow for interaction between the applications and the infrastructures beyond immature datagram delivery.
6. Pervasive Intelligence: The introduction of smart mobile devices and modern communication techniques such as unmanned aerial vehicles, virtual reality glasses, and over-the-air intelligent services are predicted to expand in acceptance. However, most of these sophisticated innovations are carried out using traditional means and are accepted to have rapid changes in the future.
7. Computation-intensive AI/ML technologies: 6G communication network will have to explore new technologies to address the potential problems such as effective usages of computation, power, and storage in the resource constraint mobile device. Thus, these new technologies will use distributed computing resources and mobile-edge devices to promote efficient ML training and interference management strategies.
8. Privacy and Security: Researchers have recently focused on developing quantum cryptography and physical layer security, but it requires appropriate research to justify their practicality in 6G communication networks. Furthermore, the potential challenges concerning privacy protection have gained massive consideration in using ML and AI in wireless communications. However, privacy protection during the system model training process is a challenging task [34] that needs to explore in detail.

10.10 Summary

The primary purpose of Chap. 10 is to recognize the proliferation of 6G communication technology and its impact on future AI/ML-based applications. We also highlight some of the key performance indicators of the 6G technology and discuss how AI/ML will empower the future 6G ecosystem. Further, we also put forward some crucial and open research problems to boost research in this domain. The information and challenges listed in this chapter will provide a new research direction to the researchers from academia and industry experts in this domain. The communications network technologies are entering a period of unprecedented change and opportunity mainly; the impact of the Network 2030 initiative on the distribution and management of network functionality will deliver a massive increase in network flexibility. These combined developments will continue the paradigm shift in how future networks are conceived, designed, built, and operated.

Further, its initiative could be the enabling network platform for other large- or small-scale initiatives. Chapter 10 also discusses ML, DL, AI, XR, AR, VR, and MR technologies for 6G. Designing efficacious communication networks in THz 6G wireless communication systems is more complicated than those at lower frequency systems because channels at the THz band are noted to be further insecure than those of lower frequencies. Therefore, 6G wireless networks will own the intelligence and capacity to encourage the most appropriate communications tactics based on channel activities sensing and signal quality estimation feedback by ML, DL, and AI technologies. In addition, to increase bandwidth and boost data rate to new dimensions, we need to use THz communication to induce 6G wireless applications such as holographic communication and digital twinning.

Moreover, THz frequencies supply access to a broader bandwidth. This wider bandwidth allows us to change how we interact with our apparatus by improving aspects like gesture recognition to assist XR-based applications, for example, the Metaverse. XR is a term including AR, VR, and MR technologies.

References

1. A. Shafie, G. N. Yang, C. Han, J. M. Jornet, M. Juntti, and T. Kurner, "Terahertz communications for 6G and beyond wireless networks: challenges, key advancements, and opportunities," *IEEE Network*, pp. 1–8, Sep. 2022.
2. W. Saad, M. Bennis, and M. Chen, "A vision of 6G wireless systems: Applications, trends, technologies, and open research problems," *IEEE Network*, vol. 34, no. 3, pp. 134–142, June 2020.
3. Y. L. Lee, D. Qin, L. C. Wang, and G. H. Sim, "6G massive radio access networks: key applications, requirements, and challenges," *IEEE Open Journal of Vehicular Technology*, vol. 2, pp. 54–66, Jan. 2021.
4. W. Jiang, B. Han, M. A. Habibi, and H. D. Schotten, "The road towards 6G: a comprehensive survey," *IEEE Open Journal of the Communications Society*, vol. 2, pp. 334–366, Feb. 2021.

5. M. Z. Chowdhury, M. Shahjalal, S. Ahmed, and Y. M. Jang, "6G wireless communication systems: applications, requirements, technologies, challenges, and research directions," *IEEE Open Journal of the Communications Society*, vol. 1, pp. 957–975, July 2020.
6. White paper: *Fundamentals of THz Technology for 6G*, Rohde&Schwarz, pp. 1–56, 2020. https://www.rohde-schwarz.com/tr/solutions/test-and-measurement/wireless-communication/cellular-standards/6g/white-paper-fundamentals-of-thz-technology-for-6g-by-rohde-schwarz-registration-_255934.html.
7. A. A. A. Boulogeorgos, E. Yaqub, M. di-Renzo, A. Alexiou, R. Desai, and R. Klinkenberg, "Machine learning: A catalyst for THz wireless networks," *Frontiers in Communication and Networks*, pp. 1–34, July 2021.
8. T. Peken, S. Adiga, R. Tandon, and T. Bose, "Deep learning for SVD and hybrid beamforming," *IEEE Transactions on Wireless Communications*, vol. 19, no. 10, pp. 6621–6642, Oct. 2020.
9. J. Mo, B. L. Ng, S. Chang, P. Huang, M. N. Kulkarni, A. Alammouri, J. C. Zhang. J. Lee, and W. J. Choi, "Beam codebook design for 5G mmWave terminals," *IEEE Access*, vol. 7, pp. 98387–98404, July 2019.
10. C. Sun, Z. Shi, and F. Jiang, "A machine learning approach for beamforming in ultra-dense network considering selfish and altruistic strategy," *IEEE Access*, vol. 8, pp. 6304–6315, Jan. 2020.
11. Z. Zhang, Y. Xiao, Z. Ma, M. Xiao, Z. Ding, X. Lei, G. K. Karagiannidis, and P. Fan, "6G wireless networks: Vision, requirements, architecture, and key technologies," *IEEE Vehicular Technology Magazine*, vol. 14, no. 3, pp. 28–41, July 2019.
12. B. Ozpoyraz, A. T. Dogukan, Y. Gevez, U. Altun, and E. Basar, "Deep learning-aided 6G wireless networks: A comprehensive survey of revolutionary PHY architectures," *IEEE Open Journal of the Communications Society, vol. 3,* pp. 1749–1809, 2022.
13. Z. Chen, C. Han, Y. Wu, L. Li, C. Huang, Z. Zhang, G. Wang, and W. Tong, "Terahertz wireless communications for 2030 and beyond- A cutting-edge frontier," *IEEE Communications Magazine*, vol. 59. no. 11, pp. 66–72, Nov. 2021.
14. M. Alsabah, M. A. Naser, B. M. Mahmood, S. H. Abdulhussain, M. R. Eissa, A. Al-Baidhani, N. K. Noordin, S. M. Sait, K. A. Al-Utaibi, and F. Hashim, "6G wireless communications networks: A comprehensive survey," *IEEE Access*, vol. 9, pp. 148191–148243, Nov. 2021.
15. L. Deng, "Three classes of deep learning architectures and their applications: A tutorial survey, " Microsoft Research, Redmond, USA. https://www.microsoft.com/en-us/research/wp-content/uploads/2016/02/Transactions-APSIPA.pdf.
16. Y. Jiao, J. J. Hall, and Y. T. Morton, "Automatic equatorial GPS amplitude scintillation detection using a machine learning algorithm," *IEEE Transactions on Aerospace and Electronic Systems*, vol. 53, no. 1, pp. 405–418, Feb. 2017.
17. T. Imai, K. Kitao and M. Inomata, "Radio propagation prediction model using convolutional neural networks by deep learning," *Proceedings of IEEE 13th European Conference on Antennas and Propagation (EuCAP)*, Krakow, Poland, 2019, pp. 1–5.
18. T. Hayashi, T. Nagao, and S. Ito, "A study on the variety and size of input data for radio propagation prediction using a deep neural network," *Proceedings of IEEE 14th European Conference on Antennas and Propagation (EuCAP)*, Copenhagen, Denmark, 2020, pp. 1–5.
19. D. Serghiou, M. Khalily, T. W. C. Brown, and R. Tafazolli, "Terahertz channel propagation phenomena, measurement techniques and modeling for 6G wireless communication applications: a survey, open challenges, and future research directions," *IEEE Communications Surveys & Tutorials*, vol. 24, no. 4, pp. 1957–1996, Sep. 2022.
20. S. Rostami and M. Maier, "The metaverse and beyond- Implementing advanced multiverse realms with smart wearables," *IEEE Access*, vol. 10, pp. 110796–110806, Oct. 2022.
21. T. Winters, "The Metaverse: prepare now for the next big thing! " Chicago, IL, USA: Independently Published, Aug. 2021.
22. S. M. Park and Y. G. Kim, "A Metaverse-taxonomy, components, applications, and open challenges," *IEEE Access*, vol. 10, pp. 4209–4251, Jan. 2022.

23. S. Liao, J. Wu, J. Li, and K. Konstantin, "Information-centric massive IoT-based ubiquitous connected VR/AR in 6G: A proposed caching consensus approach," *IEEE Internet of Things Journal*, vo. 8, no. 7, pp. 5172–5184, Oct. 2020.
24. P. Bhattacharya, D. Saraswat, A. Dave, M. Acharya, S. Tanwar, G. Sharma, and I. E. Davidson, "Coalition of 6G and blockchain in AR/VR space: challenges and future directions," *IEEE Access*, vol. 9, pp. 168455–168484, Dec. 2021.
25. W. Saad, M. Bennis, and M. Chen, "A vision of 6G wireless systems: applications, trends, technologies, and open research problems," *IEEE Network*, vol. 34, no. 3, pp. 134 – 142, June 2020.
26. M. S. Elbamby, C. Perfecto, M. Bennis, and K. Doppler, "Toward low-latency and ultra-reliable virtual reality," *IEEE Network*, vol. 32, no. 2, pp. 78–84, April 2018.
27. G. Pocovi, B. Soret, K. I. Pedersen, and P. Mogensen, "MAC layer enhancements for ultra-reliable low-latency communications in cellular networks," *Proceedings of IEEE International Conference on Communications Workshops (ICC Workshops)*, Paris, France, 21–25 May 2017, pp. 1–6.
28. A. Moldovan, P. Karunakaran, I. F. Akyildiz, and W. H. Gerstacker, "Coverage and achievable rate analysis for indoor terahertz wireless networks," *Proceedings of IEEE International Conference on Communications (ICC)*, Paris, France, 21–25 May 2017, pp. 1–7.
29. M. M. Aygün, Y. Cagriogüt, H. Baysal, and Y. Tascioglu, "Visuo-haptic mixed reality simulation using unbound handheld tools," *Applied Science*, vol. 10, no. 15, pp. 5344/1–15, July 2020.
30. Q. Zhou, D. Yu, M. N Reinoso, J. Newn, J. Goncalves, and E. Velloso, "Eyes-free target acquisition during walking in immersive mixed reality," *IEEE Transactions on Visualization and Computer Graphics*, vol. 26. no. 12, pp. 3423–3433, Dec. 2020.
31. H. Viswanathan and P. E. Mogensen, "Communications in the 6G Era," *IEEE Access*, vol. 8, pp. 57063–57074, March 2020.
32. A. Clemm, M. T. Vega, H. K. Ravuri, T. Wauters, and F. De Turck, "Toward truly immersive holographic-type communication: Challenges and solutions," *IEEE Communications Magazine*, vol. 58, no. 1, pp. 93–99, Jan. 2020.
33. https://www.itu.int/en/ITU-T/focusgroups/net2030/Documents/White_Paper.pdf.
34. https://www.itu.int/

MIX
Papier aus verantwortungsvollen Quellen
Paper from responsible sources
FSC® C105338

If you have any concerns about our products,
you can contact us on
ProductSafety@springernature.com

In case Publisher is established outside the EU,
the EU authorized representative is:
Springer Nature Customer Service Center GmbH
Europaplatz 3, 69115 Heidelberg, Germany

Printed by Libri Plureos GmbH
in Hamburg, Germany